Preface

In developing its basic radiation protection recommendations, as given in NCRP Report No. 116, *Limitation of Exposure to Ionizing Radiation* (NCRP, 1993a), the Council reiterated its acceptance of the linear-nonthreshold hypothesis for the risk-dose relationship. Specifically, "based on the hypothesis that genetic effects and some cancers may result from damage to a single cell, the Council assumes that, for radiation-protection purposes, the risk of stochastic effects is proportional to dose without threshold, throughout the range of dose and dose rates of importance in routine radiation protection. Furthermore, the probability of response (risk) is assumed, for radiation protection purposes, to accumulate linearly with dose. At higher doses received acutely, such as in accidents, more complex (nonlinear) dose-risk relationships may apply." This Report is the result of an in-depth review by NCRP Scientific Committee 1-6 of the scientific basis for this assumption, *i.e.*, the relationship between dose and risk at low doses.

Scientific Committee 1-6 sought and obtained written and oral input from several scientists in the United States who held many different views regarding the science associated with this subject and I want to thank those scientists for their frank and candid input to the Committee's work.

Since this Committee was constituted to address the scientific issues, the implications of the Committee's work for radiation protection policy will be addressed by NCRP at a later point in time.

Serving on NCRP Scientific Committee 1-6 on Linearity of Dose Response were:

Arthur C. Upton, *Chairman*
University of Medicine and Dentistry of New Jersey
Robert Wood Johnson Medical School
New Brunswick, New Jersey

Members

S. James Adelstein
Harvard Medical School
Boston, Massachusetts

Eric J. Hall
Columbia University
New York, New York

David J. Brenner
Columbia University
New York, New York

Kelly H. Clifton
University of Wisconsin
Madison, Wisconsin

Stuart C. Finch
University of Medicine and Dentistry of New Jersey
Camden, New Jersey

Howard L. Liber
Massachusetts General Hospital
Boston, Massachusetts

Robert B. Painter
University of California
San Francisco, California

R. Julian Preston
U.S. Environmental Protection Agency
Research Triangle Park, North Carolina

Roy E. Shore
New York University Medical Center
New York, New York

Advisor

Amy Kronenberg
Lawrence Berkeley National Laboratory
Berkeley, California

NCRP Secretariat

W. Roger Ney, *Consultant* (1999–2001)
Eric E. Kearsley, *Staff Scientist* (1997–1998)
William M. Beckner, *Senior Staff Scientist* (1995–1997)
Cindy L. O'Brien, *Managing Editor*

The Council wishes to express its appreciation to the Committee members for the time and effort devoted to the preparation of this Report and to the U.S. Nuclear Regulatory Commission for its financial support of this activity.

Charles B. Meinhold
President

NCRP Report No. 136

Evaluation of the Linear-Nonthreshold Dose-Response Model for Ionizing Radiation

Recommendations of the
NATIONAL COUNCIL ON RADIATION PROTECTION AND MEASUREMENTS

Issued June 4, 2001

National Council on Radiation Protection and Measurements
7910 Woodmont Avenue, Suite 800 / Bethesda, Maryland 30814

LEGAL NOTICE

This Report was prepared by the National Council on Radiation Protection and Measurements (NCRP). The Council strives to provide accurate, complete and useful information in its documents. However, neither the NCRP, the members of NCRP, other persons contributing to or assisting in the preparation of this Report, nor any person acting on the behalf of any of these parties: (a) makes any warranty or representation, express or implied, with respect to the accuracy, completeness or usefulness of the information contained in this Report, or that the use of any information, method or process disclosed in this Report may not infringe on privately owned rights; or (b) assumes any liability with respect to the use of, or for damages resulting from the use of any information, method or process disclosed in this Report, *under the Civil Rights Act of 1964, Section 701 et seq. as amended 42 U.S.C. Section 2000e et seq. (Title VII) or any other statutory or common law theory governing liability.*

Library of Congress Cataloging-in-Publication Data

Evaluation of the linear-nonthreshold dose-response model for ionizing radiation.
p. cm. — (NCRP report ; no. 136)
"June 2001."
Includes bibliographical references and index.
ISBN 0-929600-69-X
1. Radiation—Toxicology. 2. Low-level radiation—Dose-response relationship. I. National Council on Radiation Protection and Measurements. Scientific Committee 1-6 on Linearity of Dose Response. II. Series.
RA1231.R2 E935 2001
612′.01448—dc21 2001032614

[For detailed information on the availability of NCRP publications see page 273.]

Contents

1. Executive Summary

This Report presents an evaluation of the existing data on the dose-response relationships and current understanding of the health effects of low doses of ionizing radiation.[1] This reevaluation was carried out by Scientific Committee 1-6 of the National Council on Radiation Protection and Measurements (NCRP), which was charged to reassess the weight of scientific evidence for and against the linear-nonthreshold dose-response model, without reference to associated policy implications. The evaluation was prompted by the need to reassess the common use, for radiation protection purposes, of the linear-nonthreshold dose-response hypothesis in the light of new experimental and epidemiological findings, including growing evidence of adaptive responses to small doses of radiation which may enhance the capacity of cells to withstand the effects of further radiation exposure, and new evidence concerning the possible nature of neoplastic initiation.

The evaluation focuses on the mutagenic, clastogenic (chromosome-damaging), and carcinogenic effects of radiation, since these effects are generally postulated to be stochastic and to increase in frequency as linear-nonthreshold functions of radiation dose.[2] For each type of effect, the relevant theoretical, experimental and epidemiological data are considered. Furthermore, in an effort to avoid overlooking pertinent data in the evaluation, input was obtained from authorities in the field and from the scientific community at large.

The evaluation begins by considering the way in which radiation energy is deposited within cells and its implications for dose-response relationships. As is customary, the amount of radiation producing an effect is conveniently specified as the energy absorbed per unit mass in the irradiated system; *i.e.*, the dose (D). At the outset, it is noted that virtually all existing experimental and epidemiological data on the effects of sparsely ionizing [*i.e.*, low linear-energy transfer (LET)] radiation come from observations at doses far above those in

[1]In this Report, the word "dose" is frequently used in its generic sense.

[2]Publication 26 of the ICRP (1977) was the first to describe in detail that "stochastic" effects are those for which the probability of an effect occurring, rather than its severity, is regarded as a function of dose without a threshold.

which a single cell is struck, on the average, by no more than one radiation track. This means that any effects attributable to lower doses of radiation in the millisievert range can be estimated only by extrapolation, guided by radiation damage and repair models. Based on direct experimental observations involving alpha-particle microbeam experiments and theoretical considerations, it is concluded that cellular traversal by a single radiation track of any type of ionizing radiation has a non-zero probability of depositing enough energy in a critical macromolecular target, such as deoxyribonucleic acid (DNA), to injure, but not necessarily kill the cell in question. Hence, when the average number of traversals is well below one, it is concluded that the number of independently affected cells may increase as a nonthreshold function of the dose. Moreover, there is now evidence that cells in the neighborhood of those hit may also exhibit signs of radiation damage. The dose-response relationships have not been determined, but if each hit cell influences a number of surrounding cells, there could be a linear dose response until all cells are hit (Azzam *et al.*, 1998; Deshpande *et al.*, 1996; Lehnert and Goodwin, 1997; Lorimore *et al.*, 1998; Mothersill and Seymour, 1997; 1998; Nagasawa and Little, 1992).

Of the various macromolecular targets within cells that may be altered by radiation, DNA is the most critical, since genomic damage may leave a cell viable, but permanently altered. Several types of initial or primary DNA damage are known to result from ionizing irradiation, including single-strand breaks (ssbs), nucleotide base damages (bds) and loss, DNA-protein cross-links (dpcs), double-strand breaks (dsbs), and multiply-damaged sites (mds) of a type which is extremely rare in nonirradiated cells. Most such lesions in DNA are repairable to varying degrees, depending on the repair capacity of the affected cells. Dsbs and mds are induced only by ionizing radiation (and some radiomimetic chemicals) and are complex and extremely difficult substrates for DNA repair enzymes to handle; the repair of these lesions has been observed to be inaccurate where their frequencies have been amenable to measurement. Although the extent to which repair may alter their production at doses in the millisievert range remains to be determined, it is noteworthy that at higher doses all types of DNA lesions appear to be formed linearly with increasing dose and that they are induced so sparsely in the low-dose range that interactions between adjacent lesions produced by different radiation tracks are extremely rare.

Any DNA lesions that remain unrepaired, or are misrepaired, may be expressed as point mutations (resulting from nucleotide base-pair substitutions or from the insertion or deletion of small numbers of base pairs), larger deletions (involving the loss of hundreds-to-millions of base pairs), genetic recombination events (involving the exchange of sequences of base pairs between homologous chromosomes), and chromosome aberrations. Mutations of all types appear to be inducible by ionizing radiation, but their dose-response curves vary in shape, depending on the dose, the type of mutation scored, the LET and dose rate of the radiation, and the genetic background of the exposed cells. The frequency of mutations induced by a given dose of low-LET radiation has generally been observed to decrease with decreasing dose rate, implying that some premutational damage that does not accumulate too rapidly in the exposed cells can be repaired. The capacity for repair of premutational damage is also evident from the fact that prior exposure to a small "conditioning" dose of low-LET radiation may reduce the frequency with which mutations are produced by a subsequent "challenge" dose in cells of some individuals. It is noteworthy, nevertheless, that mutational changes of various types (including those types implicated in carcinogenesis) have generally been observed to be induced with linear kinetics at low-to-intermediate dose levels in human and animal cells.

The misrepair of lesions in DNA can also give rise to chromosome aberrations, the frequency of which varies markedly with the dose, dose rate, and LET of the radiation. In cells exposed to high-LET radiation, the response typically rises as a linear function of the dose, with a slope that is essentially dose-rate-independent, whereas in cells exposed to low-LET radiation the curve rises less steeply, as a linear-quadratic function of the dose after acute irradiation. At low-dose rates, the linear portion of the curve predominates and is a limiting slope at low doses. The apparent linearity of the latter dose-response relationship implies that traversal of the cell by a single low-LET radiation track may occasionally suffice to cause a nonlethal chromosome aberration, but the likelihood of such an effect would depend on the fidelity with which DNA damage is repaired at such low-dose levels.

It is noteworthy that prior exposure to a small (*e.g.*, 10 mSv) "conditioning" dose of radiation has been observed to enhance the repair of chromosome aberrations for such DNA lesions in the cells of some persons; however, the existing data imply that this type of adaptive response is not elicited in every individual, that the response lasts no more than a few hours when it does occur, that a dose of at least 5 mSv delivered at a dose rate of at least 50 mSv min^{-1} is required to elicit the response, and that the response typically reduces the aberration frequency by no more than one-half. On the basis of the existing evidence it appears likely that this adaptive response acts primarily to reduce the quadratic (two-hit) component of the dose-response curve, without changing the slope of the linear component. While the existing data do not exclude the possibility that a threshold for the induction of chromosome aberrations may exist in the millisievert dose range, there is no body of data supporting such a possibility, nor would such a threshold be consistent with current understanding of the mechanisms of chromosome aberration formation at low doses.

The significance of nonlethal mutations and chromosome aberrations is that they are implicated in the causation of cancer, a clonal disorder that may result from such changes in only one cell in the relevant organ. The types of functional genetic changes implicated thus far in carcinogenesis include the activation of oncogenes, the inactivation or loss of tumor-suppressor genes, and alterations of various other growth-regulatory genetic elements (*e.g.*, loss of apoptosis genes, mutation in DNA repair genes). The specific roles that such changes may play in the cancer process remain to be fully elucidated. However, the neoplastic transformation of cells by irradiation *in vitro*, a process which is analogous in many respects to carcinogenesis *in vivo*, typically involves a step-wise series of such genetic alterations, in the course of which the affected cells often accumulate progressively, growing numbers of mutations and/or chromosomal abnormalities, a pattern indicative of genomic instability. Although the precise nature of each step in the process remains to be elucidated in full, the frequency with which initial *in vitro* alterations are produced by ionizing radiation typically exceeds any known *in vivo* radiation-induced mutation rate by several orders of magnitude, suggesting that epigenetic changes, as well as genetic changes, are involved. Further research into the significance of *in vitro* neoplastic transformation for *in vivo* carcinogenesis is clearly needed. It is also noteworthy that susceptibility to neoplastic transformation *in vitro* varies markedly with the genetic background of the exposed cells, their stage in the cell cycle, the species and strain from which the cells were derived, and many other variables. The process is further complicated by evidence that transformed cells may release diffusible substances into the surrounding medium that enhance the transformation of neighboring cells. Not surprisingly, therefore, the dose-response curve for neoplastic transformation is complex in shape and subject to variation, depending on the particular cells and experimental conditions under investigation. Little is presently known about the shape of the curve in the low-dose domain, but evidence suggests that a small percentage of exposed cells may be transformed by only one alpha-particle traversal of the nucleus.

The dose-response relationships for carcinogenic effects of radiation have been studied most extensively in laboratory animals, in which benign and malignant neoplasms of many types have been observed to be readily inducible by large doses of radiation. The dose-response curves for such neoplasms vary widely, depending on the neoplasm in question, the genetic background, age and sex of the exposed animals, the LET and dose rate of irradiation, and other variables. In general, low-LET radiation is appreciably less tumorigenic than high-LET radiation, and its tumorigenic effectiveness is reduced at low-dose rates, whereas the tumorigenic effectiveness of high-LET radiation tends to remain relatively constant. Not every type of neoplasm is inducible, however; some types actually decrease in frequency with increasing dose, and there are others that are induced in detectable numbers only at high-dose levels, signifying the existence of effective or actual thresholds for their induction. For certain types of neoplasms, however, and for the life-shortening effects of all radiation-induced neoplasms combined, the data are consistent with (linear or linear-quadratic) nonthreshold relationships, although the data do not suffice to define the dose-response relationships unambiguously in the dose range below 0.5 Sv. The variations among neoplasms in dose-response relationships point to differences in causal mechanisms which remain to be elucidated. Nevertheless, it is clear from the existing data that tumor induction *in vivo* is a multistage process in which the initial radiation-induced alteration typically occurs at a frequency exceeding that of any known radiation-induced specific locus mutation and is followed by the activation of oncogenes, inactivation or loss of tumor-suppressor genes, and other mutations and/or chromosomal abnormalities, often associated with genomic instability in the affected cells.

Dose-dependent increases in the frequency of many, but not all, types of neoplasms are well documented in human populations as well as in laboratory animals. The dose-response relationships for such neoplasms likewise vary, depending on the type of neoplasm, the LET and dose rate of irradiation, the age, sex, and genetic background of the exposed individuals, and other variables. The data come largely from observations at relatively high doses and dose rates and do not suffice to define the shape of the dose-response curve in the millisievert dose range; however, it is noteworthy that: (1) the dose-response curve for the overall frequency of solid cancers in the atomic-bomb survivors is not inconsistent with a linear function down to a dose of 50 mSv; (2) there is evidence suggesting that prenatal exposure to a dose of only about 10 mSv of x ray may suffice to increase the subsequent risk of childhood cancer; (3) analysis of the pooled data from several large cohorts of radiation workers supports the existence of a dose-dependent excess of leukemia from occupational irradiation that is similar in magnitude to the excess observed in atomic-bomb survivors; (4) a dose of about 100 mSv to the thyroid gland in childhood significantly increases the incidence of thyroid cancer later in life; and (5) highly fractionated doses of about 10 mSv per fraction, delivered in multiple fluoroscopic examinations during the treatment of pulmonary tuberculosis (TB) with artificial pneumothorax, appear to be fully additive in their carcinogenic effects on the female breast in women exposed under the age of 50, although much less than fully additive in carcinogenic effects on the lung. At the same time, it is important to note that the rates of cancer in most populations exposed to low-level radiation have not been found to be detectably increased, and that in most cases the rates have appeared to be decreased. For example, the large pooled study of radiation worker cohorts did not show positive effect for solid tumors. In general, however, because of limitations in statistical power and the potential for confounding, low-dose epidemiological studies are of limited value in assessing dose-response relationships and have produced results with sufficiently wide confidence limits to be consistent with an increased effect, a decreased effect, or no effect.

Another factor complicating the assessment of the dose-response relationship is uncertainty about the extent to which the effects of radiation may be reduced by adaptive responses in the low-dose domain. Adaptive responses may account, at least in part, for the reduced effectiveness of low-LET radiation at low-dose rates. It is not clear, however, that such responses can be elicited by a dose of less than 1 mSv delivered at a rate of less than 0.05 Sv min^{-1}, or that the responses can increase the fidelity of DNA repair processes sufficiently to make the processes error-free. In a significant percentage of individuals, moreover, the capacity to elicit such responses appears to be lacking. The available data on adaptive responses do not suffice, therefore, to either exclude or confirm a linear-nonthreshold dose-incidence relationship for mutagenic and carcinogenic effects of radiation in the low-dose domain.

In conclusion, the weight of evidence, both experimental and theoretical, suggests that for many of the biological lesions which are precursors to cancer (such as mutations and chromosome aberrations) the possibility of a linear-nonthreshold dose-response relationship at low radiation doses cannot be excluded. The weight of epidemiological evidence, of necessity somewhat more limited, also suggests that for some types of cancer there may be no significant departure from a linear-nonthreshold relationship at low-to-intermediate doses above the dose level where statistically significant increases above background levels of radiation can be detected. The existing epidemiological data on the effects of low-level irradiation are inconclusive, however, and, in some cases, contradictory, which has prompted some observers to dispute the validity of the linear-nonthreshold dose-response model for extrapolation below the range of observations to zero dose. Although other dose-response relationships for the mutagenic and carcinogenic effects of low-level radiation cannot be excluded, no alternate dose-response relationship appears to be more plausible than the linear-nonthreshold model on the basis of present scientific knowledge.

In keeping with previous reviews by the NCRP (1980; 1993b; 1997), the Council concludes that there is no conclusive evidence on which to reject the assumption of a linear-nonthreshold dose-response relationship for many of the risks attributable to low-level ionizing radiation although additional data are needed (NCRP, 1993c). However, while many, but not all, scientific data support this assumption (NCRP, 1995), the probability of effects at very low doses such as are received from natural background (NCRP, 1987) is so small that it may never be possible to prove or disprove the validity of the linear-nonthreshold assumption.

2. Introduction

The setting of dose limits for radiation protection is presently based on the hypothesis that the mutagenic, clastogenic and carcinogenic effects of radiation are stochastic effects, the frequency of which is proportional to the radiation dose, at low (millisievert) levels of ionizing radiation exposure (ACRP, 1996; ICRP, 1991a; NCRP, 1993a; NRPB, 1995; Sinclair, 1998). Hence, although there is evidence that the magnitude of such effects may vary, depending on the LET of the radiation and dose rate of irradiation, a linear-nonthreshold dose-response model (*e.g.*, see Curve "a" in Figure 2.1) in which the dose is appropriately weighted for LET and dose rate has generally been recommended for use in estimating the risks attributable to low-level irradiation for purposes of radiation protection.

The experimental and epidemiological data on which the linear-nonthreshold model has been based have come primarily from observations at moderate-to-high levels of exposure and cannot exclude the possibility that thresholds for the mutagenic and carcinogenic effects of radiation may exist for humans in the very low (millisievert) dose domain, where quantitative data are not available. Consequently, there is a clear need to reevaluate the model periodically and to modify it, if necessary, in the light of new information.

Among the data that have prompted reexamination of the model in recent years by various national and international groups (*e.g.*, ACRP, 1996; FAS, 1995: Fry *et al.*, 1998; NRPB, 1995; OECD, 1998; UNSCEAR, 1993; 1994) is evidence that irradiation may elicit adaptive reactions in some exposed cells and organisms which can enhance their resistance to further doses of radiation (UNSCEAR, 1993). Such evidence has, in fact, been interpreted by some observers (*e.g.*, Jaworowski, 1995; Kondo, 1993; Luckey, 1991; 1994; Sugahara *et al.*, 1992) to imply that the net effects of low-level irradiation may actually be beneficial to the health of those affected ("hormesis"), although the prevailing evidence has generally been interpreted to be insufficient to support this view (*e.g.*, ACRP, 1996; NRPB, 1995; OECD, 1998; UNSCEAR, 1993; 1994; Wojcik and Shadley, 2000).

This Report reviews the extent to which existing data on the causative mechanisms and dose-response relationships for the effects of low-level ionizing radiation are, or are not, consistent with a linear-

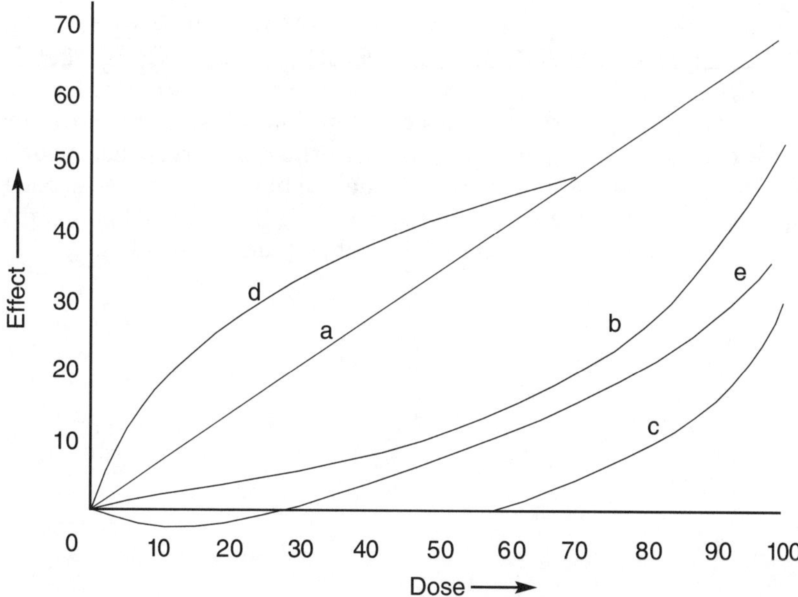

Fig. 2.1. Schematic representation of contrasting types of dose-response relationships. (a) linear-nonthreshold dose-response relationship over the entire dose range, down to zero dose; (b) linear-nonthreshold relationship only at low-to-intermediate levels of dose, above which the curve bends upward (as is characteristic of the linear-quadratic type of relationship); (c) threshold dose-response relationship, in which no effect is produced at doses below the threshold indicated on the intercept; (d) supralinear response in which the effects per unit dose at low doses exceeds that of higher doses; (e) hormetic response in which the frequency of effect is reduced at low doses and increased only at higher doses.

nonthreshold dose-response model. To this end, this Report evaluates the relevant data on the mutagenic, clastogenic and carcinogenic effects of low doses of radiation, which, as noted above, are generally classified as *stochastic* effects for purposes of radiation protection. Conversely, effects that are generally classified as deterministic (*e.g.*, teratogenic effects, impairment of fertility, and depression of immunity) are not considered herein, since effective or actual thresholds for such effects are known or presumed to exist (ICRP, 1991a).

In striving to consider relevant information, the Council solicited input from the scientific community at large, and it acknowledges with pleasure the many contributions of data and insights provided by other scientists. Owing to the vast amount of information on the effects of low-level ionizing radiation that has been published, and the fact that other in-depth reviews of the relevant dose-response relationships have appeared elsewhere (ACRP, 1996; FAS, 1995; ICRP, 1991a; NAS/NRC, 1990; NCRP, 1993a; NRPB, 1995; UNSCEAR, 1986; 1993; 1994), an exhaustive or comprehensive description of the literature was not the goal of this Report but rather a critical evaluation of the linear-nonthreshold dose-response model. The sources of all data that are cited herein have, nevertheless, been appropriately documented in the Report, and the Council has sought to leave no significant aspect of the subject unaddressed.

3. Biophysical

3.1 Energy Deposition and Its Relevance to Questions of Low-Dose Response

The initial, damaging events of ionizing radiation are qualitatively different from those of other mutagens or carcinogens. Specifically, the energy imparted, and the subsequent radiation products, such as free radicals, occur in clusters along the structural tracks of charged particles rather than in simple, uniform or random patterns.

Depending on the absorbed dose and on the type and energy of the charged particles, the resulting inhomogeneity of the microdistribution of energy deposition can be substantial. Measurements in randomly selected microscopic volumes will yield energy concentrations, or concentrations of subsequent radiation products, which deviate considerably from their average values, and these variations depend in complex ways on the sizes of the reference volumes, the magnitudes of the doses, and the types of ionizing radiations.

Although defined at each point, the absorbed dose is a macroscopic quantity because its value is unaffected by microscopic fluctuations of energy deposition. While the absorbed dose determines the average energy deposited in a specified target volume, each individual target reacts to the *actual energy*, either directly or indirectly, deposited in it rather than to the average. The relevant size of the target volume may vary in different situations; for some endpoints, it may well be that of the cell nucleus, but for others it may be smaller, or even larger than the nucleus, covering the entire cell or several cells. The characterization of energy depositions on micrometer (and smaller) scales is the field of microdosimetry (ICRU, 1983; Rossi, 1967).

3.1.1 *Track Structure*

All ionizing radiations deposit energy through ionization or excitation of the atoms and molecules in the material through which they travel. Generally speaking, most of the energy depositions are produced by secondary or higher-order electrons that are set in motion by the primary radiation, be it a photon, a neutron, or a charged particle. It is likely that the most biologically significant energy-deposition events involve ionization, whereby an electron is removed from an atom or molecule.

Because the probabilities of all the relevant interaction processes between the different radiations and the atoms and molecules of the absorbing medium can be estimated (with various degrees of realism), it is possible to simulate, on a computer, the passage of a particle (and its secondaries) as it travels through a medium (*e.g.*, Brenner and Zaider, 1984; Paretzke, 1987). A typical example is shown in Figure 3.1, which shows simulations of the passage of a variety of radiations through the periphery of a cell nucleus. Each point represents the location (projected onto two dimensions) of an ionization event, and the very localized and clustered nature of energy deposition by ionizing radiation is clearly apparent.

It is important to realize that radiation energy deposition is a stochastic process, and that no two radiation tracks will be the same. This is illustrated in Figure 3.2, which shows multiple tracks produced by protons of four different energies, each track being quite different from the others.

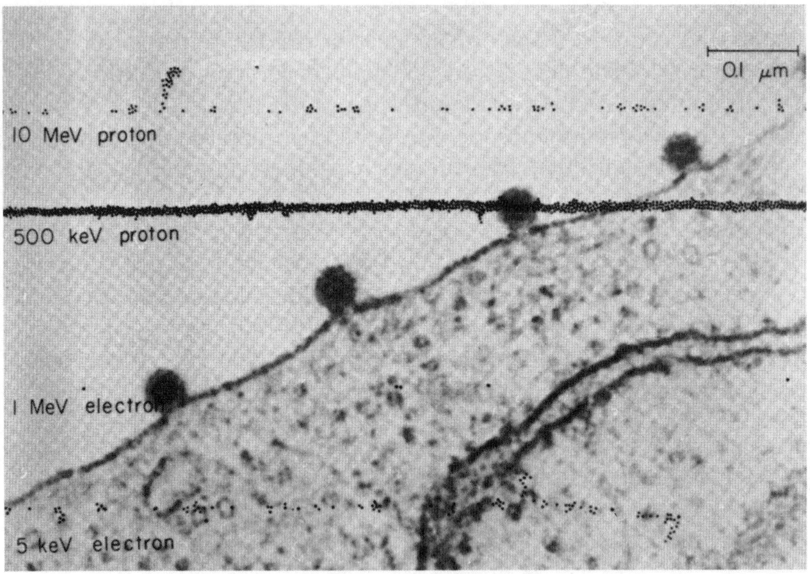

Fig. 3.1. Diagram of simulated charged-particle tracks superimposed on a micrograph of part of a mammalian cell. The viruses budding from the outer cell membrane permit an added comparison of size. In the projected track segments, which cross the figure horizontally, the dots represent ionizations. The lateral extension of the track core is somewhat enlarged in order to resolve the individual energy transfers (Kellerer, 1987).

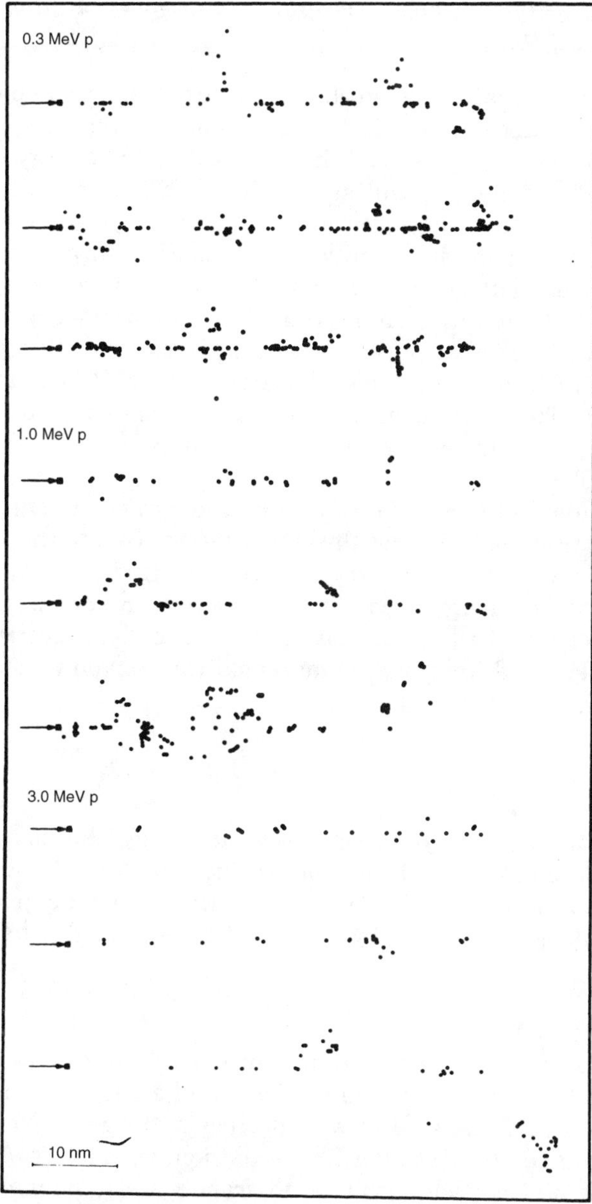

Fig. 3.2. 50 nm segments of Monte Carlo-simulated tracks of protons passing through water. The dots are the positions of individual ionizations, projected onto the x/y plane, for a particle moving in the positive x direction. Three tracks are shown for each energy, illustrating the fact that, even for the same energy, each track is quite different, because of the stochastic nature of ionizing-radiation energy deposition (Paretzke, 1987).

3.1.2 *Quantitative Characterization of Energy Deposition in Small Sites*

In order to quantify the stochastic nature of energy deposition in cellular and subcellular objects, a fundamental quantity known as the *specific energy* (z), is used. It is defined as the energy imparted to specified volumes per unit mass (ICRU, 1983), and it is measured in the same units as the absorbed dose (the average specific energy). The variation of specific energy across identical target volumes, is characterized by the distribution function $f(z;D)\mathrm{d}z$, representing the probability of depositing in a given site a specific energy between z and $z \pm \mathrm{d}z$. This distribution depends, among other things, on the dimensions of the volume under consideration and D (*i.e.*, the average value of z). The relative statistical fluctuations of z about its mean value (*i.e.*, σ_z/D) are larger for smaller volumes, smaller doses, and lower LET.

Energy deposition can be caused by the passage of one, or more than one, track of radiation through a target. Due to the relevance of single traversals to the low-dose situation, it is useful to consider the spectrum of energy depositions from single traversals, the single event spectrum [$f_1(z)$]. [Note that the dose dependent spectrum $f(z;D)$ can be calculated from $f_1(z)$ by mathematical convolution (Kellerer, 1985)]. The average of $f_1(z)$, *i.e.*:

$$z_\mathrm{F} = \int z\, f_1(z)\mathrm{d}z \tag{3.1}$$

is called the "frequency-averaged specific energy," but is simply the average specific energy deposition produced by a single traversal of a given radiation through the sensitive site. Thus, for a given D, the mean number of traversals by radiation tracks through a given target is given by:

$$n = D/z_\mathrm{F}. \tag{3.2}$$

Typical values of z_F are shown in Figure 3.3. Note that z_F increases with LET (and, indeed, z_F can be thought of as the microdosimetric correlate of LET), as well as with decreasing target site size.

Thus, a given dose of high-LET radiation, such as a dose of neutrons or alpha particles, will result from a much smaller average number of traversals than would be the case for the same dose of low-LET radiation, such as x rays. This is illustrated in Figure 3.4. Furthermore, identical cells receiving the same dose of a given type of radiation will be subject to a range of specific energy depositions [characterized by the distributions $f(z;D)$ or $f_1(z)$], because of a variety of effects such as energy straggling, track length distributions, and

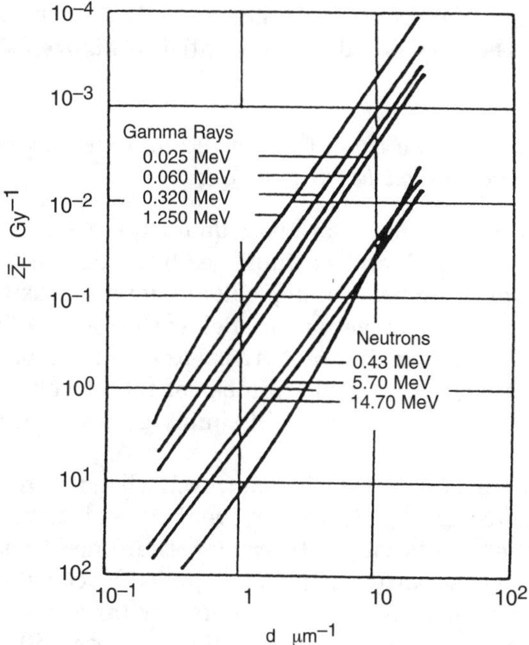

Fig. 3.3. Calculated values of the mean specific energy per event ($\bar{z}_F$) in unit density spheres of the indicated diameter (d) for gamma radiation and neutrons of different energies (ICRU, 1983).

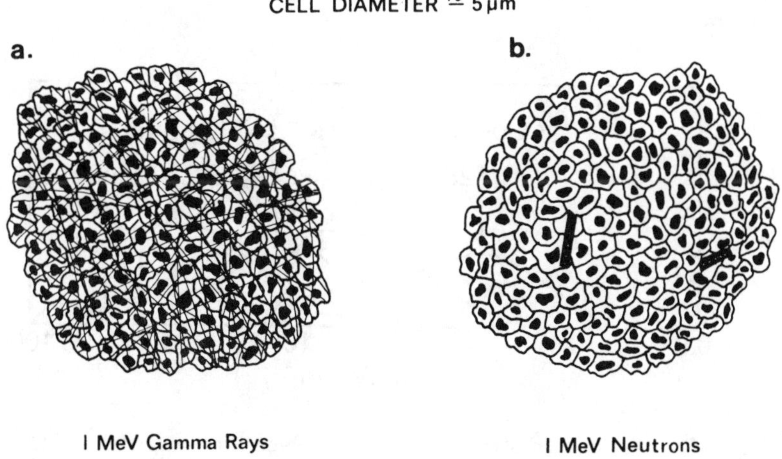

Fig. 3.4. Schematic representation of track patterns produced in about 150 cells by 10 mGy of gamma rays and 10 mGy of neutrons (Rossi, 1980).

delta ray escape (Kellerer and Chmelevsky, 1975). These distributions can often be very broad, as illustrated in Figure 3.5.

3.1.3 *Definition of Low Dose, Corresponding to an Average of One Energy Deposition Event per Target*

Based on the above considerations, a quantitative measure of what constitutes a "low dose" can be established by estimating the dose at which the average number of independent energy-deposition events experienced by a given target is one. Below this dose, effects due to the interactions between different tracks or events become progressively more infrequent, and the number of target volumes subject to this same single-event insult will simply decrease in proportion to the dose.

According to the Poisson distribution, even when the average number of independent energy deposition events in a given target is one, 26 percent of the targets will be struck more than once. Consequently, a slightly more conservative definition, applied by Goodhead (1988), corresponds to a mean number of events per target of 0.2. In this case, less than two percent of all possible targets will experience more than one event, and less than 10 percent of the hit targets will experience more than one event. This illustrates the difficulty of attempting to define a "low dose."

The dose corresponding to an average of one event per target is a measurable or calculable quantity, using microdosimetric techniques

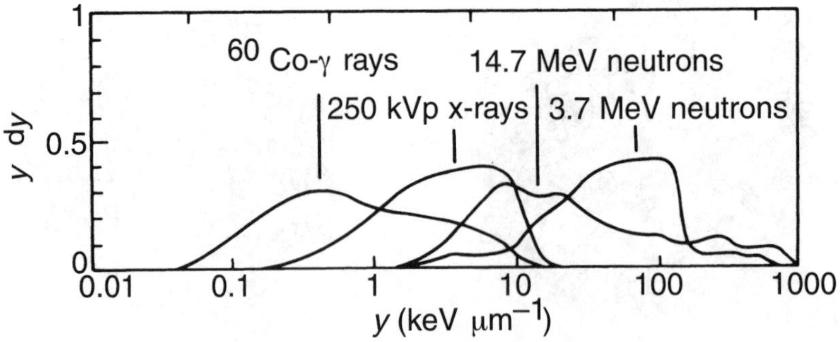

Fig. 3.5. Measured distributions in dose of specific energy (z) and lineal energy (y) for a 1 μm diameter spherical tissue region. Note the large differences in energy deposition properties of high- and low-LET radiation (Kellerer and Rossi, 1972).

(ICRU, 1983). It is the so-called "frequency-averaged specific energy per event"—the mean energy per unit mass deposited by single events in the target (z_F) (see Equation 3.1 and Figure 3.3). As discussed above, however, this quantity (and thus the definition of low dose) depends strongly on the assumed size of the sensitive target. For photons of varying energies, z_F can be estimated from measurements in different-sized targets (Kliauga and Dvorak, 1978), and corresponding measurements have been reported for different energy neutrons (Srdoc and Marino, 1996).

Appropriate target sizes for consideration are those of typical human cell nuclei (100 to 1,000 μm^3) (Altman and Katz, 1976) or, perhaps of greater relevance, the volume of nucleotides in the mammalian cell nucleus (~3 μm^3). Representative results, derived from the measured microdosimetric spectra, for spherical target volumes of 3.6 and 240 μm^3 are given in Table 3.1. Also shown are corresponding estimates for a much larger target (5,500 μm^3), designed to simulate a target consisting of a cluster of cells, each of which is able to communicate damage to other cells, thus possibly comprising a large effective target. Results are given for ^{60}Co gamma rays (1.25 MeV) and x rays (25 kVp, typical of those used in mammography). The data in Table 3.1 can be compared with the frequency of energy-deposition events in target cells due to natural background radiation. For example, an average background dose of ~0.1 mGy per month of sparsely ionizing (low-LET) radiation would result in about 1 in 10 target nuclei being subjected to an energy deposition event. Similarly, based on the results of the BEIR VI report of the Committee on the Biological Effects of Ionizing Radiations

TABLE 3.1—*Possible definitions of "low dose": The dose below which the average number of energy deposition events occurring within autonomous targets of diameter (d) is less than unity.*[a,b]

	Target Volume		
Radiation Type	3.6 μm^3 (d = 1.9 μm) (nucleotides)	240 μm^3 (d = 7.7 μm) (nucleus)	5,500 μm^3 (d = 22 μm) (cluster of cells)
1.25 MeV gamma rays	15 mGy	0.9 mGy	0.1 mGy
25 kVp x rays	100 mGy	4.5 mGy	0.5 mGy

[a]In order to apply a more conservative definition of "low dose," corresponding to <0.2 traversals per spherical target, the doses tabulated should be divided by five.

[b]For the purposes of this Report, low doses are in the range 0 to 10 mSv, moderate doses from >10 to 100 mSv, and high doses >100 mSv.

(NAS/NRC, 1999) about 1 in 2,500 target cell nuclei in the bronchial epithelium would be traversed by an alpha particle each month in an individual living in an "average" radon home.

3.2 Implications of Energy-Deposition Patterns for Independent Cellular Effects at Low Doses

General correlations have been found between the detailed spatial and temporal properties of the initial physical features of radiation energy deposition and the likelihood of final biological consequences (Brenner and Ward, 1992; Goodhead, 1994; Goodhead and Brenner, 1983). These correlations persist despite the sequence of physical, chemical and biological events that process the initial damage. Details of the initial energy-deposition conditions provide insight into the critical features of the most relevant early biological damage and subsequent repair. Ionizing radiations produce many different possible clusters of spatially adjacent damage, and analysis of track structures from different types of radiation has shown that clustered DNA damage of severity at least comparable to, if not greater than simple dsbs (see Section 4) can be assumed to occur at biologically relevant frequencies with *all* ionizing radiations, and at any dose (Brenner and Ward, 1992; Goodhead, 1994). In other words, such clustered damage can be expected to be produced by a *single track* of ionizing radiation, with a probability that increases as the ionization density increases, but is non-zero even for x and gamma rays.

One general conclusion which follows from the stochastic nature of ionizing radiation energy deposition in small sites is that, for small absorbed doses (defined, see above, as when the average number of energy deposition events in the target is appreciably less than one), the average effect on independent targets is always proportional to dose. (Note here that, as discussed earlier, what constitutes an "independent target" may vary with endpoints. The targets could be cells, nuclei, substructures of nuclei, or even clusters of cells, as illustrated in Table 3.1). Such a linear relation between observed effect and dose must be expected at low doses regardless of the dependence of cellular effect on specific energy. This linear relation is due to the fact that, even at very low doses, finite amounts of energy are deposited in a target when this target is struck by a charged particle. The energy deposited in such single events does not depend on the dose, and so the effect in those targets which are traversed by a charged particle does not change with decreasing dose. At low doses, the only change which occurs with decreasing dose is a decrease in the proportion of targets which are subject to a single energy deposition. This can be treated quantitatively (*e.g.*, Goodhead, 1988; Kellerer and Rossi, 1975), and, as discussed above, microdosimetry can supply information regarding the range of doses in which the statement applies for different radiation qualities and for different target sizes. A schematic illustration of these concepts is given in Figure 3.6.

A possible objection to this conclusion would be that the effects of a single energy deposition event in the appropriate target might be zero, while a positive effect might be produced after multiple energy depositions. However, this hypothesis appears inconsistent with both microdosimetric and biologic evidence: First microdosimetrically, both for sparsely-ionizing and densely-ionizing radiation there is a broad distribution of the spectrum of specific energy produced in single events (see Figure 3.5). Consequently, there is always a finite probability, although it may be extremely small, that the same amount of energy deposited in two events can also be deposited in one event. In fact, recent track structure calculations have demonstrated that a single low-energy electron from an x-ray or photon interaction can produce double-strand DNA breaks and more complex clustered

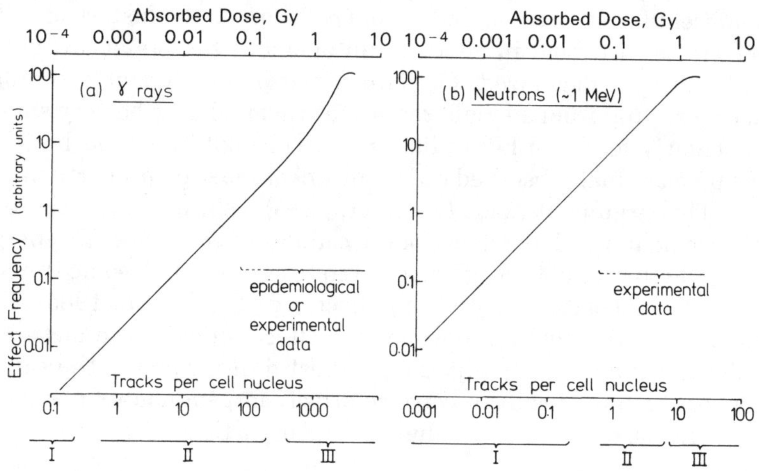

Fig. 3.6. Schematic dose-response curves for low- and high-LET radiations. The mean number of tracks correspond to 8 μm diameter spherical nuclei. Region I corresponds to "definite" single-track action on individual cells, corresponding to less than 0.2 events per cell nucleus. Region II corresponds to low doses, where single-track action on individual cells may still dominate. Region III corresponds to the region where multi-event actions will dominate (Goodhead, 1988). The brackets indicate the range of doses/tracks where epidemiological or radiobiological data are available.

DNA damage (Brenner and Ward, 1992; Goodhead, 1994). Second, biologically, there is now clear evidence from the new generation of single-particle microbeams that, at least at high-LET, traversal of a cellular nucleus by a single radiation track may produce measurable biological effects (Hei *et al.*, 1997).

As an example, let us suppose that a target cell (such as a cycling stem cell) needed to receive severe damage before it would be a candidate to be the initial clone ultimately leading to a cancer. For the sake of argument, let us assume that the severe event would be the induction of four DNA dsbs in a stem cell (the probability of which is about 7×10^{-4} at 10 mGy of x rays). Then even at this low whole-body dose of x rays, millions of target cells would be expected to receive such damage, and would thus be candidates for the subsequent stochastic processes leading to a cancer. In other words, because of the large number of relevant cellular targets in the body, no matter how unlikely the initial event at very low doses there will always be a finite number of relevant target cells subject to that initial event. This represents the initial conditions for the series of stochastic processes leading, with a finite probability proportional to the number of initially-damaged cells, to cancer induction.

These arguments suggest that in the action of ionizing radiation on subcellular structures, individual cells, or small clusters of cells, there is no threshold in effect on autonomous targets at low doses as the dose is decreased. Of course, the probability of a cellular effect resulting from a single radiation traversal may be very small, particularly for low-LET radiation, but ultimately in the limiting case of very small absorbed doses the effect must be proportional to dose. This argument holds (Kellerer, 1985) whether or not there is a threshold in the dependence of the cellular effect on specific energy z (*i.e.*, whether or not any given cell requires more than some threshold energy deposition to show an effect, *e.g.*, Bond *et al.*, 1985). The absence of a threshold in dose for the effect is due to the fact that, as discussed above, even at the smallest doses, some of the target cells, or clusters of cells, receive relatively large amounts of energy when they are struck by a single charged particle.

3.3 Implications of Energy-Deposition Patterns for Carcinogenic Effects of Radiation

The microdosimetric arguments outlined above apply: (1) to situations when the dose is sufficiently low that multiple energy deposition events in the appropriate target are rare and (2) to endpoints which result directly from the effects of energy deposition in single targets. Thus, for example, application of this argument to the endpoint of cellular mitotic death is likely to be appropriate, in that this type of killing of a given cell does appear to be largely independent of effects on other cells. Even in this simple situation, however, the appropriate target size is not clearly understood, and thus the dose below which linearity would hold is not clear, though the absence of a threshold in dose would follow independent of the target size.

Application of this argument to complex endpoints such as radiation-induced carcinogenesis is, however, more uncertain. Based on these biophysical considerations about the shape of the dose-response relation for low-dose radiation-induced carcinogenesis, conclusions can be drawn if: (1) radiogenic cancer induction is causally related to radiation-induced damage in a single cell and (2) the ways in which other cells or cell systems subsequently modify the probability that any given initially radiation-damaged cell becomes the clonal origin of a cancer do not vary with dose in a nonlinear fashion.

3.3.1 *Evidence Regarding the Clonality of Tumors*

The use of molecular-genetic approaches to the study of the monoclonality of tumors has strongly complemented the traditional approaches to this question, and the evidence that the great majority of cancers are of monoclonal origin is increasingly strong (Wainscoat and Fey, 1990; Worsham *et al.*, 1996). The main approaches to the question of human tumor clonality have been either through analyses of somatic mutations or X-chromosome inactivation in solid tumors, both of which we briefly discuss, although other approaches that have been used to establish clonality in hemopoietic neoplasms (Arnold *et al.*, 1983; Cleary *et al.*, 1988; Levy *et al.*, 1977; Minden *et al.*, 1985) have reached similar conclusions.

In studying somatic mutations, many human tumors have been shown cytogenetically to have consistent, nonrandom chromosomal aberrations, such as the classic Philadelphia chromosome in chronic myeloid leukemia (Heim and Mitelman, 1995; Nowell and Hungerford, 1960). In fact, even when other chromosomal aberrations vary, consistent marker chromosomes often indicate the presence of subclones, rather than independent clones (Nowell, 1976). More recently, molecular genetic techniques have been used to assess clonality through assessment of loss of heterozygosity at specific chromosomal loci (*e.g.*, Abeln *et al.*, 1994; Jacobs *et al.*, 1992; Miyao, 1993), through *p53* mutational analysis (*e.g.*, Jacobs *et al.*, 1992;

Kupryjanczyk *et al.*, 1996), or through DNA "fingerprinting" (Fey *et al.*, 1988).

Inactivation or methylation patterns of X-chromosome genes can be used for the detection of clonality of tumors in females who are heterozygous for a specific X-chromosome linked polymorphism. Much of the early work was done by Fialkow (1976; 1984) and colleagues, with the *G6PD* isoenzyme system (*e.g.*, Fialkow, 1976; 1984), who showed that the great majority of tumors examined were monoclonal, including carcinoma of the breast, colon, uterine cervix, ovaries, and many hematological cancers. This approach is, of course, limited in that it is restricted to women who are heterozygous for the gene for *G6PD*, and by the 1980s, the RFLP technique had been developed to examine differential methylation of various X-linked genes (Vogelstein *et al.*, 1984; 1987). These methods were applicable to about 50 percent of women. More recently, techniques based on the polymerase chain reaction (PCR) have been applied to the problem (*e.g.*, Gilliland *et al.*, 1991). For example, Noguchi *et al.* (1992) used PCR to show that DNA samples from widely separated sites with breast tumors exhibited inactivation of the same X chromosomes in each tumor from the same individual, which is strongly suggestive of a monoclonal origin.

Of course, DNA-based methods for determining clonality of tumors rely on the assumption that a cell population is homogeneous with respect to a particular marker. By the time a tumor is detected, however, it is very likely to have undergone extensive genetic changes, and so selection of subclones might have occurred, and assessment of clonality at that time might not then reflect the earliest events in tumorigenesis. A tumor might have originated, for example, from many cells, with the progeny of one of these cells (bearing the marker) eventually having outgrown all the others (Alexander, 1985).

3.3.2 *Relationship Between Initially-Damaged Cells and Tumorigenic Cells*

There is much less evidence regarding the question of whether cells or cell systems can modify the probability that any given initially radiation-damaged cell becomes the monoclonal origin of a cancer, in a manner which is nonlinear with dose. That radiation-damaged cells can be inactivated is clear from, for example, studies of apoptotic responses (Schwartz *et al.*, 1995). Presently we do not have sufficient data to distinguish whether these modifying effects are linear or nonlinear with dose. Nonlinearity could occur, for example, if a small number of damaged cells were to be inactivated with greater efficiency per cell than a larger number of damaged cells.

The discussion given above has tacitly assumed that a cell which is the origin of a radiation-induced monoclonal tumor contains damage directly produced by energy deposition in that cell. This may not be the case in that the original tumor cell may be the progeny of one exposed to an energy deposition (delayed instability), or may have been adjacent to a target that received an initial energy deposition (bystander effect).

If delayed instability is indeed an important mechanism in radiation carcinogenesis, then the relationship between the yield of initially damaged cells and that of subsequent unstable cells needs to be assessed. The current, fairly limited, evidence suggests that this relationship is linear at doses below ~1 Gy (*e.g.*, Limoli *et al.*, 1999). Similarly, if the original tumor cell was often an initially unirradiated "bystander" cell, it would be important to assess the relationship between the yield of irradiated damaged cells and nonirradiated damaged cells.

3.4 Conclusions

Microdosimetric considerations regarding the structure of energy deposition by different radiations can give some insight into the conditions under which low-dose linearity would be likely.

Specifically, it is possible (see Table 3.1) to define, for particular radiation types and particular assumed target site sizes, the doses below which multiple energy deposition events in the given targets would be rare. At such low doses, linearity of dose response would be expected if the effect were produced autonomously in individual targets, *i.e.*, independently of each other. These targets could be subnuclear structures, cell nuclei, cells, or even clusters of cells.

With regard to radiation-induced oncogenesis, the evidence for a monoclonal (single cell) origin for most cancers is highly convincing which, given the above considerations, may be considered to constitute a *prima facie* argument in support of low-dose linearity. However, the appropriate target size or sizes for the initial radiation-induced damage in the cell is not known, so the dose below which linearity would be expected is also unknown. Nevertheless, these arguments imply that the expected dose response is linear without threshold at low-to-intermediate doses. Yet, it is conceivable that other cells or cell systems subsequently modify the probability that any given initially radiation-damaged cell becomes the clonal origin of a cancer, in a manner which is nonlinear with dose, although, no conclusive evidence that such processes do or do not occur is currently available.

3.5 Research Needs

The central issues discussed in this Section relate to the biological effects produced by the traversal of single cells or cell nuclei by single radiation tracks. Consequently, the primary research needs in this area include the development of:

1. techniques for irradiating single cells by single tracks of radiation (although microbeams are beginning to become available, they require further development, in particular in terms of extending their spatial precision and their LET range to both higher and lower values);
2. sufficiently sensitive cell assay systems for quantitatively detecting low levels of biological damage in single cells, particularly for endpoints relevant to oncogenic transformation in human cells; and
3. techniques for determining whether and/or how the effects of damage to cells surrounding a radiation-altered (or otherwise initiated) cell may modify the dose response.

4. Deoxyribonucleic Acid Repair and Processing after Low Doses and Low-Dose Rates of Ionizing Radiation

Exposure to ionizing radiation induces damage of various kinds into the genetic material DNA of all organisms, including humans. These broad categories of damage are ssbs, dsbs, bds, dpcs, and mds (Figure 4.1). Several processes have evolved that counteract these DNA damages. The best known of these and easiest to appreciate are those that repair the damaged DNA *in situ*. However, other processes aid in abolishing the deleterious effects of DNA damage. One is radiation-induced apoptosis; cells with damage that could ultimately result in mutation are triggered to undergo programmed death, thereby removing them from the tissue in which they might otherwise become transformed to a precancerous state. Another process (or set of processes) causes irradiated cells to pause at one or other of the cell-cycle checkpoints ($G_1 \rightarrow S$, in S, $G_2 \rightarrow M$), allowing more time for the cells to repair damage or placing them in permanent arrest. Most mature animal cells are in G_0 and must enter the mytotic cycle to become a risk for health effects. If these cells are damaged and recruited back into cycle, they would add to the potentially carcinogenic pool of cells. The role of these processes in mutagenesis and carcinogenesis after low dose and low-dose-rate irradiation is discussed below.

4.1 Ionizing Radiation-Induced Deoxyribonucleic Acid Lesions and Their Repair

4.1.1 *Single-Strand Breaks (Including Deoxyribose Damage)*

There is about one ssb induced per cell per milligray of low-LET radiation (Ward, 1988). Two-thirds or more of the total ionizing

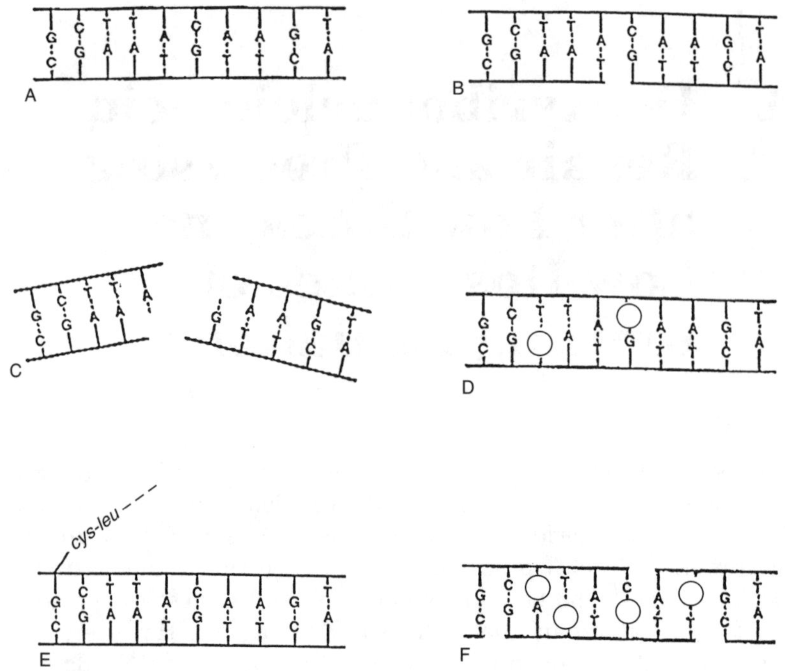

Fig. 4.1. Illustrations of the various categories of DNA damage. (A) Two-dimensional representation of the normal DNA double helix. The base pairs carrying the genetic code are complementary (*i.e.*, adenine pairs with thymine, guanine pairs with cytosine). (B) Single-strand breaks. Strand breaks are accompanied by the loss of a base. (C) Double-strand breaks. (D) Base damages. (E) DNA-protein cross-links. The "cys-leu—" indicates the first amino acid, cysteine and leucine, in a protein covalently linked to the sugar phosphate backbone of DNA. (F) Multiply-damaged sites, *i.e.*, any combination of other damage categories in a local region of the DNA (Hall, 1994).

radiation-induced DNA damage in cellular DNA is caused by indirect action (radiation-induced water radicals migrating to the DNA) (Roots and Okada, 1975) and one of the main targets of these radicals is the sugar moiety of DNA (Schulte-Frohlinde and von Sonntag, 1990). Consequently, most ssbs are formed by a series of reactions following the initial lesion in deoxyribose; base loss accompanies ssbs formation. The resulting breaks are rarely, if ever, of the simple 3′OH-5′PO_4 type that can be sealed in one step by ligase; "dirty" end groups, such as glycolate, are formed and must be enzymatically modified before base insertion and final sealing of the breaks can occur. All cells repair ssbs rapidly and completely, with half times of 3 to 5 min, and this repair is almost error-free (Ward *et al.*, 1985).

4.1.2 *Base Damage and Loss*

All four bases in cellular DNA are chemically modified by ionizing radiation; most of this damage is oxidative. The yield is slightly higher than that for ssbs. Undamaged bases are also released, generally accompanying ssbs formation. Damaged bases are repaired rapidly by redundant enzymatically catalyzed reactions that first remove the damaged base, incise the DNA on one or the other side of the apurinic or apyrimidinic site, excise the deoxyribose phosphate, fill in the resulting gap and ligate (Demple and Harrison, 1994; Figure 4.2). In most cells, this repair occurs at two levels: the

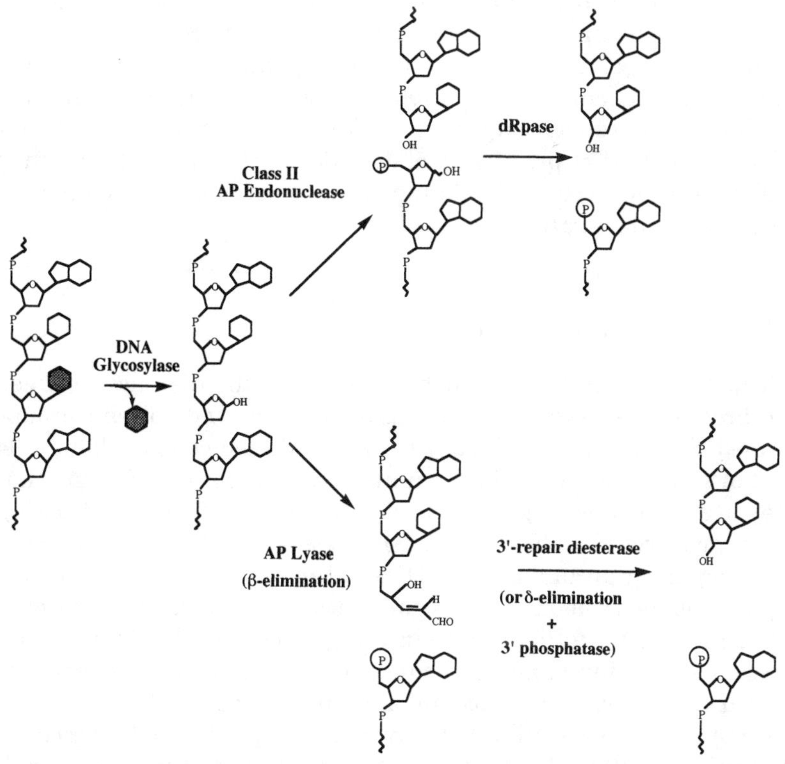

Fig. 4.2. Base excision repair pathways. In both pathways a glycosylase excises the damaged base, leaving an abasic site. The upper pathway uses an AP endonuclease that allows deoxyribophosphodiesterase to remove the sugar phosphate; after the patch is filled ligase can act directly on the consequent substrate without further modification. The lower pathway leaves a substrate that requires further processing of the 3′ end group before ligase can act (Demple and Harrison, 1994).

transcribed strands of expressed genes are rapidly repaired [transcription-coupled repair (TCR)], and the rest of the genome is repaired more slowly (global repair). The probable importance of TCR is implied by the fact that humans with the genetic disease, Cockayne's syndrome, are deficient in TCR of oxidative bds and are hypersensitive to ionizing radiation (Cooper *et al.*, 1997), although they are not at increased risk for cancer. The repair of bds in normal cells is considered to be very accurate.

4.1.3 *Deoxyribonucleic Acid-Protein Cross-Links*

Dpcs are formed in much lower yields than are other radiation-induced lesions and are poorly characterized. Nuclear matrix proteins seem to be the major proteins that are induced to bind to DNA by radiation (Oleinick *et al.*, 1990). The exact chemical nature and mechanisms of repair of dpcs are not known, but their relatively low yields indicate that processing of dpcs does not cause a deviation of the dose-response curve for mutagenesis or carcinogenesis after low doses or low-dose rates.

4.1.4 *Double-Strand Breaks*

There are many kinds of dsbs, varying in the distance between the breaks on the two DNA strands and the kinds of end groups formed (Painter, 1981). Their yield in irradiated cells is about 0.04 that of ssbs and they are induced linearly with dose, indicating that they are formed by single tracks of ionizing radiation (Ward, 1990). Dsbs can be repaired by two basic processes: homologous recombination, requiring an undamaged DNA strand as a participant in the repair and nonhomologous recombination, which actually is end-to-end rejoining following "trimming" of the ends of the break (Figure 4.3). Homologous recombination, an accurate process, appears to be relatively uncommon in mammalian cells, and is carried out by proteins similar to the *rad 51* gene product of *S. cerevisiae* (Petrini *et al.*, 1997). It should be noted that dsbs do stimulate homologous recombination, which can, *via* gene conversion, cause loss of heterozygosity. Therefore, even this "accurate" process can cause cells that are heterozygous for a defective DNA processing gene to produce progeny that are homozygous for that gene. Nonhomologous recombination is relatively inaccurate and probably accounts for many of the premutagenic lesions induced in the DNA of human cells by ionizing radiation. DNA-dependent protein kinase and the

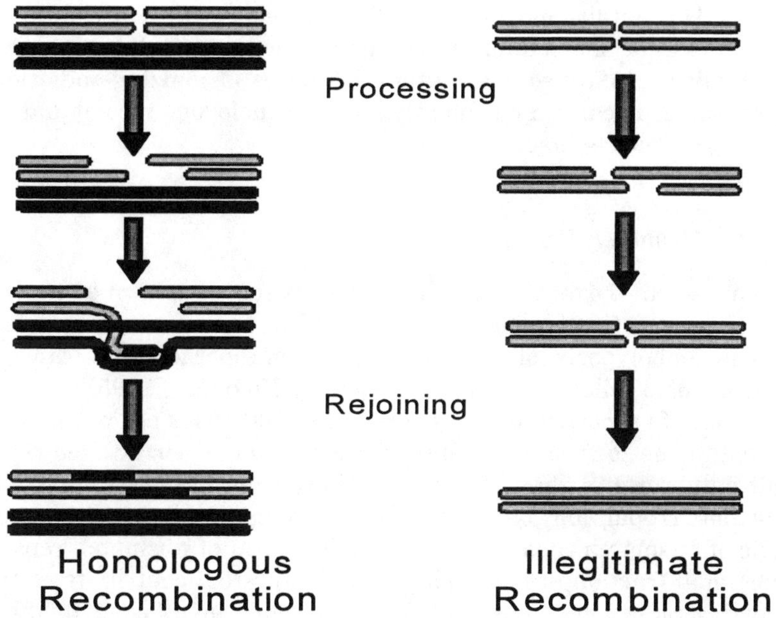

Fig. 4.3. Dsbs repair *via* homologous and nonhomologous recombination. In homologous recombination the exposed 3′ end invades the homologous duplex (usually the sister chromatid), so that the complementary strand acts as a template for gap filling. The breakage of the other strand and subsequent exchanges are not shown. In illegitimate recombination no template exists to guide gap filling (Petrini *et al.*, 1997).

Ku proteins participate in this repair process (Jeggo *et al.*, 1995). A protein complex, which includes hMre11 and hRad50 (homologues to proteins involved in the illegitimate repair of dsbs in *S. cerevisiae*) and p95, the product of the *NBS1* gene [also called nibrin (Varon *et al.*, 1998)], is also involved in the repair of dsbs in human cells (Carney *et al.*, 1998).

4.1.5 *Multiply-Damaged Sites*

Mds are complex lesions wherein ssbs or dsbs lie in close apposition to other (base and/or sugar) DNA damage; they have not yet been isolated or chemically characterized. They are caused by single tracks, and theoretical models suggest that they are induced by two to five radicals in a few base-pair stretch of DNA (Brenner and Ward, 1995). Ultimately, mds are complicated lesions that are difficult substrates for the cellular repair machinery to attack (see review by Wallace, 1998), so that, after high doses of ionizing radiation, their repair is carried out mainly by nonhomologous recombination and is, therefore, inaccurate.

4.1.6 *Mismatch Repair*

This repair system corrects base pair mismatches formed during semi-conservative synthesis of DNA and has been found to be defective in nonpolyposis colorectal cancer, in some sporadic colon cancers and in some other cancers (see review by Kolodner, 1995). Leadon *et al.* (1996) reported long-patch repair (>100 bases per patch compared to one to three for standard bds repair) in x-irradiated cells, but only those at the G_1/S border. This raises the possibility that mismatch repair may be involved because long patches are characteristic of mismatch repair. Even if this is a form of mismatch repair, the small fraction of bds repair that this system contributes after ionizing radiation suggests that it is probably insufficient to affect significantly the shape of the dose-response curve at low doses and low-dose rates.

4.1.7 *Effects of Linear-Energy Transfer*

All kinds of DNA damage are formed linearly in cellular DNA with a dose in the low-dose range (<1 mGy). However, the damage is induced so sparsely that there are no significant interactions between adjacent lesions or between the repair of these lesions. As the LET of radiation increases, more energy is dissipated along individual tracks so that (1) within a track, interaction of damaged sites increases, which results in higher frequencies of dsbs and mds along the track; and (2) between tracks, the probability of interaction decreases.

4.1.8 *Spontaneous Deoxyribonucleic Acid Damage*

It is probable that the amount of "spontaneous" damage to DNA is large compared to that from low doses of ionizing radiation, but the estimates for the extent of this damage vary widely. For instance, Beckman and Ames (1997) estimated that the steady-state level of oxidative bds alterations per human cell is about 150,000, but more recently the same group lowered this value to 24,000 (Helbock *et al.*, 1998). Even this represents a large burden, equivalent to the damage induced by about 2.5 Gy h^{-1} of ionizing radiation, assuming a mean repair time of 1 h.

Exposure of mammalian cells to hydrogen peroxide at 0 °C produces large yields of ssbs (Ward, 1995) and bds (Blakely *et al.*, 1990). However, no cell killing (Ward *et al.*, 1985) or mutations (Bradley and Erickson, 1981) are induced, which suggests that only accurately repaired, single-strand DNA damage is induced by this treatment. Ward (1995) conjectures that spontaneous DNA damage also consists almost exclusively of single strand damage. Assuming this is so, the formation of dsbs by adjacent ssbs in unirradiated cells must be extremely rare, given the vanishingly low probability of two such events occurring so close together within the 5 min, or less, required for repair of ssbs. Dsbs occur in unirradiated cells but they are formed and sealed enzymatically as part of normal processes such as DNA replication, crossing over, and antibody production. Dsbs induced by ionizing radiation, in contrast, have unusual end groups, often accompanied by bds and sugar damage, and are formed randomly in the genome in all exposed cells (Ward, 1990). Thus, the dsbs (and mds) induced by ionizing radiation (and some radiomimetic chemicals) are peculiarly difficult substrates for the cell to cope with and are thought by most radiobiologists to be the lesions that endow ionizing radiation with its uniquely toxic effects (Ward, 1995).

4.2 Cell-Cycle Checkpoints

After many kinds of insults to the cellular machinery, including ionizing radiation-induced DNA damage, cell progression is delayed at several points in the cell cycle. Cells with abnormal *p53*, a protein kinase that regulates transcription, fail to stop at the $G_1 \rightarrow S$ boundary after irradiation, whereas cells with normal *p53* often delay there (Bates and Vousden, 1996). (Other transcription factors are also involved.) This delay in G_1 presumably allows cells more time for repair of DNA before replication begins. When cells in S phase are irradiated, DNA synthesis is inhibited in a dose-dependent manner. This is at least partially mediated by a protein kinase gene, *ATM* (Enoch and Norbury, 1995). Cells from ataxia-telangiectasia (AT) patients have deficient *ATM*, and DNA synthesis is inhibited much less in irradiated AT cells than in normal cells. Thus, an apparent function of *ATM* is to shut down ongoing DNA replication and allow damage to be repaired before DNA replication is resumed. For the $G_2 \rightarrow M$ block (mitotic delay), irradiated cells move more slowly into mitosis than do unirradiated cells; presumably this delay allows more time for repair and thereby reduces the frequency of chromosomal aberrations. Again, *p53* has an important role (Bates and Vousden, 1996); the Ras gene product and other regulatory proteins are also involved (Blank *et al.*, 1997).

4.3 Programmed Cell Death (Apoptosis)

Apoptosis, a process of programmed death that has evolved to remove cells that are unnecessary or potentially dangerous, occurs in many kinds of cells that have been exposed to ionizing radiation. Apoptosis is closely linked to cell proliferation. Often the presence or absence of a single gene product determines whether a cell is blocked at a cell-cycle checkpoint, dies *via* apoptosis, or continues unabated through the cell cycle. Gene *p53*, "the guardian of the genome," is intimately involved in the initiation of apoptosis, as are *Rb*, another tumor suppressor gene, the proto-oncogene Myc, and Bcl-2 and its family members (Blank *et al.*, 1997). Once initiated, the processes that bring about the final degradation of cellular macromolecules are mediated by caspases, which are specific proteinases that exist in an inactive form in the cell until an apoptotic signal initiates their activation (Nicholson and Thornberry, 1997).

4.4 Impact of Cell-Cycle Checkpoints and Apoptosis on the Dose Response for Deoxyribonucleic Acid Repair at Low-Dose Rates

Although there is no evidence to suggest that cell-cycle checkpoints or apoptosis will affect the dose response after low doses of ionizing radiation delivered acutely, the situation may be different at low-dose rates. For instance, at one low-dose rate a single hit in a cell's DNA may induce a delay in G_1 that prevents the cell from entering S phase [there are data that suggest a single dsb can induce G_1 arrest (Huang *et al.*, 1996)], so that the next hit occurs while the cell is still in G_1. At a lower dose rate, however, the G_1 arrest may be dissipated and when the second hit occurs the cell may be S phase. The result, in terms of repair, subsequent mutation and cell transformation, may be different for the two situations. Similarly, for induction of apoptosis, two different dose rates should be considered.

At the higher dose rate, two hits will be close enough together to signal for an apoptotic response, but at the lower dose rate they will not. Ohnishi *et al.* (1999) reported that even more complicated events may occur during low-dose-rate irradiation. They found that the level of *p53* rose as usual during the first 10 h of chronic irradiation. However, the *p53* level then fell to lower than normal for the next 40 h; even a large acute radiation dose failed to increase *p53* levels during this time. Such an abrogation of the *p53* response in cells that are normally *p53* responsive suggests that the pathways leading to cell-cycle delays and/or apoptosis may become inactive after a few hours exposure to low-dose-rate irradiation. Only continued research on the dose requirements for cell-cycle delays and apoptosis will reveal if they can affect the dose response at low-dose rates.

4.5 The Adaptive Response

When cells are first exposed to a low dose of ionizing radiation and later to a higher dose, the measured damage after the second dose is sometimes less than if the cells were exposed to the higher dose alone. This adaptive response was first reported for ionizing radiation-induced chromosomal aberrations in human lymphocytes by Olivieri *et al.* (1984), and many papers have since appeared, measuring endpoints from bds (Le *et al.*, 1998) to carcinogenesis (reviewed in UNSCEAR, 1994). This response indicates an ameliorating effect of low doses and suggests the induction of a repair system. However, the adaptive response, by its very nature, requires a first dose whose effects are often unknown. At least 5 mGy is necessary to evoke the protective effect of the first dose in lymphocytes (Shadley and Wiencke, 1989). This dose, while not causing a statistically significant increase in chromosomal aberrations among the usual 100 to 200 cells scored for cytogenetic studies, does induce DNA damage in every exposed cell, and many cells contain a dsb or md. When radiation is delivered acutely there can be an effect of the adaptive response on the dose-response curve only if the cell can induce adaptation soon enough to enhance the repair of the damage which a single dose induces. However, there is a much better possibility for an effect at low-dose rates. When the time between successive ionizing events is less than the time that the protective effects of the first dose persist (and if a single track can induce adaptation), the measured damage would be expected to be less severe than that caused by a single dose. On the other hand, when the time between successive ionizing events exceeds the time for protection induced by the first dose, the dose response should revert to that shown for single doses. These kinds of effects have not been observed for carcinogenesis, but it is possible that the dose rates used thus far have been inappropriate for their demonstration. In conditions where the dose rate varies, as is probably the case for many workers in the radiation industries, the effects of the adaptive response could be very complex.

4.6 Summary

It is impossible to know whether DNA damage-modifying processes affect the shapes of dose-response curves in the millisievert dose range, because there are insufficient data for DNA processing at very low doses, and results from higher doses may not always be representative of what happens at low doses. For example, after high doses of ionizing radiation many dsbs are repaired by the inaccurate nonhomologous recombination pathway. However, because homologous recombination does occur in mammalian cells, it could be argued that this accurate pathway is the principal or only one used for repair of dsbs after low doses. If so, the repair of dsbs and mds would be very accurate, and a threshold could exist. Mutagenesis studies, reviewed in the next section, suggest that inaccurate repair does occur after low doses and low-dose rates. Indeed, recent studies suggest that radiation damage that occurs only in the cytoplasm can result in nuclear DNA lesions (the "bystander effect"); this phenomenon can occur either *via* diffusion of reactive oxygen species from cytoplasm to nucleus, or from one cell to another *via* gap junctions or the intercellular milieu [reviewed by Grosovsky (1999), also see Nagasawa and Little (1999)]. Therefore it seems probable that exposure of humans to very low doses and low-dose rates of ionizing radiation does result in permanent alterations in DNA sequences. Many steps are necessary between these events and acquisition of the cancerous state. However it is not known if this progression occurs differently after low-dose irradiation than it does after higher, known carcinogenic doses.

4.7 Research Needs

The main unresolved problems for determining if DNA processing at low doses and very low-dose-rates affects the LNT hypothesis revolve around whether DNA is repaired completely and accurately under these conditions. If all DNA damage is repaired accurately after low doses, a threshold would be assured. In the last few years, very sensitive techniques for detecting DNA damage and repair in human cells have been developed; for example, the "comet assay," for ssbs (Malyapa *et al.*, 1998; Singh *et al.*, 1995) and an ultrasensitive method for assessing bds (Le *et al.*, 1998); however, none of these new methods can measure the fidelity of repair.

It is most important to find out about the repair of dsbs and mds after very low doses. Because these lesions are repaired primarily by the inaccurate nonhomologous recombination pathway, they are the main candidates for the deleterious effects of ionizing radiation. And yet it is clear that accurate nonhomologous recombination does occur in somatic cells; such recombination occurs frequently between sister chromatids in S and G_2 cells, although recombination between homologous chromosomes is relatively rare. Technologies for measuring dsb repair have become more and more sensitive; they should be used with cells that are mutant in homologous and/or nonhomologous recombination pathways to generate data on the mechanisms of dsb repair after low radiation doses.

Experimental techniques should also be developed to determine the effects of low doses on apoptosis, cell-cycle delays, and the adaptive response; the few data available now are insufficient to generalize to scenarios that are important for carcinogenesis.

5. Mutagenesis

5.1 Introduction

Ionizing radiation induces virtually all classes of mutation in exposed cells (NAS/NRC, 1990; UNSCEAR, 1988). Point mutations consist of base pair substitutions and small insertions or deletions of one to several base pairs. These are generally produced by radiation less frequently than other types of mutants, but nevertheless they are induced to statistically significant levels. Larger deletions of hundreds to millions of base pairs also are induced readily by ionizing radiation. Finally, genetic recombination events between homologous chromosomes also can be induced. These disparate classes of mutation are not likely to arise by a single mechanism. Theoretically, a mutation may arise from one "hit," or energy depositing event in the genetic target (producing, for example, a single DNA lesion); such a mutation would be expected to exhibit a linear dose response without a threshold. Alternatively, a mutation may arise from two or more "hits" (*e.g.*, a dsb or other form of multiple damage). In this case, a mutation still may exhibit a linear dose response if a single track or particle induces the damage. At higher doses, a quadratic dose dependence would be expected when damage from two discrete tracks or particles interact. Consequently, it is reasonable to postulate that there will be different dose-response relationships among the different mutational classes. If so, the overall dose-response curve will equal the sum of the components, and the relative frequencies of each will dictate the final shape.

5.2 Potential Mechanisms of Mutagenesis

5.2.1 *Replication Errors*

One mechanism by which point mutations may arise is from damaged bases that are misreplicated. Replication errors can arise from a single damaged base. Unless there are saturable levels of radical scavengers or of DNA repair pathways for bds, this sort of DNA damage should form as a linear function of dose, and thus any mutations resulting in this fashion would also arise as a linear function.

5.2.2 *Mutations Arising During Repair*

Misrepair of DNA lesions is another likely mechanism for mutation formation. Generally, simple bds should be repaired with high fidelity, since there is an intact template on the opposing strand; thus it seems unlikely that point mutations would develop from errors during such repair. On the other hand, regions with multiple bds could be a better substrate for misrepair and the subsequent development of mutation.

DNA dsbs are a much greater challenge for the cell to repair, since there is no intact template readily available. One mechanism for repair of dsbs is by intrachromosomal recombination (often called nonhomologous recombination, see Figure 4.3); this restores structural integrity to the DNA, but at the expense of the genetic fidelity, since the intervening sequence is lost or altered. Such large-scale deletions are thought to arise from one or more dsbs. If in fact two or more "hits" are required, one would imagine that the dose response for deletions would contain a nonlinear (quadratic or higher) component. Deletions generated from multiple hits still could develop linearly, with multiple damage from single tracks or particles. Multilocus deletions on autosomal chromosomes would result in loss of heterozygosity of all deleted alleles; as a consequence, recessive mutations within the region may be expressed.

Intermolecular recombination (gene conversion or strand exchange between homologous chromosomes; often termed homologous recombination, see Figure 4.3) also can function to repair damage; in such a case, the sequence information in the sister chromatid or homologous chromosome serves as a template for repair synthesis. This is not misrepair, as damage should be repaired faithfully. However, if the homologue is utilized, the process leads to loss of homozygosity of all genes distal to the crossover point, and thus it will unmask any recessive alleles by replacing the original wild-type sequence with the defective copy. It is unclear whether interchromosomal recombination would require multiple strand breaks to be initiated.

5.3 Dose-Response Studies with Low Linear-Energy Transfer Radiation

5.3.1 *Human in Vivo*

Dose-response data for mutagenesis in humans are scanty, especially in human germ cells, because the relevant controlled experiments cannot be performed. Nevertheless, it is clear that ionizing radiation is mutagenic, although there are disparate suggestions as to the doses of ionizing radiation at which a significant increase in mutation frequency can be measured. It is noteworthy, however, that mutations have been observed to be induced in the cells of some organisms at doses below 10 mGy (*e.g.*, Sparrow *et al.*, 1972).

Mutation frequencies at the hypoxanthine phosphioribosyl traniferase (HPRT) locus in peripheral blood T-lymphocytes among atomic-bomb survivors were elevated (Hirai *et al.*, 1995). This is especially true when mutation frequencies were examined in those survivors who exhibited elevated levels of chromosome aberrations (Hakoda *et al.*, 1988). Significant dose-related increases in mutations at the glycophorin A locus also were seen in this set of 39 exposed survivors (Langlois *et al.*, 1993). Data were fitted to linear curves but there was considerable scatter in the data among the individuals assigned the same nominal dose.

On the other hand, in extensive studies looking for mutations inherited by the offspring of atomic-bomb survivors (utilizing altered electrophoretic mobility or activity of a series of 30 proteins as the endpoint), there were no significant differences between exposed and control groups (Neel *et al.*, 1988).

Similarly, victims of the Chernobyl accident exhibited increased variant frequencies at the glycophorin-A locus. These somatic mutations were the result of large-scale alterations. Once again, data were fitted to a linear dose response, but the scatter was very large (Jensen *et al.*, 1995). Da Cruz *et al.* (1997) and colleagues have studied the population exposed to ^{137}Cs radiation in the accident in Goiana, Brazil in 1987. For the HPRT locus, they estimated a doubling dose of 1.73 $\pm$ 0.47 Gy.

Cancer patients who have received radiotherapy exhibit increased mutation frequencies at the HPRT locus (*e.g.*, Caggana *et al.*, 1991; 1992; Messing and Bradley, 1985; Saia-Trepat *et al.*, 1990). Generally these studies are not amenable to dose-response analysis, since total doses received vary only within a restricted range, the doses are fractionated and not usually delivered to the whole body, and mutation frequencies are not determined until several months after treatment. In one exception, Nicklas *et al.* (1990) studied patients before and after treatment with radioimmunotherapy with ^{131}I. Mean mutation frequencies were $11.5 \pm 5.1 \times 10^{-6}$ pre and $27.8 \pm 16.1 \times 10^{-6}$ post. Known quantities of ^{131}I ranged from 310 to 5,600 MBq, and the mutation frequency data fit well to either a linear or linear-quadratic relationship. Analysis of individual mutants indicated that there was an excess of deletions, which is consistent with induction by radiation.

There are some reports of very low doses of ionizing radiation inducing somatic mutation at the HPRT locus. These are surprising results, in that the doses involved are well below what would be expected to yield a measurable increase in a small population. In one study, a group of 13 to 14 radiation technicians who were exposed largely to gamma rays, received an average dose of 2.2 mSv in the six months prior to the assay (Messing *et al.*, 1989). These technicians in 1986 had an average mutation frequency at the HPRT locus of $12.8 \pm 8.6 \times 10^{-6}$ versus $9.5 \pm 8.2 \times 10^{-6}$ in the controls; in 1984, the corresponding frequencies were $7.2 \pm 5.9 \times 10^{-6}$ versus $3.1 \pm 1.9 \times 10^{-6}$ in the controls; and these differences between the two time points were ascribed to laboratory procedures. The authors' conclusion was that the 2.2 mSv average dose yielded a 50 to 100 percent increase in mutant frequency. In a second study, in 10 nuclear medicine patients, a total dose to the lymphocytes of 10 to 50 mGy from ^{99m}Tc increased HPRT mutation frequency from $2.09 \pm 3.18 \times 10^{-6}$ to $7.62 \pm 20.7 \times 10^{-6}$ (Seifert *et al.*, 1987). In a third report, Seifert *et al.* (1993) examined factory workers with potential exposure to external gamma and internal tritium radiation. There were two peak exposure periods, the first at 48 to 68 weeks before sampling of blood cells, and the second at 20 to 34 weeks before sampling. There were no overall differences in mutation frequency between exposed and unexposed workers. However, the authors claimed that there were higher mutation frequencies associated with the earlier exposure, all due to the external gamma dose of 20 to 40 mSv. It is not clear from the data presented in the manuscript how this conclusion was reached.

Similar studies failed to reproduce these results. Kelsey *et al.* (1991a) showed that there were no significant differences in either mutation frequency or chromosome aberration frequency among 24 patients before or after injection with ^{201}Tl. Similarly, there was no effect of ^{99m}Tc on either endpoint in 17 patients (Kelsey *et al.*, 1991b).

Khaidakov *et al.* (1997) reported that Russian cosmonauts exposed in earth orbit to 4 to 127 mGy had HPRT-mutant frequencies that were 2.4 to 5-fold higher than age-corrected values for healthy, unexposed subjects. This finding could be an indication that mutagenic effects occur at very low doses. However, the group studied consisted of only five individuals, and a control group was not available. In addition, although the samples were obtained in 1991, the exposures occurred between 1969 and 1990.

5.3.2 *Animal in Vivo*

In vivo studies in mice have produced data much more amenable to quantitative analysis. Linear dose-response curves for genetic alterations in a series of gene targets have been observed for both acute and chronic dose-rate protocols for mouse spermatogonia exposed to low-LET radiation using the specific locus assay (*e.g.*, Russell and Kelly, 1982a; 1982b; Searle and Beechey, 1974). A dose-rate effect was seen down to approximately 8 mGy min^{-1}. Acute treatments were three times more mutagenic than the low-dose rates (Figure 5.1). These results strongly suggest that single radiation

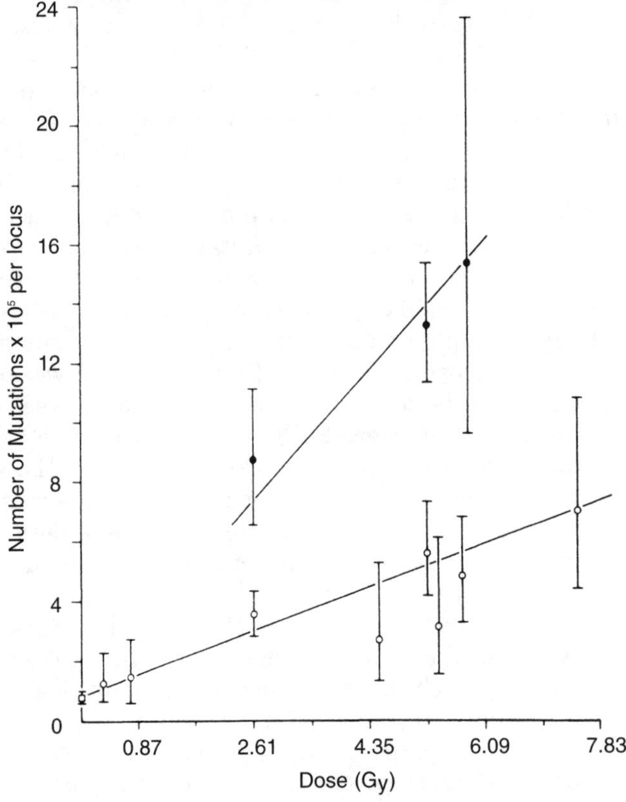

Fig. 5.1. Straight lines of best fit, by the method of maximum likelihood, for specific-locus data obtained for mouse spermatogonia under chronic (○) and acute (●) exposure conditions. Chronic experiments include all of those with unfractionated exposures at air kerma[3] rates of 7 mGy min^{-1} and lower. Acute experiments include all of those with unfractionated exposures up to an air kerma of 5.8 Gy with air kerma rates from 630 to 780 mGy min^{-1}. The 90 percent confidence intervals (CI) of data points are shown. Data points at each dose are combined in the figure but are kept separate in the computations (Russell and Kelly, 1982a).

[3]The figure was originally published with the abscissa labeled "Dose (R)." The units were assumed to be roentgens and converted to air kerma.

tracks can produce mutations, and thus predict that the induction of specific-locus mutations should follow a linear dose response without a threshold. The dose-rate data show no difference between these tracks delivered at a rate of about 10 min^{-1} or 10 d^{-1}. This suggests that the type of repair that is responsible for this dose-rate effect is effective only for relatively high concentrations of damage, but it does not prevent other processes from modifying the response at lower total numbers of tracks.

Lorenz *et al.* (1994) irradiated mice with ^{137}Cs, then isolated spleens and determined the HPRT mutation frequency in the T-lymphocytes. Both a dose and dose-rate dependence was observed. After acute irradiation (0.5 Gy min^{-1}), a linear-quadratic response was observed for mutation frequencies scored 8 to 10 weeks after treatment. For the lower dose rates (1 Gy d^{-1}, 1 Gy $week^{-1}$), a linear dose response was obtained.

A new assay, the mouse pink-eye system (Schiestl *et al.*, 1994), is an *in vivo* reversion assay which detects DNA deletions, the alteration detected being the loss of an existing intragenic duplication. X-ray-induced deletions thus detected were fitted linearly between 0.01 and 1 Gy, providing another example of a specific type of mutation being induced with linear dose relationship.

5.3.3 *Mammalian Cells in Vitro*

Numerous *in vitro* dose-response studies using mammalian cells have explored the effects of differing genetic backgrounds, using different genetic loci, and examining the characteristics of purported inducible repair systems.

5.3.3.1 *Assays at the Hypoxanthine Phosphioribosyl Traniferase Locus.* The most commonly used genetic locus for mutation studies in mammalian cells is the HPRT locus. It is easily selected for with purine analogs such as 6-thioguanine. It is X-linked and thus hemizygous, or functionally so, in all mammalian species (this is advantageous in that there is no second compensating allele to mask phenotypic changes after a mutation, but disadvantageous for two reasons: (1) mutational mechanisms involving homologous chromosomes cannot be studied, and (2) very large deletion events may include an essential gene that will result in cell death).

In some human cell systems, the assay gives a linear dose response with no apparent threshold. These cell systems include human fibroblasts (*e.g.*, Cox and Masson, 1979) and human lymphoblasts (Liber *et al.*, 1983). Generally, there are few data below 200 mGy, so thresholds cannot be ruled out. However, Grosovsky and Little (1985) performed a fractionated experiment in which lymphoblast cells were treated daily with 10 to 100 mGy of acute x rays. The final mutation frequency observed was equal to that seen for a single acute exposure, suggesting that the increments were strictly additive, and that a dose as low as 10 mGy was effective at inducing mutation. On the other hand, in other human cell systems, notably peripheral blood T-lymphocytes treated *in vitro*, either purely quadratic or linear-quadratic responses have been observed (Sanderson *et al.*, 1984; Vijayalaxmi and Evans, 1984). In the majority of rodent studies, the dose response is nonlinear, generally fitting a linear-quadratic relationship best. Within the same laboratory group, human fibroblast versus rodent V79 cells have maintained this linear versus linear-quadratic trend (Cox and Masson, 1979; Thacker*et al.*, 1979).

In Chinese hamster ovary cells, Nagasawa and Little (1993) amplified the exons of HPRT with the PCR to characterize mutants induced by acute x irradiation. They showed that the overall mutation rate fit a linear-quadratic equation best. However, they also found that point mutations and also partial deletion mutations were generated with a linear dose response; only the total gene deletions arose nonlinearly. The total gene deletions, which dominated, determined the nonlinear nature of the dose-response curve as a whole. These data are shown in Figure 5.2.

5.3.3.2 *Assays at Other Genetic Loci.* There have been studies at the heterozygous thymidine kinase (Tk1) locus in both human and mouse cell systems. Mutations at this locus can arise by all of the same pathways as at HPRT, but in addition, can arise from mechanisms involving the homologous chromosome, and by very large intrachromosomal deletions. These latter mechanisms lead to loss of heterozygosity of the autosomal locus; such alterations could be frequent occurrences in most cancers as they can unmask recessive genes in the oncogenic pathway.

In L5178Y mouse cells (heterozygous for Tk1 and mutant for *p53*), the dose-response curve for mutations of the Tk locus was reported to be nonlinear (Nakamura and Okada, 1981; Nakamura *et al.*, 1982). However, later experiments with a different subclone indicated a linear response (Moore *et al.*, 1986). In addition, Moore *et al.* (1986) reported that at the Tk locus in L5178Y mouse lymphoma cells, gamma rays induced both large-colony (thought to arise from small intragenic alterations) and small-colony (thought to arise from large

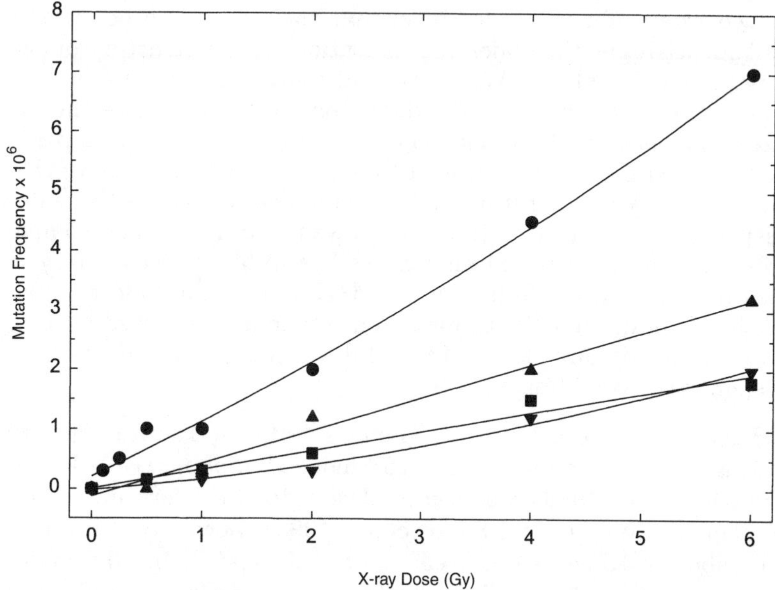

Fig. 5.2. Mutagenicity of x rays at the HPRT locus in CHO cells. Cells are treated with the indicated dose of x rays, and HPRT mutants were characterized with the polymerase chain reaction. The symbols indicate overall frequency of mutations (●), frequency of total deletions (▼), frequency of partial deletions (▲), and frequency of point mutations (■) (Nagasawa and Little, 1993).[4]

multi-locus alterations including both deletions and recombination events) mutants with linear kinetics.

In Tk6 human lymphoblast cells, thymidine kinase mutants also arise mainly from loss of heterozyosity resulting from a combination of deletions and recombination events. Interestingly, the best-fit curve is linear (Konig and Kiefer, 1988), suggesting that these large-scale genetic rearrangements can arise through damage from a single track.

Studies with low-LET radiation at the dihydrofolate reductase locus, where one mechanism by which mutants can arise is by gene amplification, have shown that the dose response is nonlinear in EMT-9 mouse cells (Hahn *et al.*, 1990). However, in L5178Y mouse lymphoma cells, the dose response was linear (Nakamura *et al.*, 1982).

Waldren and colleagues developed a human/hamster hybrid cell system (A_L) which contains a single human chromosome 11 that is almost entirely dispensable (Waldren *et al.*, 1986). Mutation frequencies seen in the human chromosome in the cells of this assay are up

[4]Nagasawa, H. and Little, J.B. (1993). Personal communication (Harvard University School of Public Health, Boston).

to two orders of magnitude higher than those seen in other systems. Not surprisingly, the majority of mutations screened arise from very large-scale alterations. What is somewhat surprising is that x rays were more mutagenic at lower doses than at higher doses; the average slope between 0 and 0.55 Gy was 3.45×10^{-3} mutants per cell per gray, while the slope from 1 to 6 Gy was approximately 0.75×10^{-3}. There was no hint of a positive quadratic term in these dose-response data. Recently, this system was modified so that complete loss of the human chromosome, as well as stable translocations with loss of the 11p arm, could be detected (Kraemer and Waldren, 1997). With this additional class, mutation frequencies induced by x rays were an additional factor of 12-fold higher; however, the dose response was still linear.

5.3.3.3 *Dose-Rate Effects.* The effect of dose rate on radiation mutagenesis is complex. In L5178Y mouse lymphoma cells Furuno-Fukushi *et al.* (1988) reported no dose-rate effect for mutagenesis. Supporting this finding, Evans *et al.* (1990) showed that there was only about a 15 percent decrease in the slope of the mutation dose-response curve at HPRT for low-dose-rate exposure (0.02 Gy h^{-1} compared to acute at 53 Gy h^{-1}). However, at the autosomal Tk locus, the lower dose rate was significantly less mutagenic in radiation-resistant strains, but about equally effective in radiation sensitive strains. In another study with mouse lymphoma cells, lowering the dose rate from 500 to 8 mGy min^{-1} changed the shape of the dose response at the HPRT locus from linear-quadratic to linear. At the dihydrofolate reductase locus in these same cells, the 8 mGy min^{-1} dose rate was less mutagenic by a factor of two, but the dose response was linear at both dose rates (Nakamura and Okada, 1981).

In V79 hamster cells, Crompton *et al.* (1985; 1990) reported that lowering the dose rate from 4 Gy min^{-1} to 50 mGy h^{-1} decreased the mutagenic efficiency of gamma rays at the HPRT locus; however, decreasing the dose rate still further to 8 mGy h^{-1} led to a dramatic increase in mutagenic efficiency, to at least several-fold higher than the acute treatment. In these experiments all dose-response curves were nonlinear.

In human Tk6 lymphoblast cells (where the acute dose response is linear), several early experiments showed no evidence of a dose-rate effect during chronic exposure to either gamma radiation (Konig and Kiefer, 1988) or beta particles from tritiated water (Liber *et al.*, 1985; Tabocchini *et al.*, 1989). More recently, however, Amundson and Chen (1996) reported on low-dose rates in these cells. As with V79 cells, the effect was an inverse one, with the lowest dose rate of 27 mGy h^{-1} producing more mutation than the acute dose protocol. One explanation is that this inverse dose-rate effect is caused by correlated variations in sensitivity to both killing and mutagenesis, across the cell cycle.

In the human/hamster A_L hybrid system, a low-dose rate of 0.5 mGy min^{-1} was as effective at inducing S1 mutation as the high-dose rate of 2 Gy min^{-1} (Waldren *et al.*, 1992).

5.3.3.4 *Effect of Genetic Background.* The genetic status of the cell may determine the nature of its response to ionizing radiation.

Human cells that are heterozygous for a mutation in the retinoblastoma gene have the same linear dose-response curve for mutagenesis after treatment with gamma rays as do normal fibroblasts (Wang *et al.*, 1986).

As described above, Tk6 human lymphoblast cells which have a wild-type *p53* gene exhibit a linear dose response after treatment with x rays. On the other hand, the closely related Tk1 human lymphoblast cells with mutant *p53* are considerably more mutable by x rays, and exhibit a nonlinear dose response after acute irradiation (Amundson *et al.*, 1993; Xia *et al.*, 1995). Similarly, over-expression of Bcl-2 or Bcl-xL, which suppress apoptosis, also alters the shape of the dose-response curve in Tk6 (Cherbonnel-Lasserre *et al.*, 1996). The *p53* mutant cells also are more mutable by high-LET radiation (Amundson *et al.*, 1996). Recently, L5178Y mouse lymphoma cells were found to be mutant at the *p53* locus (Storer *et al.*, 1997). However, it is not certain whether all of the various strains of this frequently used cell line carry this mutation.

In human cancer, increased mutability of cells already harboring mutation in *p53* or with dysregulated Bcl-2 or Bcl-xL could be important if radiation acted to induce a late step in carcinogenesis.

5.3.3.5 *Inducible Systems.* Any DNA repair systems that are inducible could impact on mutation dose-response relationships; mutagenesis could be increased if the induced system was error-prone, or decreased if it were error-free. Clearly, the dose dependence and other characteristics of induction of such repair, including the type of damage that leads to induction, as well as whether induction was a graded response or an "all or none," would be critically important for understanding its effect on mutation at low doses.

5.3.3.5.1 *Genomic instability.* Hamster cells surviving a high dose of low-LET radiation (on the order of 10 Gy) sometimes exhibit a persistently elevated mutation rate at the HPRT locus. This prolonged genomic instability is evident in about 10 percent of the survivors and lasts 30 generations or more (Chang and Little, 1992). The molecular spectra of the mutants arising in these experiments are the sort generally associated with spontaneous events rather than radiation induced, in that the majority are point mutations (Little *et al.*, 1997). Similar findings have been reported in human lymphoblast cells (Grosovsky *et al.*, 1996). So far, there is no plausible mechanism to explain these observations. Furthermore, there is no indication of whether this sort of genetic instability might be induced at low radiation doses, or what dose-response relationships it might follow. However, if such a response were induced at low doses, it is possible that radiation carcinogenesis could be due at least in part to such indirectly induced mutations. If that were true, one might not be able to use the dose-response relationship discussed above for directly-induced mutations to reach conclusions about carcinogenesis. If not, one might overestimate the risk of carcinogenesis at low doses.

5.3.3.5.2 *Adaptive response.* Exposure *in vitro* to low doses of tritiated-thymidine, x or gamma rays (approximately 10 mGy) can cause lymphoid cells to adapt so that subsequent doses of 3 Gy delivered 5 to 24 h later are only about half as mutagenic (Kelsey *et al.*, 1991c; Rigaud *et al.*, 1993; Sanderson and Morley, 1986; Sasaki, 1995; Zhou *et al.*, 1991; 1994). Adaptation is reported also to protect selectively against the formation of deletions (Rigaud *et al.*, 1993; 1995; Ueno *et al.*, 1996; Zhou *et al.*, 1994). There is inter-individual variability in the expression of the adaptive response; in one cytogenetic study 14/18 people exhibited a response (Bosi and Olivieri, 1989). At least for cytogenetic endpoints, there is a dose below which an adaptive response is not observed (Shadley and Wiencke, 1989). This of course suggests that the response, which is generally thought to reflect the induction of a repair system, requires a certain level of damage before it is induced; it follows from this that multiple radiation tracks are likely to be needed to induce the adaptive response in an exposed cell. Thus it is improbable that the adaptive response is operational very much of the time in any particular cell receiving background levels of low-LET radiation. Even so, individual cells would rarely receive a second discrete dose of radiation within the short time period during which the adaptive response apparently remains effective (*i.e.*, 24 to 48 h). Thus, unless the response has an effect on a single radiation dose (it has not been demonstrated whether the adaptive response is implemented quickly enough to have an impact on the inducing dose itself), it seems unlikely that it will have an important role in protecting cells from radiation damage.

5.4 Dose-Response Studies with High Linear-Energy Transfer Radiation

Animal studies with the mouse specific-locus assay have shown that fission neutrons mutate spermatogonia without evidence of a dose-rate dependence, at least not at total doses below 1 Gy. On the other hand, the relative mutagenicity of neutrons for oocytes was reduced at low-dose rates (Batchelor *et al.*, 1969).

Numerous experiments have been done on mammalian cells *in vitro*. Generally the resulting dose-response relationships are linear, especially in human cells (*e.g.*, Cox and Masson, 1979; Kranert *et al.*, 1990; Kronenberg and Little, 1989a; 1989b; Metting *et al.*, 1992; Nakamura *et al.*, 1982; Stoll *et al.*, 1995; Tsuboi *et al.*, 1992). Kronenberg (1991) reported an inverse dose-rate effect for neutrons; dose rates of $<14\ \mu Gy\ min^{-1}$ were about three times more mutagenic than higher rates. Both the high-dose-rate and low-dose-rate data had linear dose-response relationships. In rodent cells, the curves have sometimes been observed to be linear, and sometimes

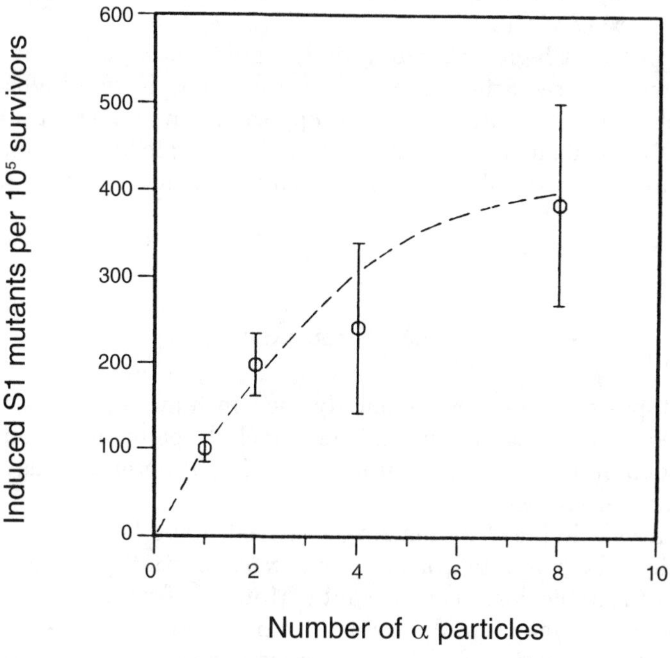

Fig. 5.3. Induced mutants at the S1 locus per 10^5 surviving A_L cells irradiated with an exact number of alpha-particle traversals at 90 keV μm^{-1}. Induced mutant yield equals total mutant yield minus background incidence. The background mutant fraction in A_L cells used in these experiments averaged 45 per 10^5 survivors. Background mutant fraction was subtracted from treated points. Data were pooled from three experiments, and the curves fitted using the least-square method (bars represent ± SEM) (Hei *et al.*, 1997).

curvilinear. In some studies, significant increases in mutation rates were produced by what was likely to be a single particle traversing the target chromosome (Kronenberg, 1991; Kronenberg and Little, 1989a; Kronenberg *et al.*, 1995).

The recent development of a microbeam facility at the Radiological Research Accelerator Facility of Columbia University (Randers-Pherson *et al.*, 1995) has allowed mutagenesis studies to be done in cells traversed by known numbers of alpha particles. Hei *et al.* (1997) have shown that a single alpha particle induces approximately a three-fold increase in the frequency of mutations at the S1 locus in the A_L cell system (Figure 5.3). This suggests that, at least for alpha particles, there is no observable threshold for this type of mutation.

Molecular analyses have revealed that the majority of mutants induced by high-LET particles are large-scale in nature (*e.g.*, Bao *et al.*, 1995; Chaudhry *et al.*, 1996; Kronenberg and Little, 1989b; Kronenberg *et al.*, 1995).

The fact that high-LET radiation is significantly more mutagenic than low-LET radiation suggests that mds may have a high likelihood of producing mutations. This supports the hypothesis that multiple DNA lesions are more mutagenic than single lesions, and may explain why low-LET dose-response curves sometimes have a quadratic component.

5.5 Summary

All types of mutations commonly seen in human cancers can be induced by ionizing radiation. These include point mutations and small deletions, as well as mutations arising at heterozygous loci by loss of heterozygosity.

Mutations induced by ionizing radiation have been observed in somatic cells irradiated *in vivo*, but *in vitro* studies have yielded the most reliable dose-response data. Mutation frequencies increase either with linear, quadratic or linear-quadratic dose-response curves; the shapes of the dose-response curves vary depending on the LET of the radiation and the dose rate, the nature of the genetic endpoint and thus the class of mutation being examined, and the genetic background of the cell. For high-LET radiation, mutation induction is linearly related to dose; for low-LET radiation, linear and linear-quadratic responses have been found. There is no direct evidence for a threshold in either case. Therefore, if radiation-induced cancer results directly from the induction of mutations involved in the oncogenic pathway, the data reported do not support the existence of a threshold. There is an intriguing corollary hypothesis, namely that radiation may induce a persistent state of genomic instability, and that this could be a contributory mechanism for radiation-induced carcinogenesis; however, at this time, nothing is known about the kinetics of such an effect. However, if carcinogenesis involves additional steps which depend on the microenvironment of the initiated cell (see Section 3), a nonlinear response cannot be ruled out.

Under some experimental conditions the adaptive response can protect cells against the mutagenic effect of ionizing radiation; however, based on the time course involved, the response seems unlikely to have a significant impact on radiation effects in human populations.

5.6 Research Needs

Dose-response studies in human populations exposed to radiation should be pursued; this will be especially important when good dose estimates can be obtained. At present these types of studies will be most useful if they involve a wide range of doses. It would be difficult to detect significant changes in mutation frequency for low-dose exposures using current methods, so new technologies for making quantitative measurements of mutation frequencies in humans will need to be developed. One example of such a new technology would be molecular markers specific to radiation exposure. Such new approaches would also be invaluable for investigating mutagenesis *in vitro*.

More studies on dose-response relationships for the different mutational classes are warranted. Experimental variables that will be of interest include irradiation conditions and genetic background of the cell.

More studies on the phenomenon of radiation-induced genomic instability are required. Questions to be investigated include the dose response, irradiation conditions, initial genetic background of the cell, subsequent genetic background of the apparent mutator cells, and the relation of this phenomenon to cancer. Detailed studies on hypermutability of the unstable variants also will be of great interest.

For the adaptive response, it will be important to determine whether it is effective for a single radiation dose. The mechanisms of induction and of repair need to be determined. The reasons for individual variations should be examined.

6. Chromosome Aberrations Induced by Low Doses and Low-Dose Rates of Ionizing Radiation

Section 4 provides a description of the different types of ionizing radiation-induced DNA damage and the current status of what is known about its repair. The broad categories of DNA damage are ssbs, dsbs, bds or modifications, dpcs and mds, all defined and illustrated in Section 4. The various modes of repair (excision, ligation or recombination) have a finite probability of making errors. Of course, excision repair is relatively error-free and nonhomologuous end-joining is more error-prone. The errors can either be failure to repair or inaccurate repair (misrepair). In this Section the role of these DNA repair errors is considered, together with DNA replication errors in the formation of chromosome aberrations, and how knowledge of the mechanism of formation of aberrations might be used in defining the shape of the dose-response curve for aberrations at low doses and low-dose rates of ionizing radiation. In addition, there is a discussion of the relevance of chromosome aberration analysis for predicting frequencies of specific tumor types, and thereby its utility for predicting the shape of cancer dose-response curves at low doses and low-dose rates of ionizing radiation. For the purpose of the discussion in this Section, a low dose is considered to be less than 50 mGy. A low-dose rate is considered to be one that requires a total time to deliver the dose that is long compared to the repair time for the different DNA lesions, *e.g.*, greater than 6 h.

6.1 Misrepair, Misreplication, and Chromosome Aberration Formation

The types of chromosome aberrations induced by ionizing radiation differ, depending on the stage of the cell cycle in which they are formed. The consequences for cell killing and inherited genetic alterations depend on the types of aberrations, as illustrated by the brief discussion that follows. A complete listing of all types of chromosome aberrations can be found in the comprehensive review by Savage (1979).

6.1.1 *Chromosome-Type Aberrations*

Chromosome aberrations induced by ionizing radiation in G_1 chromosomes or unreplicated chromosomal regions in the S-phase are of the chromosome-type, *i.e.*, both chromatids are involved in the aberration. These chromosome-type aberrations result from radiation-induced DNA damage to an unreplicated G_1 or S-phase chromosome that consists of a single DNA double helix. Failure to repair a dsb, for example, will produce a terminal deletion, whereas incorrect repair can lead to interchanges (dicentrics, reciprocal translocations, or rings), inversions, or interstitial deletions. An aberration induced in G_1 is replicated in the S-phase, resulting in the observed involvement of both chromosome arms in the aberration. Thus, chromosome-type aberrations induced in G_1 are a consequence of errors of DNA repair and can result from single DNA lesions (terminal deletions) or the interactions of two or more lesions (interchanges, inversions, interstitial deletions), as depicted in Figure 6.1.

6.1.2 *Chromatid-Type Aberrations*

Chromosome aberrations induced by ionizing radiation in G_2 of the cell cycle or in replicated regions of S-phase chromosomes are of the chromatid-type, *i.e.*, only one chromatid of a chromosome is involved in an aberration. (An exception to this classification might appear to be the isochromatid deletions, but the most plausible mode for their formation involves independent breaks at nonhomologous regions of the chromosome, *i.e.*, each break is a chromatid-type break.) As with chromosome-type aberrations, chromatid-type aberrations are terminal deletions, interchanges (inter- or intrachromosomal), interstitial deletions, and a range of other intrachromosomal rearrangements (Savage, 1979). Chromatid-type aberrations can be formed by errors of DNA repair, either by a failure to repair or by incorrect repair. In addition, and in contrast to chromosome-type aberrations, chromatid-type aberrations can be formed by errors of DNA replication caused by a damaged template.

Thus, chromosomal aberrations can be produced by errors of DNA repair or DNA replication, and so considerations of the kinetics of DNA repair for specific types of DNA damage are important for determining which mode of action is more likely to occur.

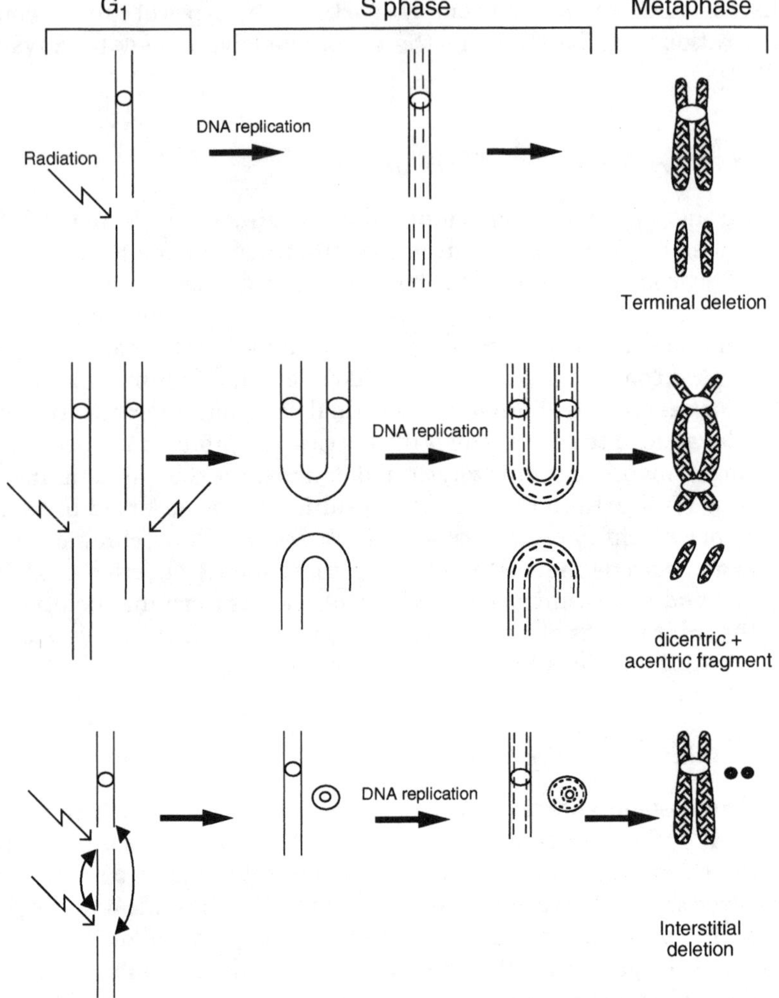

Fig. 6.1. Diagrammatic representation of the conversion of radiation-induced DNA damage into a chromosome-type aberration (terminal deletion, dicentric, or interstitial deletion) observable at metaphase.

6.1.3 *Mechanisms of Formation of Chromosome Aberrations*

Errors in DNA repair or replication that can ultimately lead to chromosome aberrations are described in Section 4. It remains important to establish the types of DNA damage that are involved and how they can be misrepaired or misreplicated. It is necessary ultimately for both halves of the DNA double helix to be involved in the formation of a chromosome aberration; these induced double-strand interactions can be from DNA dsbs induced directly or from dsbs induced indirectly through the repair of bds and possibly through the misrepair of ssbs. This latter is rather unlikely given the high probability of correct rejoining of a ssb, from considerations of chromosomal (DNA) geometry, and the very rapid repair of ssbs that necessarily limits the time available for interactions. Given the involvement of different types of DNA damage in the formation of chromosomal aberrations, it is necessary to consider effects of low- and high-LET radiations separately since different frequencies of the different classes of DNA damage are induced by these two general radiation qualities.

6.1.3.1 *Low Linear-Energy Transfer Radiations.* The types of DNA lesions induced by ionizing radiation and their modes of repair are described in Section 4. It is quite straightforward to perceive how directly-induced dsbs can lead to chromosome alterations: unrepaired dsbs would result in deletions and misrepaired dsbs in inter- or intrachromosomal exchanges. This view has been formalized in the breakage first model for aberration formation, originally proposed by Sax (1938). This model predicts that the dose-response curve for low-LET-induced inter- and intrachanges that require two breaks (produced by one or two tracks) is nonlinear, and follows the general formula:

$$Y = \alpha D + \beta D^2 \tag{6.1}$$

Whereas the dose-response curve for terminal deletions that require only one break for their formation is linear:

$$Y = \alpha D \tag{6.2}$$

Despite the fact that the breakage first hypothesis and the sole involvement of dsbs in chromosome aberration formation has simplicity, it fails to explain some of the available data, and does not consider a high proportion of the induced DNA damage, most notably alterations to DNA bases.

The model described by Revell (1974) proposes that all chromosome aberrations, including deletions, are produced from an exchange process, essentially complete for interchanges and intrachanges and incomplete for deletions. In its original concept the exchanges were derived from interactions of lesions. In molecular terms, these lesions can be dsbs, base alterations, or mds, and that the interaction to form an aberration can occur at the time of DNA repair (G_1, G_2, and much of S-phase of the cell cycle) and of DNA replication (S-phase). The relative involvement of repair and replication in the S-phase in the formation of aberrations will depend on the time between damage induction and replication of a particular genomic region. That is, damage induced some hours before replication will be repaired prior to replication, and as a corollary, damage induced in S post-replication will not be subject to replication errors.

On the basis of the published literature, it seems that a combination of the two models (breakage-first and exchange) is probably correct. The study by Duncan and Evans (1983) perhaps best represents this view. They showed that a proportion (about 20 percent) of chromatid deletions induced by bleomycin (which is radiomimetic) were associated with a sister chromatid exchange. Revell's exchange hypothesis predicts that 40 percent of chromatid deletions would be associated with sister chromatid exchange, whereas the breakage-first model predicts that none would be. It is important to know the specific modes of formation of chromosome aberrations, since they affect the shapes of dose-response curves at low doses or dose rates.

Additional support for deletions arising as products of incomplete exchange events is provided by the studies of Brewen and Brock (1968). They showed that almost all chromosome-type terminal deletions, induced in G_1 by x rays, were the result of incomplete exchange formation. The consequence of this is that terminal deletions, like interchanges, can be produced by one or two tracks, and fit the relationship $Y = \alpha D + \beta D^2$. For low-LET radiations on the basis of the preceding discussion, each type of aberration can be formed by a one-track or a two-track mechanism, and each will have a linear-quadratic dose-response curve overall.

An exception to this is suggested from a recent study by Griffin *et al.* (1996) which assessed the efficiency of 1.5 keV aluminum x rays in inducing complex chromosome aberrations (requiring three or more interacting lesions for their formation). The authors suggested that damaged DNA might interact with undamaged DNA to produce some of the aberrations. This is in contrast to the data of Cornforth (1990), who concluded that a one-hit exchange probably did not occur, although it could not be ruled out at low doses. The importance of a one-hit exchange process to the shape of the dose-response curve at high doses and exchange yields at low doses is readily apparent, and the question of its likelihood, especially at low doses, requires further study.

Another component of the mechanism of formation of chromosome aberrations that is pertinent to the consideration of low-dose linearity is the magnitude of the rejoining distance, namely, the distance apart that two radiation-induced lesions can be in order for their interaction to produce an aberration. This information can be combined with energy deposition as discussed in Section 3, to predict the probability of aberrations over the low-dose range. The estimates of rejoining distance have been made in several ways and quite a wide range of values has been estimated. Virsik *et al.* (1980) using data for the induction of chromosome aberrations by ultrasoft x rays concluded that exchange aberrations could be produced by local energy events within a distance of less than 7 nm. Schmid and Bauchinger (1975) used a microdosimetric approach for 15 MeV neutrons to estimate a rejoining distance of at least 1.8 μm. More recently, Savage (1996) used an empirical approach to estimate that high- or low-LET radiation-induced chromosome aberrations could result from dsbs produced within a well defined shell several hundred nanometers from each break. It would appear that these rejoining distances are consistent with linear dose-response curves at low exposure levels.

6.1.3.2 *High Linear-Energy Transfer Radiations.* The DNA damage induced by high-LET radiations has, in general, been less well characterized than for low-LET radiations. For the present discussion, it is sufficient to classify the damage in a way similar to that for low-LET radiations; that is, as DNA strand breaks, bds and mds. However, the relative frequencies of the various types is different for the two types of radiation qualities, with most significantly a much higher proportion of mds with high-LET radiations (Ward, 1990).

The chromosome aberration types induced by high-LET radiations are the same as those induced by low-LET radiations, although some more complex exchanges are observed after exposures from high-LET radiations. Thus, it would appear that similar processes of misrepair and misreplication occur. Double-strand gaps are most likely the predominant lesion involved in chromosome aberration formation by misrepair of high-LET-induced lesions (Goodhead, 1995). This is deduced from the fact that the dose-response curve is linear, with each aberration being the result of a single ionization track. The greater effectiveness of high-LET radiations in inducing chromosome aberrations is probably a reflection of an increased yield of double-strand DNA damage per unit of dose and a greater probability of misrepair of an mds than of a single dsb (Goodhead, 1994).

6.1.4 *Dose-Response Curves: Acute and Chronic Exposures*

6.1.4.1 *Low Linear-Energy Transfer Radiations.* As described above (Section 6.1.3.1), the dose response for all types of chromosome aberrations is described by $Y = \alpha D + \beta D^2$. Preston (1992) has argued that this dose response can be viewed not only as a combination of one-track (αD) and two track (βD^2) events but that the two components are representative of different types of DNA damages: the linear being from dsbs and the dose-squared from bds and/or mds. However, Monte Carlo simulations of DNA strand breaks using sophisticated track structure models have demonstrated that a single low-LET particle can produce the full range of DNA damage, including mds (Brenner and Ward, 1992; Goodhead, 1994). Nevertheless, at low doses, the major mechanism in the formation of aberrations is misrepair of dsbs. Similarly, for low-dose rates, the linear curve obtained for aberrations is likely to be representative of charged particle interactions that result in dsbs. The aberration frequencies at low doses will be effectively identical following chronic or acute exposures.

Thus, for considerations of the shape of the dose-response curve for chromosome aberrations induced by low-LET radiations, the kinetics and fidelity of repair of dsbs are the appropriate parameters. The problem remains that existing methods for the assessment of the induction and repair of DNA dsbs are applicable only at relatively high doses, and predicting the response over the low-dose range is difficult (see Section 4).

6.1.4.2 *High Linear-Energy Transfer Radiations.* The dose-response curves for all aberration types for high-LET radiation exposures have been observed to be linear over the dose ranges tested, *e.g.*, 3.5 to 244 mGy (Pohl-Ruling *et al.*, 1986), and they fit the equation:

$$Y = \alpha D \tag{6.3}$$

This relationship is readily explained by assuming that one ionization track produces the DNA lesions required for the different aberration types. Using this basic assumption of aberration formation, the dose-response curve for low-dose rates will also be linear and equivalent to that for acute exposures. There is no mechanistic evidence to suggest that the dose-response curve would deviate from linearity at low doses.

6.2 Distribution of Aberrations Within and Among Cells

The nature of the distribution of chromosome aberrations among cells and within individual cells provides a broad picture of the distribution among cells of DNA damage that can be converted into chromosome aberrations and/or the distribution of probabilities of repair errors. It is not abundantly clear how the various distributions will influence the shape of the dose-response curve at low levels of exposure, but, as discussed below, they would probably not reduce it to a sublinear form.

6.2.1 *Intercellular Distributions of Chromosome Aberrations*

With low-LET radiation exposures, the frequency of chromosome aberrations in each irradiated cell can be described by a Poisson distribution function (*i.e.*, they are randomly distributed over a broad dose range). There is no apparent deviation up to doses of about 200 mGy although deviations from a Poisson distribution are increasingly difficult to establish as the aberration frequency decreases. The situation has been somewhat complicated by new observations using fluorescence *in situ* hybridization (FISH) indicating that complex aberrations (*i.e.*, those involving more than two chromosomes) can be produced by x rays at doses of radiation of about 1 Gy (Simpson and Savage, 1996). These aberrations cannot be observed by conventional staining procedures. The inclusion of complex aberrations could cause the distribution of total aberrations to become overdispersed (*i.e.*, to have a higher proportion of cells with multiple chromosome aberrations than predicted by a Poisson distribution). At low doses (10 to 50 mGy) of low-LET radiation, such effects would be expected to be minimal except perhaps for a slight increase in the slope of the dose-response curve.

With high-LET radiations the aberrations (and chromosome breaks observed by stimulating premature chromosome condensation) are over dispersed. This is expected, based on the magnitude of energy deposition and consequent DNA damage produced by high-LET interactions. At low doses this will have little influence on the shape of the dose-response curve since the frequency of multiply-damaged cells is still very low. Recent data from Griffin *et al.* (1995), however, have shown quite high frequencies of complex aberrations with ^{238}Pu alpha particles, using FISH at doses of 0.4 to 1 Gy.

The number of chromosome aberrations in each cell at low doses or low-dose rates for both high- and low-LET radiations will have little impact on the shape of the dose-response curve. That is, there will be zero or one aberration per cell, with the relative proportions of each being dependent upon dose.

6.2.2 *Inter- and Intrachromosomal Distribution of Chromosome Aberrations*

There is evidence that specific regions of chromosomes vary in their susceptibility to spontaneous and induced aberrations. In surveys of the naturally occurring breakpoints in healthy unirradiated donors, chromatid deletions occur most frequently at three sites, 3p14.3, 16q23, and Xq27 (Aurias, 1993). In addition, recurrent rearrangements are observed that involve breaks at the T-cell antigen receptor loci on chromosomes 7 and 14 in phytohemagglutinin stimulated lymphocytes (Wallace *et al.*, 1984; Welch *et al.*, 1975). Koduru and Chaganti (1988) reported that congenital chromosome breakage was clustered within light bands in Giemsa-banded chromosome preparations. Chromosome-specific fragile sites and cellular oncogene sites were also involved more frequently than a random prediction.

For radiation-induced chromosome aberrations, there is generally a good agreement between the observed distribution of aberrations among chromosomes and the estimated distribution based on relative DNA content. There are, however, some reported deviations from this general rule. For example, Kamada and Tanaka (1983) reported that the distribution of breakpoints in T-lymphocyte chromosomes from atomic-bomb survivors was nonrandom. A significantly higher incidence of breakpoints than expected for a random distribution was found for 22q1, 14q3, 5q3, 21q2, and 18p1. Bone-marrow cells had some clusters of breakpoints leading to deletions of the long arm of chromosome 22. Some of these breakpoints are similar to, or the same as, those associated with human leukemias. The same authors also noted that there were fewer breaks in chromosomes 1, 2 and X than predicted from a random distribution. In contrast, Ohtaki (1992) reported that chromosome 1 was involved at a significantly higher frequency than predicted in the lymphocytes of Hiroshima atomic-bomb survivors. Bouffler *et al.* (1991) showed that mouse chromosome 2 was more frequently involved than would be predicted on a random basis for the induction of chromosome aberrations by x rays. A deletion in chromosome 2 is associated with murine myeloid leukemogenesis, and the breakpoints are at (or near) interstitial telomere-like sequences.

As is the case with background aberrations, radiation-induced aberrations are preferentially located in light bands or at the junction of light and dark bands obtained by Giemsa-banding methodology (Savage, 1977). The reason for this localization has not been identified. It is known that interstitial telomere-like sequences are hot spots for recombination, leading to enhanced chromosome breakpoints in these regions (reviewed in Preston, 1997). There is some evidence showing that repair of radiation-induced DNA damage is also not random throughout the genome. Regions of DNA that are transcriptionally active in a particular cell type are repaired more rapidly than silent or nontranscribed regions (Leadon and Cooper, 1993). The more slowly repaired genomic regions could be hot spots for aberrations produced by errors of DNA replication and the rapidly repaired regions could be more susceptible to the formation of aberrations by errors of DNA repair.

Thus, there are both inter- and intrachromosomal deviations from a random distribution of chromosome breakpoints. The occurrence of a higher frequency of aberrations at fragile sites and recombination breakpoints at loci involved in tumor formation suggests that any influence these might have on the dose-response curve for tumors would be to produce a higher estimated tumor response than would be estimated from simply using total aberration dose-response curves, with aberrations being used as a surrogate for cancer.

6.3 Uncertainties in Shapes of Dose-Response Curves at Low Doses

6.3.1 *Nonlinear and Threshold Responses*

As noted in Section 6.1.3.1, the dose-response curve for the majority of chromosomal aberrations is of the form:

$$Y = \alpha D + \beta D^2. \tag{6.4}$$

This curve shows a continually changing slope at high exposures but is essentially linear at low exposures. This low-dose relationship is based on the assumption that the relevant mechanisms and probabilities of errors of repair or replication are the same as those at exposures above 50 mGy. Unfortunately such assumptions are based on extrapolation rather than direct measurement. However, it is important to note that lowering the dose rate changes the low-LET aberration dose-response curve from quadratic to increasingly more linear (see, for example, Zaichkina *et al.*, 1997). At some limiting dose rate the curve becomes linear and does not appear to change in shape or magnitude with further decreases in dose rate (reviewed in Bender *et al.,* 1988). The magnitude of the low-dose and low-dose-rate responses will be defined by α. If α is small and β is large, the chromosome aberration dose response approaches $Y = \beta D^2$. In this case, the slope at low exposures will be greatly reduced compared to that with a higher α. The variation in α among lymphocytes of different species has been nicely exemplified by Brewen *et al.* (1973) who showed that α could vary from 10.3×10^{-4} in the marmoset, through 5.6×10^{-4} in humans, to about zero for the wallaby, where clearly the best fit was to $Y = \beta D^2$ as compared to a linear-quadratic or linear one. In the case of the wallaby, the response at the lower exposures (0.5 and 1 Gy) was much lower than for the other species. If there are situations in humans for particular cell types or exposure conditions where $Y = \beta D^2$ for chromosome aberrations, then an argument could be proposed for a cancer response considerably lower than that for the more typical $Y = \alpha D + \beta D^2$ exposure, given the proviso of an association between chromosome aberration responses and those for cancer. At this time there does not appear to be any human chromosome aberration dose response that is purely dose-squared.

A threshold response for radiation-induced chromosome aberrations would be one that has no increase in aberrations until some specific amount of dose, or DNA damage, is obtained. The point of departure could be linear or nonlinear. In general, responses at low exposures are predicted rather than being obtained from direct observation. However, there is one set of data that is very pertinent to this discussion. In a 10-laboratory collaboration (Pohl-Ruling *et al.*, 1983), unstable chromosome aberrations were analyzed in human peripheral lymphocytes irradiated *in vitro* with 250 kVp x rays at doses of 4, 10, 20, 30, 50, 100 or 300 mGy. The reported results were (1) that the frequencies of all types of aberrations at 4 mGy are below the control values, (2) there was not a significant increase in dicentrics up to 20 mGy or up to 50 mGy for terminal deletions, (3) the frequencies of total aberrations were not significantly different at 10, 20 and 30 mGy, and (4) if the data over the whole dose range were fit to $Y = c + \alpha D + \beta D^2$ (where c is the control value), the frequencies at 10, 20, 30 and 50 mGy are below the fitted curve, and the data were clearly nonlinear. The authors concluded that the latter observations could be explained by the increased overall effectiveness of DNA repair mechanisms at doses below 50 mGy. The fact that 10 laboratories participated in this study led others to question the impact of interlaboratory variation not being adequately handled by the statistical analysis, for which standard errors of the mean based on the total number of cells analyzed was used (Lloyd *et al.*, 1992). Also, the control value was significantly higher than the generally observed one. Lloyd *et al.* (1992) in a six-laboratory collaborative study, showed that interlaboratory and interdonor effects introduced statistical uncertainties such that below 20 mGy the lymphocyte aberration assay could not distinguish between a linear or a threshold model, although the dose-response curve was linear down to 20 mGy.

In a similar collaborative effort with 14 laboratories (Pohl-Ruling *et al.*, 1986) the investigators analyzed chromosome aberrations in human peripheral lymphocytes exposed *in vitro* to 14.8 MeV D-T neutrons at doses of 3.5 to 244 mGy. The dose-response curves for dicentrics, terminal deletions, and total aberrations fitted well to $Y = \alpha D$, although it was noted that the data were insufficient to rule out a threshold response below 1 mGy. These data for x rays and neutrons are important in the discussion for linear no threshold responses versus threshold responses at low doses. Despite the quantity of work involved it would appear to be greatly advantageous to conduct a repeat study, or to analyze the data more completely.

Recent data by Tucker *et al.* (1998) provide support for a linear dose-response curve for chromosome translocations assayed using FISH. Mice were exposed to daily doses of 6.4, 18.5, or 55 mGy of ^{137}Cs gamma rays for 21, 42 or 63 d. The dose-response curve for reciprocal translocations assessed in peripheral lymphocytes was linear over the accumulated dose range, but with a lower alpha value than that for acute exposures. This approach provides a method for assessing cytogenetic responses at very low exposure (0.5 mGy suggested by the authors) for each fraction of an extended exposure, allowing for further consideration of the shape of the dose-response curve at very low exposures.

6.3.2 *Effect of Adaptive Response*

An adaptive response for chromosome aberrations has been quite extensively discussed (UNSCEAR, 1994; Vijayalaxmi *et al.*, 1995; Wolff, 1996). In a typical case where an adaptive response has been observed, an adaptive or priming dose of x rays, around 10 mGy, is followed by a challenge dose of 1.5 Gy. However, whether or not an adaptive response is observed varies from study to study and even among individuals from a single study (Bauchinger *et al.*, 1989; Sankaranarayanan *et al.*, 1989; Vijayalaxmi *et al.*, 1995). In fact, synergistic responses have been reported for human lymphocytes exposed to x-ray priming doses of 20 mGy (Olivieri *et al.*, 1994). A thorough discussion of this variability has been presented by UNSCEAR (1994). A general observation is that there is an approximately two-fold reduction in chromosome aberration frequency for achromatic lesions and deletions under adaptive conditions. Thus, the aberration frequencies at doses below 1.5 Gy might be reduced by a factor of two, leading to a nonlinear response but of reduced magnitude compared to a single acute exposure. It is conceivable that the observed adaptive response merely reduces the two-track component of the dose-response curve, without changing the slope at low doses.

There is no evidence to support the contention that an adaptive response can reduce the frequency of chromosome aberrations to zero, thereby leading to a threshold response at low doses (see review in UNSCEAR, 1994). The possible threshold response described in Section 6.3.1 for low-dose x rays and chromosome aberrations is more likely to involve inherent repair effectiveness rather than adaptive effects. It is perhaps important to note that under conditions of chronic exposures, the dose-response curve for chromosome aberrations is linear and the slope is as predicted from the acute, nonlinear response. It is predicted that a true adaptive response should occur at very low-dose rates, but there appears to be no evidence for it. Thus, there remains an ambiguity about an adaptive response for radiation-induced chromosome aberration induction.

6.3.3 *Efficiency of Deoxyribonucleic Acid Repair*

The shape of the dose-response curve for chromosome aberrations is either a reflection of misrepair (*e.g.*, dicentrics, reciprocal translocations, rings, interstitial deletions, and inversions) or of incomplete repair (terminal deletions). The dose-response curve will thus be influenced by the efficiency of DNA repair over the dose range of study. For example, a repair process that depends on dose, is responsive, rapid and complete but error-prone, will result in an increasing slope for dicentrics, reciprocal translocations, rings, and inversions. On the other hand, a repair process that depends on dose but is slow and inefficient at completing repair, could result in an increase in slope for deletions with dose. Either of these repair scenarios will result in an increased slope with dose, but would not result in a threshold response for chromosome aberrations at low doses.

In contrast to the error-prone repair scenario discussed in the preceding paragraph, it is pertinent to consider the effect of a low-dose, error-free pathway. The assessment of repair of radiation-induced DNA damage is almost exclusively at rather high exposure levels as a consequence of the overall insensitivity of the methods used. Given the high efficiency of DNA repair processes at these high-dose levels, it might be reasonable to hypothesize that at very low doses the repair of dsbs by recombinational repair or bds by excision repair is error-free or sufficiently close to it that a threshold or effective threshold obtains. There is no direct experimental evidence for this at present, and it is unlikely that any will be readily forthcoming.

6.3.4 *Inducibility of Deoxyribonucleic Acid Repair and Cell-Cycle Checkpoints*

If DNA repair is induced at low doses of radiation, there could be an enhancement in the rate of DNA repair of induced damage over that of unexposed cells. Provided that the induced repair is not specifically error-prone, the enhanced repair could reduce chromosome aberration yields, particularly in cells exposed during DNA synthesis. A reduction in aberration yields in other cell-cycle stages could occur if the time that the damage remains unrepaired enhances the probability of producing a chromosome aberration, and if the more rapid repair is not more error-prone. This reduction in the yield of chromosome aberrations produced by errors of DNA replication on a damaged DNA template could be even greater if a G_1/S cell-cycle checkpoint is induced by low exposure levels, or low levels of DNA damage, thereby providing more time for repair prior to replication. There is limited evidence for an induced checkpoint response at low levels of induced DNA damage, since most such studies are conducted at rather high exposure levels. However, Huang *et al.* (1996) showed that *p53*-dependent arrest can be obtained by nuclear injection of linearized plasmid DNA, circular DNA with a large gap or single-stranded circular phagemid. The authors suggested that the arrest mechanism in normal human fibroblasts can be activated by a very few dsbs, and that only one may be sufficient based on a statistical argument. If, on the other hand, a cell-cycle checkpoint is not induced at low doses, then more aberrations are predicted to be produced by unrepaired DNA damage during replication.

From the preceding discussion on the effects of inducible DNA repair and cell-cycle checkpoints on chromosome aberration frequencies, it can be seen that scenarios can be developed for supralinear, linear, sublinear and perhaps even threshold dose responses at low levels of exposure. Unfortunately, defining the most likely scenario would seem to be impossible at this time, without the influx of additional, pertinent data.

6.3.5 *Genomic Instability*

Genomic instability, a condition in which affected cells accumulate genetic alterations more rapidly than do normal cells, is a common characteristic of neoplastic progression in premalignant and malignant cells (Nowell, 1976). More recently, delayed genomic instability has been observed several cell generations after exposure in the progeny of irradiated rodent and human cells in culture and *in vivo* (reviewed by Kronenberg, 1994; Little, 1998: Wright, 1998). For example, new nonclonal chromosome aberrations were observed in 40 to 60 percent of mitoses in mouse marrow stem cell cultures scored several cell generations after alpha irradiation (Kadhim *et al.*, 1992). Similarly, delayed instability was observed in ~10 percent of surviving gamma-irradiated human T-lymphocytes, a frequency at least three orders of magnitude greater than expected for specific locus mutations (Lambert *et al.*, 1998). Interest in delayed genomic instability was stimulated by the observations that the frequencies both of radiogenic transformation of cultured cells *in vitro* (Little, 1994) and of radiogenic initiation of cancer per susceptible cell in rodents *in vivo* (Clifton, 1990; Kamiya *et al.*, 1995) are orders of magnitude greater than the frequencies of acutely-induced radiogenic mutations in specific loci. Instability is expressed in a variety of ways in irradiated cell cultures, *e.g.*, in delayed clonal heterogeneity, delayed gene mutations, delayed chromosomal aberrations, reproductive death, reduced plating efficiency, giant cell formation, cell fusion, lowered attachment ability, and morphological neoplastic transformation (Morgan *et al.*, 1996). Insofar as genomic instability may play an important role early in radiogenic cancer induction, it is of importance to consider current knowledge of the physical and biological factors that influence its development and thereby may influence radiation dose-cancer response relationships.

The abnormalities that develop several cell generations after irradiation are predominantly chromatid aberrations that arise in the S or G_2 phases of the cell cycle. This is significant in that such chromatid aberrations must have arisen in the same cell cycle as the one in which they are observed. If observed at subsequent cell divisions, they are derived chromosome-type aberrations (Savage, 1979). For example, delayed chromatid abnormalities were most commonly observed among initially normal human marrow stem cells after exposure to alpha particles (Kadhim *et al.*, 1994), among cultured gamma-irradiated BALB/c mouse mammary epithelial cells (Ponnaiya *et al.*, 1997a), and among gamma- or neutron-irradiated immortal human breast cells (Ponnaiya *et al.*, 1997b). These were not noted to occur in specific chromosomes. In contrast, the delayed chromosomal instability observed in human fibroblasts 20 to 25 culture passages after accelerated heavy ion irradiation were predominantly dicentrics (chromosome-type aberrations induced in G_1) involving telomeric regions of chromosome arms 13p and q, and less frequently, 1p, 16p, and 16q. These nonrandom rearrangements developed in parallel with the appearance of clones with unbalanced karyotypes to which they may have been causally related (Martins *et al.*, 1993). No such specificity was observed among cultured V79 Chinese hamster cells after irradiation with x rays, neutrons, or alpha particles (Trott *et al.*, 1998) nor generally in other reports. Most commonly, a single unstable clone may include cells with one or more new and different karyotypic aberration(s).

The majority of mutations directly induced and observed soon after x irradiation in the HPRT locus of cultured Chinese hamster ovary cells were total or partial deletions, involving loss of one or more whole exon(s). In contrast, mutations that arose several cell cycles after irradiation, like those arising spontaneously in the absence of known radiation exposure, were point mutations involving changes in but 100 to 150 base pairs (Little, 1994; 1998). Mutations in the tumor suppressor *p53* gene have been observed in a variety of human neoplasms including breast cancer. Three distinct single codon *p53* mutants were isolated from one mammary epithelial cell clone derived from a BALB/c mouse that had been irradiated with 1 Gy gamma rays. These mutations were acquired several cell generations after irradiation and the estimated mutation frequency per cell was $\sim 10^{-2}$ Gy^{-1}. Cells bearing *p53* mutations were preneoplastic, forming tumors several months after transplantation in syngeneic recipients. Sublines of some *p53* mutant cells also acquired overexpression of c-Myc and were neoplastic, giving rise to tumors within weeks after grafting (Selvanayagam *et al.*, 1995; Ullrich and Ponnaiya, 1998).

Induction of delayed genomic instability is radiation type-dependent, occurring more frequently and/or earlier following exposures of high-LET than to low-LET radiations (Kadhim *et al.*, 1998; Trott *et al.*, 1998). The frequency also depends on the biological material and the endpoint examined. For example, delayed reproductive cell death, micronucleus formation, dicentric chromosome formation, and apoptosis tended to follow similar patterns but with different magnitudes of effect and/or timing in x-irradiated V79 hamster cells. There was a near-linear increase with dose to 3 to 6 Gy x irradiation followed by a plateau at higher doses (Trott *et al.*, 1998). Mean frequencies of chromatid aberrations were determined ~25 population doublings after plating of BALB/c mammary epithelial cells gamma-irradiated *in vitro* or *in vivo*. In the latter case, the cells were removed from the donor mice and plated 24 h or four weeks after irradiation. Chromatid aberrations in all three groups of cells increased near linearly to ~0.25 Gy. The increase in aberrations slowed to a plateau thereafter up to the largest dose of 3 Gy (Figure 6.2) (Ullrich and Davis, 1999).

Delayed chromosomal instability was examined in GM10115 hamster cells that contained a single copy of human chromosome 4 and 20 to 24 hamster chromosomes (Limoli *et al.*, 1999). Clones from x-irradiated progenitors were scored as positive if they contained three or more cells with aberrations that involved the human chromosome as demonstrated by FISH. In some experiments, the cells were treated before irradiation to induce substitution of 25 to 60 percent of thymidine in their DNA with BrdU (bromodeoxyuridine). The

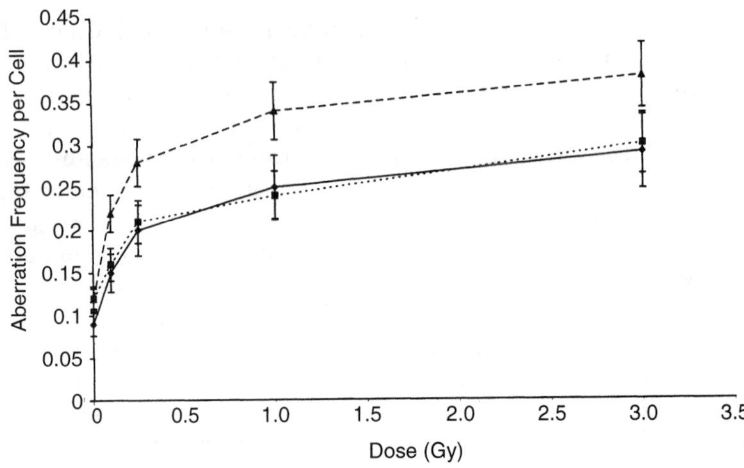

Fig. 6.2. Dose response for ^{137}Cs gamma-ray induced delayed cytogenetic instability in BALB/c mouse mammary epithelial cells. Mean aberrations per cell at 25 population doublings after plating. Cells were irradiated *in vitro* and plated immediately (▲ —) or irradiated *in vivo* in living mice and removed and plated either 24 h after irradiation (■ · · ·) or four weeks after irradiation (▲ – – –). Aberrations frequencies were determined from analyses of approximately 70 metaphases from each of five mice per data point. Vertical bars are standard errors of the means (Ullrich and Davis, 1999).

radiation dose-chromosome anomaly response was roughly linear over the dose range of 1 to 10 Gy for cells without BrdU substitution; a maximum of ~30 percent of the clones contained cells with aberrations after 10 Gy (Figure 6.3). Delayed aberration frequencies after doses <0.5 Gy were not distinguishable from those in clones from unirradiated cells. In clones of irradiated cells with incorporated BrdU, the dose-delayed aberration response rose rapidly after 0.5 to 2.5 Gy and plateaued at ~30 percent of the clones at higher doses (Limoli *et al.*, 1999). The physicochemical composition of the DNA thus strongly influenced the induction of delayed genomic instability.

No effect of dose rate on induction of delayed genomic instability by gamma rays was observed either in GM10115 cells as above (1 to 10 Gy at 0.092 Gy min^{-1} versus 17.45 Gy min^{-1}) (Limoli *et al.*, 1999) or in primary human T-lymphocytes (3 Gy at 45 Gy h^{-1} versus 0.024 Gy h^{-1}) (Lambert *et al.*, 1998). In contrast, division of a total dose of gamma rays into 25 daily fractions of 0.01 Gy each greatly reduced delayed chromatid instability in BALB/c mouse mammary epithelial cells *in vivo* below that seen after a single exposure to 0.25 Gy (Figure 6.4) (Ullrich and Davis, 1999).

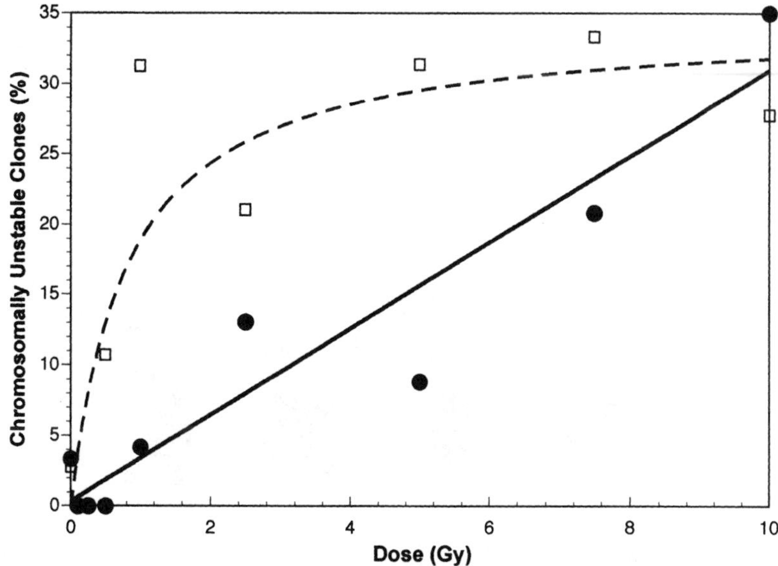

Fig. 6.3. Dose-response relationships for induced chromosomal instability in BrdU-substituted (□, – – –) and unsubstituted (●, —) clonal progeny of x-irradiated GM10115 hamster cells that contained a single human chromosome 4 and 20 to 24 hamster chromosomes. Aberrations that were scored all involved human chromosome 4 as recognized by FISH analysis. 25 to 66 percent of the DNA thymidine had been replaced in the BrdU-substituted cells before irradiation (Limoli *et al.*, 1999).

Susceptibility to induction of instability is dependent on the genetic background of the cells, and radiation-induced genomic instability is transmissible *in vivo* (Wright, 1998). Susceptibility of mouse mammary epithelial cells to development of delayed chromatid aberrations following gamma irradiation is a dominantly inherited characteristic. Mammary cells from BALB/c mice are susceptible to radiation-induced delayed karyotypic abnormalities as well as to radiogenic transformation *in vitro* and radiogenic cancer induction *in vivo*. Mammary cells of C57BL/6 mice are resistant to all three radiation-induced effects, as are cells from F1 hybrids between the two strains (Figure 6.5) (Ullrich and Ponnaiya, 1998). Comparable results were obtained with bone-marrow cells from mice of the CBA/H strain, the closely related DBA/2 strain, and the more distantly related C57BL/6 strain following irradiation with 0.5 Gy alpha particles. Although there were increases in chromosomal aberrations in the clonal progeny of the irradiated cells above those in nonirradiated cells from all three strains, C57BL/6 mouse cells were

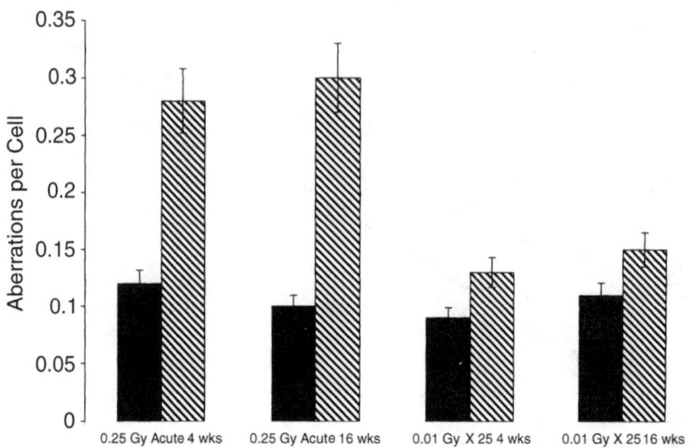

Fig. 6.4. Effect of fractionation on radiation-induced delayed cytogenetic instability. Mean aberration frequencies per cell were evaluated at 25 population doublings after plating. BALB/c mouse mammary cells were ^{137}Cs gamma irradiated *in vivo* with 0.25 Gy either as a single acute dose or in 25 daily fractions of 0.01 Gy each. The mammary cells were then collected from the mice at 4 or 16 weeks after irradiation and plated for culture and analysis. Data from sham-irradiated cells (solid bars) and irradiated cells (hatched bars) are means ± SEM of analyses of approximately 70 metaphases from each of five mice per data bar (Ullrich and Davis, 1999).

significantly less affected than cells from the other strains (Watson *et al.*, 1997). This resistance was inherited dominantly in cells from F1 hybrids of the C57BL/6 mice crossed with either of the two other strains. It is worthy of note that CBA/H mice are susceptible to radiation-induced myeloid leukemia; C57BL/6 mice are resistant to this neoplasm. Cells from some humans appear to also be resistant to induction of genomic instability. For example 0.25, 0.5 or 1 Gy alpha-particle irradiations produced delayed karyotypic aberrations, predominantly chromatid anomalies, in human marrow cell samples from but two of four normal individuals (Kadhim *et al.*, 1994).

A distinctive aspect of radiation-induced genomic instability is the "bystander effect." For example, the frequencies of mouse marrow cells that developed clones that expressed delayed chromosomal instability after alpha irradiation exceeded the number of surviving cells that would be expected to have been traversed by one or more particle tracks at the fluences investigated (Kadhim *et al.*, 1992).

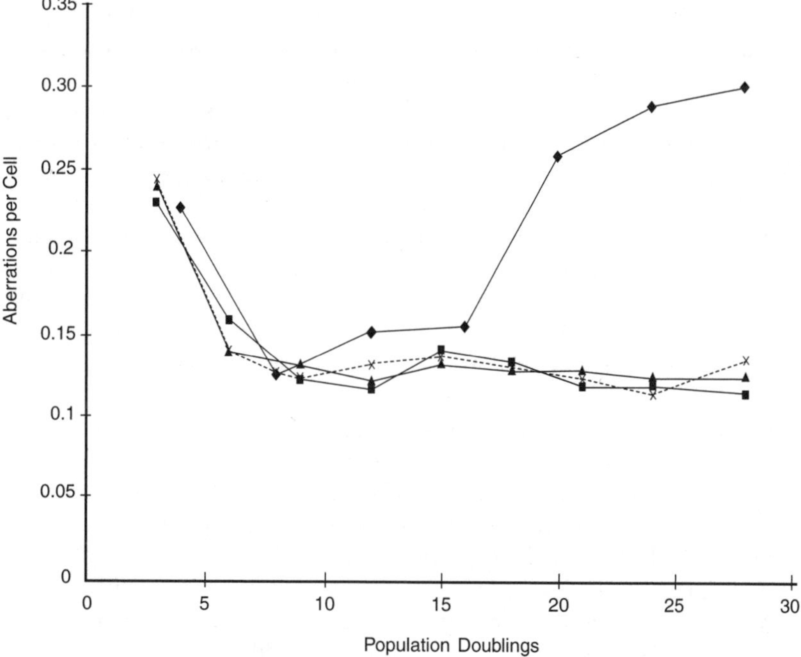

Fig. 6.5. Effect of genetic background on yields of delayed chromatid aberrations in the clonal progeny of 3 Gy gamma-irradiated primary mouse mammary epithelial cells as a function of population doublings after exposure. Cells from mammary cancer-susceptible BALB/c mice (solid diamonds, upper line), from mammary cancer resistant C57BL/6 mice (solid squares, lower line), from B6CF1 hybrids (solid triangles, lower line) and from CB6F1 hybrid mice (single "x"s, dashed line) (Ullrich and Ponnaiya, 1998).

Mouse marrow cells were subsequently exposed to 1 Gy of alpha particles with and without a grid shield interposed between the source and the cell sample. In the absence of the grid, 20 percent of the surviving cells would have been traversed by one or more alpha particles; with the grid in place the expected proportion of surviving cells that suffered at least one traversal was three percent. Despite the marked radiation-related differences in cell survival and in cells suffering particle traversals, 22, 21 and 7 percent of the clones from cells irradiated without the grid, cells irradiated with the grid, and unirradiated cells, respectively, expressed delayed chromosomal instability (Lorimore *et al.*, 1998). The data suggest that factors(s) induced in the cells that had been traversed by one or more particles were transmitted to unirradiated neighboring cells in which delayed instability subsequently developed.

Loeb (1994) has observed that a mutation in any one of more than 20 known genes can dramatically increase the rate of mutations throughout the *E. coli* genome. In particular he notes that a mutator phenotype results from a mutation in any one of three mismatch repair genes or other genes associated with maintenance of genetic stability in *E. coli*, and that there are homologous genes in human tissues. For example, hereditary nonpolyposis colorectal cancer is associated with an acquired mutation in one of the normal copies of the homologous mismatch repair genes hMSH2 or hMLH1 in an individual who has inherited a mutant hMSH2 or hMLH1 allele, respectively. Loeb postulates that a mutation in one of these human genes associated with maintenance of genomic stability may serve as an early step in cancer induction and that mutations in other genome-stabilizing genes may serve similar roles. However, for the very high frequencies of radiogenic genomic instabilities discussed herein to be attributable to directly induced radiogenic mutations would require: (1) that the genes in question be orders of magnitude more mutable than are the genes tested to date, or (2) that there be orders of magnitude more of such genes in human and rodent genomes than have been recognized, or (3) that induction of a condition of high susceptibility to mutation preceded the specific mutational change.

Alternatively, oxidative stress including aberrant oxyradical metabolism has been suggested as a mechanism of induction of genomic instability (Lorimore *et al.*, 1998; Watson *et al.*, 1997; Wright, 1998), and several other mechanisms have been proposed (Chen *et al.*, 1998; Morgan *et al.*, 1996). In the absence of more direct experimental support, these all remain speculative, however.

Recent results with differing biological systems and endpoints in several different laboratories have strengthened the postulated linkage between induced delayed genomic instability and radiation-induced cancer initiation and promotion. Although it has not been mechanistically established, it is an attractive hypothesis on several grounds. If genomic instability is ultimately shown to play a common role in radiation carcinogenesis, it is likely to significantly influence the shapes and natures of radiation dose-cancer response relationships. For example, the effects of pre-existing mutation-sensitive cells on the dose-response curve for radiation-induced genetic alterations and/or cancer would be to increase the slope of the curve rather than to decrease it. The size of any sensitive subpopulation would dictate the magnitude of the change, and possibly the dose range over which there would be such an increase in slope. There does not appear to be a scenario whereby the slope of the response curve would be decreased other than by selective killing of sensitive cells at low doses. The latter would be effective only in cases in which sensitive cells comprise a significant fraction of the entire cell population.

It is of importance to note that a causal relationship between the various measures of radiation-induced genomic instability and carcinogenesis has not been established, and thus the use of data on such induced genomic instability in cancer risk assessment is premature. However, research results in the area of radiation-induced genomic instability merit continuing attention.

6.4 Association Between Chromosomal Changes and Cancer

The fact that cancer is a multistep process and that all tumors contain genetic alterations and exhibit variable degrees of chromosome instability has been known for some time (Nowell, 1976). In recent years, with the advent first of chromosome banding techniques, and then FISH and a whole range of molecular techniques for identifying genomic rearrangements, more detail on specific genetic alterations involved in or associated with the formation of a tumor has been obtained. In fact, the genetic alterations associated with specific steps in the multi-step process of formation of some tumors have been described (Fearon and Vogelstein, 1990; Sandberg, 1993). The chromosomal changes associated with particular tumor types have been comprehensively classified in Heim and Mitelman (1995) and Mitelman (1994). The development of biologically-based dose-response models initially for chemicals, but subsequently for radiation provide a basis for incorporating mechanistic data into risk assessments (Kai *et al.*, 1997).

The general mode of formation of tumors is to activate oncogenes and/or inactivate tumor suppressor genes, resulting in altered cell proliferative capacity or alterations in other cellular "housekeeping" processes. These genetic alterations can be produced by point mutation, deletion, translocation, mitotic recombination, amplification, or non-disjunction, leading to aneuploidy. Two examples will serve to illustrate this point.

A child born with one mutant allele of the recessive gene for retinoblastoma (RB1), frequently a deletion at chromosome 13q14, needs to lose only the one copy of the wild-type allele in a single retinoblast cell in order for a tumor to form during development of the retina (Knudson *et al.*, 1975). This occurs with a frequency approaching 100 percent, given that there are several million retinal cells in each eye. Other steps might also be required and occur at relatively high frequency as a consequence of RB1 loss. The loss of the wild-type allele can arise by nondisjunction, nondisjunction followed by chromosome reduplication (leading to a diploid karyotype), mitotic recombination (reciprocal translocation between homologues), deletion that includes the RB1 locus, gene silencing, or mutation of the RB1 locus (point mutation or small deletion) (Cavenee *et al.*, 1983). Given that loss of the wild-type allele at the RB1 locus is rate limiting, it can be predicted that the dose-response curve for retinoblastomas induced by a specific carcinogen can be represented by a dose-response function that is a combination of the function for point mutations, non-disjunction and interstitial deletions. Knowing the spectrum of the various classes of alterations in a selectable gene induced by the agent under study will provide a reasonable estimate of the specific contributions of each class to overall tumor induction. In simple terms, point mutations (including small deletions) require one DNA lesion for their formation, and will be linear with dose for high- and low-LET radiations. Large, interstitial deletions would generally require two lesions for their formation and are predicted to fit a linear-quadratic curve for low-LET radiations, and a linear curve for high-LET radiations. Non-disjunction is nonlinear with dose (Dulout and Natarajan, 1987; Eastmond and Pinkel, 1990), probably having a threshold at low exposure levels for low- and possibly for high-LET radiations. Thus, the information developed for genetic alterations can be used in a qualitative way to describe the shape of the dose-response curve for a specific tumor type at low exposure levels, beyond the point of observation for tumors themselves. Similar arguments can be followed for loss of the wild-type allele of other tumor suppressor genes.

A second example is that of the role of reciprocal translocations in the development of hematopoietic tumors. A frequent and often apparently singular alteration in leukemias and lymphomas is a reciprocal translocation. The breakpoints are often adjacent to or within a proto-oncogene or within or adjacent to an immunoglobulin locus (B-cell cancers) or a T-cell antigen receptor locus (T-cell cancers) (Rabbitts, 1994). Two outcomes arise from such translocations. In the first, a fusion gene (and fusion protein) can be produced that has a new function, as is the case in chronic myeloid leukemia involving a t(9:22). In this case part of the *BCR* gene on chromosome 22 is brought into juxtaposition by the translocation with part of the ABL proto-oncogene from chromosome 9. The resulting chimeric gene codes for a fusion protein that has acquired an autophosphorylation activity in contrast to the role of ABL alone as a tyrosine-specific protein kinase (Heisterkamp *et al.*, 1993). In fact, high doses of x or gamma rays can induce leukemia-associated fusion genes in hematopoietic cell lines (Deininger *et al.*, 1998; Ito *et al.*, 1993). A second outcome is that a proto-oncogene (typically) is translocated adjacent to a strong immunoglobulin or T-cell antigen receptor promoter that serves to activate the normally silent proto-oncogene. An example is the t(14:18) that is involved in human follicular B-cell lymphomas. The breakpoint on chromosome 14 is within the J_H region of the IgH locus and is part of the V(D)J recombination process of active Ig genes (Showe and Croce, 1987). The chromosome 18 breakpoint is proximate to or within the Bcl-2 proto-oncogene, resulting in placement of Bcl-2 adjacent to powerful enhancer elements of the IgH locus. Many other similar examples have been described in Rabbitts (1994).

Since reciprocal translocations appear to be the rate-limiting step for many leukemias and lymphomas, data on the dose-response curve for translocation induction by ionizing radiations can be used to develop dose-response curves for these tumor types at doses below those at which direct observation is possible. The dose-response curve for translocations is linear-quadratic for low-LET radiations and linear for high-LET radiations. Thus, the respective dose-response curves for leukemias and lymphomas that involve translocations will be linear for high-LET radiations and linear-quadratic for low-LET radiations. This prediction has been tested in an initial way by Mendelsohn (1995) for tumor data obtained for the survivors of the atomic bombs in Hiroshima and Nagasaki. The overall relative risk (RR) for leukemias is linear-quadratic with dose and quantitatively similar to the RR for chromosome aberrations in lymphocytes of the same population. The response at low doses is predicted to be linear or indistinguishable from linear. However, it should be noted that combining all forms of leukemia, as was done by Mendelsohn (1995), can be misleading, given the different modes of their formation.

Thus, given that there is a strong association between the production of specific chromosome alterations and chromosome alterations as a consequence of genomic instability and tumor formation, it is proposed that an appropriate assay for clastogenicity is a good predictor of carcinogenicity for ionizing radiations, and further, perhaps of the shape of the dose-response curve at low exposures.

Knowledge of the mechanism of induction of a particular tumor type, especially a determination of the specific alteration(s) that defines the rate-limiting step(s), or the initiating step, might be used to predict the shape of the dose-response curve at low levels of exposure. As described above, the dose-response curve for point mutations is linear for high- and low-LET radiations, and that for interstitial deletions or translocations is linear-quadratic for low-LET radiations and linear for high-LET radiations. It is reasonable also to use predictions of the frequency of induction of the *specific* genetic alterations involved in this rate-limiting step to provide a quantitative estimate of cancer risk at low exposures.

6.5 Biological Dosimetry for Chromosome Aberrations

6.5.1 *Acute Exposures*

Over the past 30 y there have been a large number of reports of the use of chromosome aberration analysis for estimating biological dose (Bender *et al.*, 1988). The great majority of these involve one or very few individuals since they are radiation accident cases. The exceptions are atomic-bomb survivors, individuals associated with the Chernobyl accident, two smaller populations accidentally exposed to fallout radiation from a nuclear detonation at Bikini Atoll (43 Rongelap Islanders and 14 Japanese fisherman), and individuals treated with x irradiation for ankylosing spondylitis.

The Bikini Atoll study populations received large doses, ranging from 0.7 to 1.5 Gy for the Rongelap Islanders (Lisco and Conard, 1967) and 1.7 to 6.9 Gy for the Japanese fisherman (Ishihara and Kumatori, 1983). A linear response provided the better fit for both study groups; however, since the blood samples were taken 10 or more years after exposure and the aberrations assessed were the unstable ones (dicentrics, rings and acentric fragments), the frequencies were low, and only very limited evidence for the dose-response curve could be assessed.

Chromosome aberrations were also assessed in peripheral lymphocytes of individuals with ankylosing spondylitis who were exposed to single doses or a course of 10 treatments of partial body x irradiation (Buckton, 1983). The dose-response curve for dicentrics and rings following single doses was linear-quadratic and that for accumulated exposures of 1.5 Gy was linear. Again the doses were quite high and any predictions of the responses at low doses could be made only by high-to-low-dose extrapolations.

There is a broad range of published papers on the potential cytogenetic effects of exposure to radiations as a consequence of the Chernobyl accident. These involve investigations on individuals who served as liquidators at the Chernobyl site, inhabitants (children and adults) of surrounding areas, and individuals at distant sites potentially impacted by radiation contamination. The majority of these are lacking information on exposure and, in fact, the chromosome aberration frequencies in peripheral lymphocytes are used as a biological dosimeter. Thus, dose-response information is almost completely absent. The study of Semov *et al.* (1994) does, however, provide some information. Unstable chromosome aberrations were assayed in the peripheral lymphocytes of 31 cleanup workers of the Chernobyl accident who were placed, based upon physical dosimetry, in an exposure category of 120 to 300 mGy. Linear dose-response curves were reported for dicentrics and for all unstable chromosome-type aberrations. The coefficients were 1.4 ± 0.4 per 100 cells Gy^{-1} for dicentrics and 7.2 ± 1.2 per 100 cells Gy^{-1} for all unstable aberrations. These values are similar to the linear coefficient for *in vitro* dose-response curves measured using human lymphocytes exposed to low-LET radiation (Bender *et al.*, 1988).

A considerable amount of effort has been expended on the analysis of chromosome aberrations in atomic-bomb survivors, extending over more than 30 y. The initial analyses were for unstable chromosome aberrations, because of availability of techniques. Subsequent studies have been performed for stable and unstable aberrations. The advantage of studies with stable aberrations is that the frequency of these aberrations declines relatively little with time after exposure, since they are not cell lethal and aberrant precursor cells can repopulate the peripheral lymphocyte pool. However, since the studies were started some 20 y after the exposure, information on dose-response curves is still for residual aberrations and not for induced frequencies (Awa *et al.*, 1978). This is much less of a concern for stable aberrations, as noted, but not negligible when attempting to assess effects at low doses by extrapolation. The assignment of the subjects into broad dose groupings precludes an accurate assessment, by observation or by extrapolation to the low-dose region, of the shape of the dose-response curve at low doses. A reassessment using narrower dose groups is needed for describing the dose-response curve at low doses, provided that the number of individuals in each dose group allows for appropriate sensitivity of analysis.

More recent data have been collected for stable aberrations (largely reciprocal translocations and inversions) using FISH methodology (Lucas *et al.*, 1992) or conventional staining or G-banding (Stram *et al.*, 1993). Lucas *et al.* (1992) showed that the translocation frequencies (measured with specific chromosome painting probes and estimated by expansion to a whole genome response) for 20 atomic-bomb survivors were similar to those obtained for first division lymphocytes irradiated *in vitro*. Thus, the dose-response curve was linear-quadratic with essentially a low-dose linear component. Stram *et al.* (1993) provide a more complete statistical analysis of data on stable chromosome aberrations in 1,703 individuals exposed to atomic-bomb radiation in Hiroshima and Nagasaki and analyzed between 1968 and 1985. They calculated an increase in aberration frequency of 0.08 Sv^{-1} at low doses in Hiroshima samples and 0.126 Sv^{-1} in Nagasaki samples. The chromosome data overall fit a linear-quadratic dose relationship with a greater curvature for Nagasaki samples. The coefficients of the slope were similar to those obtained for acute exposures *in vitro*. It remains difficult to assert that there is or is not linearity at low doses, as opposed to a threshold, since there is an extrapolation over a broad dose range to obtain information in the low-dose range. A conclusion about low-dose linearity is particularly difficult for the Nagasaki data, given the curvature of the dose response, and the relatively limited data for doses (DS86 bone marrow) below 0.25 Gy (Stram *et al.*, 1993).

6.5.2 *Chronic Exposures*

A number of studies have been conducted on populations exposed chronically to ionizing radiations either occupationally or environmentally. The occupational exposure studies are more readily interpreted since the relevant estimates of dose are more accurate in such studies.

Evans *et al.* (1979) reported on chromosome aberration frequencies in 197 nuclear dockyard workers who were followed over a 10 y period. These workers were exposed to mixed neutron-gamma radiation during the refueling of nuclear reactors, with most exposures being below 50 mSv y^{-1}. Aberration frequency was linear as a function of dose and was influenced by age and time of blood sampling after exposure. The aberration frequency for this chronic occupational exposure was $2.32 \pm 1.01 \times 10^{-6}$ Gy^{-1}, which is similar to the mean value of the alpha coefficient of $2.5 \pm 1.1 \times 10^{-6}$ Gy^{-1} in the linear terms of the dose response that was derived from four other large studies summarized in Bender *et al.* (1988).

Lloyd *et al.* (1980) studied aberration frequencies in peripheral lymphocytes of 146 radiation workers from United Kingdom nuclear establishments. The authors employed a half-life of 3 y to weigh individual increments of dose, and they obtained a linear dose-response curve for dicentrics, with a coefficient of $2.22 \pm 0.94 \times 10^{-6}$ Gy^{-1}, which was similar to that observed by Evans *et al.* (1979).

A number of investigators have observed chromosome aberration rates to be elevated in persons residing in areas of high natural background radiation (*e.g.*, Wang *et al.*, 1990a). A number of additional studies have been conducted on populations exposed to high background radiation. These have been summarized by Pohl-Ruling and Fischer (1983). The major problem with these studies, with the exception perhaps of the one conducted on individuals from "the Radon Spa" in Bad-Gastein, Austria, is that it is very difficult to estimate dose, since exposure can be both external and internal. In general, dose-response curves for chromosome aberrations were linear. In the Bad-Gastein study, individuals were exposed externally to gamma rays and internally to alpha radiation from ^{222}Rn and its decay products. At low-dose levels and low-dose rates, largely the result of x-ray external radiation, the dose-response curve for total chromosome aberrations was linear, up to annual doses of 3 mGy, with a plateau for additional dose increments (Pohl-Ruling and Fischer, 1983).

6.5.3 *Evidence for Threshold and/or Linearity in Dose Response*

As noted in the two previous sections on cytogenetic population monitoring following acute exposures (Section 6.5.1) and chronic exposures (Section 6.5.2) with samples taken shortly after or at long times after exposure, the majority of studies report linear dose-response relationships in the lower dose range with the coefficient being quite similar to the alpha coefficient of the *in vitro* linear-quadratic dose-response curves. The only departure from this conclusion was a study of chronic exposure (Pohl-Ruling and Fischer, 1983) that showed an enhanced but linear response at very low doses and dose rates for mixed gamma and alpha exposures. Thus, the majority of biological dosimetry studies, given the problems of accurate dose assessment, provide broad support for a linear response at low-dose levels.

6.5.4 *Implications for Dose Response for Carcinogenic Effects*

In Section 6.4 the association of chromosome alterations with cancer was discussed. It is apparent that there is an association between cell transformation and tumors and altered karyotypes, and that specific chromosomal changes (deletions, translocations and aneuploidies) have frequently been shown to be present in tumors. The utility of chromosome aberration data for predicting cancer dose-response curves at low doses requires that the role of any particular chromosomal change in cancer induction be known, and that such a change be rate limiting, in whole or part, in order that it can be used in a predictive way. Given the fact that the great majority of biological dosimetry studies have been conducted using chromosome aberration types that are lethal to the cell (large deletions, dicentrics and rings), their interpretation in terms of cancer endpoints has to be quite limited. If it is assumed that the dose-response relationships of induction of nonlethal (symmetrical) chromosome aberrations are similar to those for lethal (asymmetrical) ones for which there are data (Lucas *et al.*, 1996), then some general conclusions can be proposed. The fact that all dose-response curves obtained with biological dosimetry studies are linear or linear-quadratic, whether the exposure is acute or chronic, suggests that low-dose responses for cancer are linear at low doses, although the slope of this linear curve will be reduced relative to the linear extrapolation from the lowest point of tumor observation.

6.6 Summary and Conclusions

The available data demonstrate that the majority, if not all, of the different classes of chromosome aberrations require two separate DNA lesions (dsbs, base alterations, cross-links, or complex lesions) for their formation. These two lesions can be produced by a single track of either a high- or low-LET radiation or by two independent tracks of low-LET radiation. However, for higher acute doses of low-LET radiation, multiple tracks within small target volumes increase the probability of multiple DNA lesions or mds. The modes of formation of aberrations involve errors of repair in unreplicated and replicated DNA and errors of replication on a damaged DNA template during normal semi-conservative replication. The consequences of these modes of formation of aberrations are that the dose-response curves for the different classes are linear-quadratic with acute low-LET exposures and are linear with acute, high-LET exposures; with chronic exposures, the curves for low-LET approach a limiting linear slope (x) even at high-dose levels; whereas with high-LET radiations, the dose-response curves for chronic exposures are similar or identical to those for acute exposures. In addition, the slope for the linear response with chronic exposures is the same as that for the linear component of the linear-quadratic curve for acute low-LET exposures. Thus, target theory would predict low-dose linearity for chromosome aberrations for all exposure conditions and radiation types, with differences in relative effectiveness for different conditions and radiation types. The problem with this interpretation is that the fidelity of DNA repair at low levels of exposure has not been assessed, and since radiation-induced aberrations are produced most frequently by errors of DNA repair, the shape of the dose-response curve at very low doses is open to some interpretation. If DNA repair is error-free at low-dose levels, an effective threshold would be predicted. This view of low-dose linearity is supported to some extent by population monitoring studies for acute and chronic exposures to low-LET radiations (Section 6.5), with the proviso that exposure assessments are generally not highly accurate.

There is increasing evidence that chromosome alterations are not only present in tumors as a consequence of genomic instability, but that some of these specific alterations are involved directly in tumor formation. If this is indeed the case, the shape of dose-response curves for chromosome aberrations provides an insight into the shape of dose-response curves for cancer. Such an interpretation would be markedly strengthened if the cancer dose-response curve at low doses could be specifically modeled from data on a particular chromosome aberration involved in tumor formation. Suffice it to say that qualitative characteristics of the dose-response curve for a given cancer, inasmuch as the tumor broadly involves chromosome changes in its formation, can be derived from pertinent chromosome aberration data. Additional data are needed to warrant a dose-response curve for aberration induction that is not linear at low doses. Predictions can be made for threshold responses, but the existing data do not support or refute them.

6.7 Research Needs

The preceding section on the interpretation of chromosome aberration data *per se* and in the context of dose-response curves for cancer serves to identify the areas of research that are needed in order to reduce uncertainty about the dose-response relationships for aberrations and their use in cancer risk assessment.

The mechanism of formation of chromosome alterations, both structural and numerical, following radiation exposure needs to be studied in detail. The relationship between the different classes of radiation-induced DNA damage and the fidelity of their repair in the formation of chromosome aberrations needs to be studied extensively. In particular, it is of importance to develop methods that will allow for the understanding of the processes at low exposure levels. While some pertinent information can be obtained for higher, acute exposures for use in extrapolating to describe responses at low exposures, this must be regarded as an interim scenario. There is also a clear need to establish the mechanisms of induction of chromosomal alterations at low-dose rates for comparison with those operating with acute exposures.

The mechanisms underlying cellular responses to radiation that lead to alterations in overall sensitivity to chromosome aberration induction need to be investigated. These include: establishing the nature of inducible DNA repair processes, especially at low doses and dose rates; the induction of cell-cycle checkpoints at low doses and dose rates; the induction of genomic instability, in general, but subsequently at low doses and dose rates; the universality of an adaptive response, especially for chronic exposures; and the effect of cancer susceptible genotypes on chromosome aberration induction at low doses and dose rates.

The molecular basis of radiation-induced tumors needs to be more clearly understood for animal model systems and, wherever possible, for humans. In particular for the present discussion, there is a need to further elucidate the role of specific chromosome structural and numerical alterations in the production of tumors following high- and low-LET radiations. For the purposes of estimating cancer risk at low radiation doses, there is a need to establish which specific steps in the multistage cancer model for particular tumor types are involved in describing the form of the tumor dose-response curve. In addition, the dose-response curves for specific chromosomal alterations involved in radiation-induced tumors need to be established. The data developed for chromosomal alterations need to be combined with data for other biomarkers of tumor response for deriving biologically based tumor response models.

If more accurate dose assessments can be obtained for human populations exposed to radiation, then valuable information can be generated on responses at low levels of chronic exposures for chromosome aberration classes pertinent to tumor induction. FISH methods now allow transmissible aberrations to be assessed, in contrast to most earlier studies where asymmetrical aberrations were the ones analyzable by conventional staining procedures. G-banding was only rarely used for the analysis of symmetrical (transmissible) aberrations. In the unlikely event that significant population exposures occur, it is essential that high-quality analyses of exposure and chromosome aberrations be conducted in order to predict cancer outcomes. Such assessments have been made for chemical exposures, with a preliminary indication of a good correlation (Bonassi *et al.*, 1995; Hagmar *et al.*, 1994).

7. Oncogenic Transformation *in Vitro* and Genomic Instability

7.1 Dose-Response Relationships

Extensive data are available on the oncogenic transformation of rodent cells, especially C3H10T½ cells and Syrian hamster embryo cells, exposed to both high- and low-LET radiations. In general, dose-response relationships for oncogenic transformations are more complex than for chromosome aberriation or mutations. Remarkable features of the data include the following:

1. The high incidence of oncogenic transformation, around 10^{-3} at 1 Gy, is much too high to be attributed to a conventional mutational event.
2. The transforming event can be identified morphologically so that large-scale experiments can be readily scored entirely *in vitro*. Because of this, extensive quantitative data are available with these systems. Ingenious experiments have been devised that indicate that oncogenic transformation in the petri dish, a surrogate for carcinogenesis *in vivo*, is a multi-step process with the first event having a high probability (too high to be a mutation) and the second event having a fixed but low probability per cell division. This was inferred from experiments with C3H10T½ cells in which it was found that if successive dilutions were made so that fewer and fewer cells were seeded per dish requiring more and more cell divisions before confluence was reached, the same number of transformed foci per dish were still observed (Figure 7.1), *i.e.*, the more divisions a cell went through before confluence, the more likely it was to be transformed. It is this observation that also leads to the inference that a very large proportion (essentially all) of the original population received the first event.

In general, dose-response relationships for human cells are far fewer in number. Extensive data are available on a hybrid cell line (HeLa—normal fibroblasts), in which the conversion to a malignant state can be recognized by changes in surface markers (Figure 7.2).

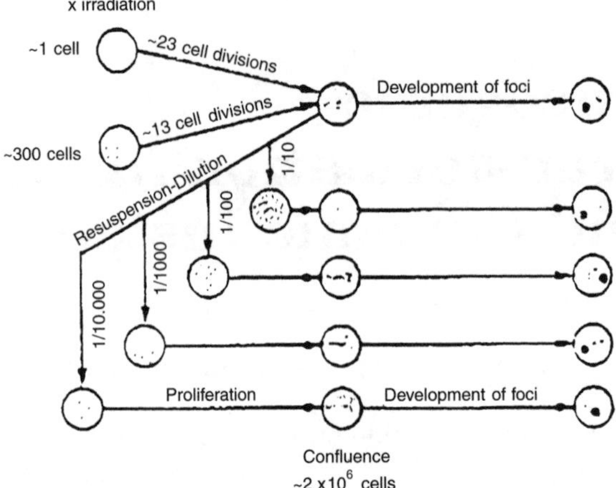

Fig. 7.1. Schematic diagram of dilution experiments with C3H10T½ cells which imply that the first step in oncogenic transformation by radiation is a common event, while the second step occurs with a probability related to the number of cell divisions before confluence is reached (Kennedy, 1985). Since more and more transformations are observed as more cell divisions are allowed before confluence, implies that a large proportion of the original cell population received the first event.

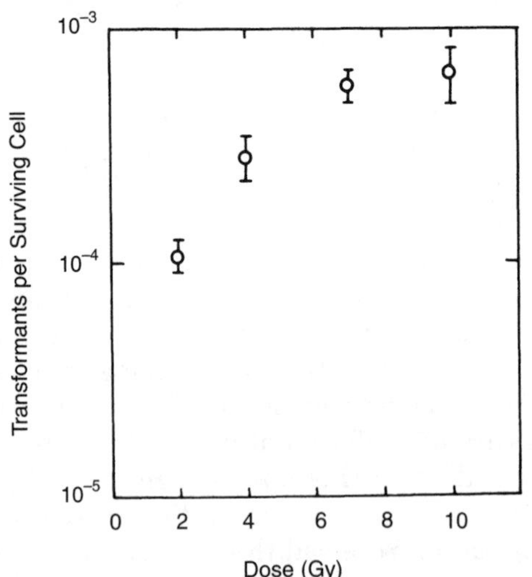

Fig. 7.2. Gamma-radiation-induced oncogenic transformation of the human hybrid cell line CGL1 to the form which expresses a 75 kDa membrane phosphorprotein, a transformation which correlates with the expression of tumorigenicity (Redpath *et al.*, 1987).

Oncogenic transformation for this cell line involves the deletion of a tumor-suppressor gene on chromosome 11, and other changes on chromosome 14 (Mendonca *et al.*, 1998; Redpath *et al.*, 1987). The dose-response data for oncogenic transformation of these cells are almost identical both qualitatively and quantitatively to those for C3H10T½ mouse fibroblasts.

No one has yet succeeded in transforming primary human cells with ionizing radiation. Several groups have shown transformation of cells of human origin immortalized with various viruses. Point estimates of the transformation frequency are available for immortalized human fibroblasts and epithelial cells, which are in the range of 10^{-5} to 10^{-7} for radiation doses of less than 1 Gy (Hei *et al.*, 1994; Rhim and Dritschilo, 1991; Thraves *et al.*, 1990) which is two orders of magnitude lower than for rodent cells (Figure 7.3). In most instances, the conversion to malignancy must be identified by tumorigenesis in animals, since morphological changes that can be seen in a petri dish are not an early event in human cells (Hei *et al.*, 1994; Namba *et al.*, 1986; Willey *et al.*, 1991). Because of the size and difficulty of the experiments involved, dose-response relationships are not available for human cells, even for immortalized human cells.

If a very approximate estimate is made of the number of target cells at risk in the female breast, the incidence of radiation-induced breast cancer in the Japanese atomic-bomb survivors translates into a transformation frequency per cell of about 10^{-15} Gy^{-1}. It should be no surprise, therefore, that it is difficult to transform normal human cells unless they are immortalized.

7.2 Shape of the Dose-Response Relationship for Oncogenic Transformation

The first dose-response relationship for oncogenic transformation *in vitro* was obtained by Borek and Hall (1973) using Golden hamster embryo cells and is illustrated in Figure 7.4. Dose-response relationships for C3H10T½ cells are more numerous; examples of x-ray dose-response curves from three laboratories are shown in Figures 7.5, and 7.6, for doses up to 12 Gy. When the transformation frequency per surviving cell is plotted as a function of dose, the frequency rises as an apparently linear function of dose until it reaches a plateau of about 10^{-2} to 10^{-3} at a dose of about 6 Gy. If the data are expressed in terms of transformation frequency per initial cell at risk, the frequency rises rapidly to a peak of about 10^{-4} at about 3 Gy and subsequently falls off to parallel the cell survival curve (Figure 7.6).

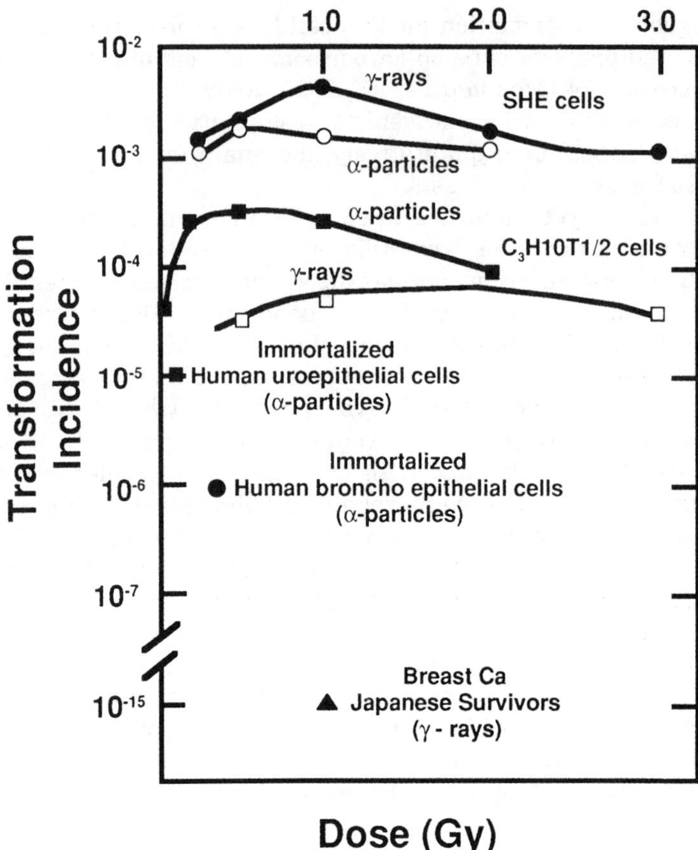

Fig. 7.3. Summary from data in the literature of the approximate incidence of oncogenic transformation as a function of radiation dose for various cellular systems. It is intended simply to illustrate that rodent cells are much more sensitive to radiation-induced oncogenesis than even immortalized human cells. The figure for breast cells in the Japanese survivors is a rough calculation based on the incidence of breast cancer for a given dose and an estimate of the number of breast cells at risk.

In the absence of radiation, these *in vitro* assays have a transformation frequency of about 10^{-4}.

The lowest dose at which a significant excess can be seen depends on (1) the radiosensitivity of the endpoint, (2) the size of the study (number of dishes), and (3) the spontaneous background level of the endpoint. Hamster embryo cells are more sensitive to the induction of oncogenic transformations and the background level is extremely low, so that a dose as low as 0.01 Gy of x rays can be detected

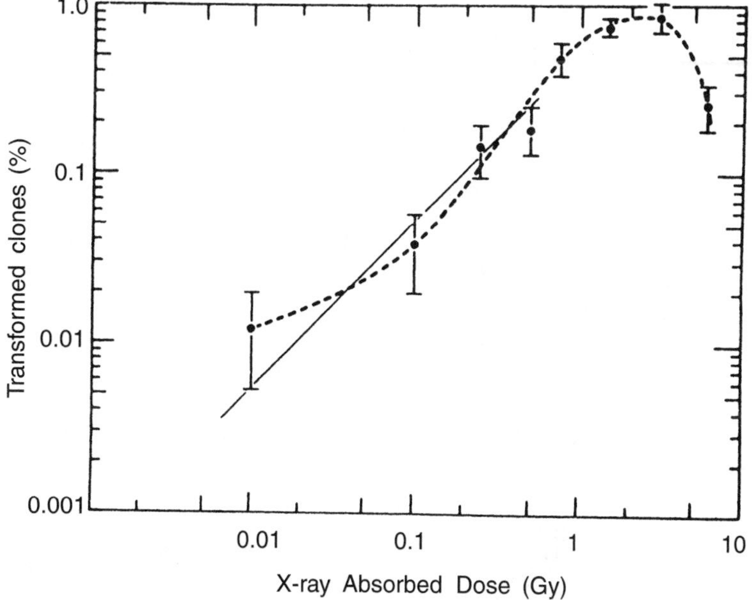

Fig. 7.4. Incidence of hamster embryo cell transformation following exposure *in vitro* to x irradiation. For doses at which more than one experiment was performed, the data were pooled; the mean value, together with the standard deviation, is plotted in the figure. The broken line is drawn by eye to the mean data points; the solid line has a slope of +1 and passes through the error bars of each datum point (Borek and Hall, 1973).

(Figure 7.4). By contrast, the lowest dose used in experiments with C3H10T½ cells, even when hundreds of dishes are used, is 0.1 Gy because of the high background level of transformation characteristic of this cell line (Figure 7.7).

A few attempts have been made to investigate the shape of the dose-response relationship at doses below 1 Gy, either of x rays or of more densely ionizing radiations (Bettega *et al.*, 1985; Miller and Hall, 1978; Miller *et al.*, 1979; 1982). Such experiments are fraught with difficulties, and a large number of dishes must be used to detect the increase in transformation incidence above the background level. There is some evidence of a structure to the curve, *i.e.*, the initial portion is not an extrapolation of the higher dose region (Figures 7.7 and 7.8). While such data are a strong function of the exact experimental technique used, they do suggest that caution must be used when extrapolating from intermediate to low doses. The complex shape of the dose-response relationship when efforts are made to examine the shape critically at low doses is most likely due to the

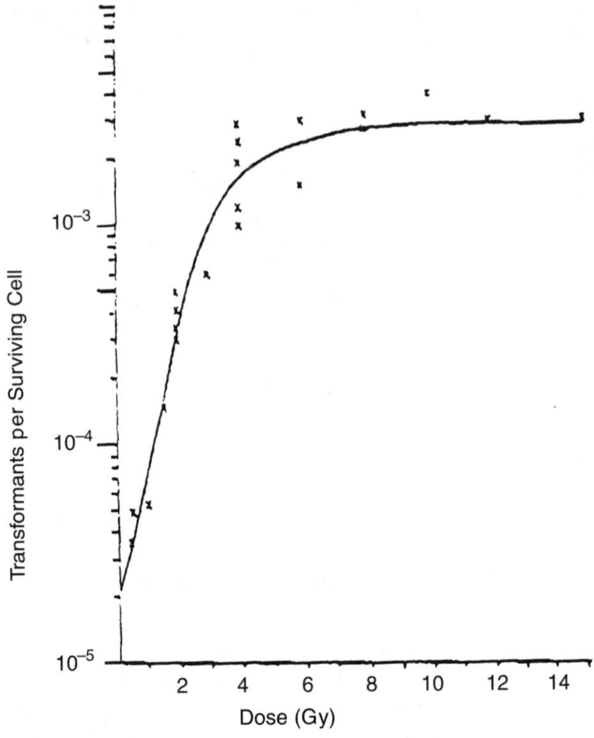

Fig. 7.5. X-radiation-induced oncogenic transformation in exponentially growing C3H10T½ cells *in vitro*. Each point represents data from 20 to 100 dishes. At low doses with low associated transformation frequencies, more dishes were necessary in order to obtain a total number of transformants comparable to that observed at higher doses (Terzaghi and Little, 1976).

variation of sensitivity to oncogenic transformation with phase of the cell cycle. Two laboratories have shown that, in the case of x or gamma rays, there is a marked variation in sensitivity through the cell cycle, with a narrow window of marked sensitivity in G_2/M (Cao *et al*., 1992; Miller *et al*., 1992). This is illustrated in Figure 7.9. This variation in sensitivity would lead to a very steep initial shape at very low doses as the most sensitive cells are transformed, followed by a less steep dose-response relationship for the majority of the population. It would also account for the variability in results between different laboratories and the dependence of transformation incidence on the precise experimental conditions, since small changes in technique can alter the proportion of cells in or close to mitosis.

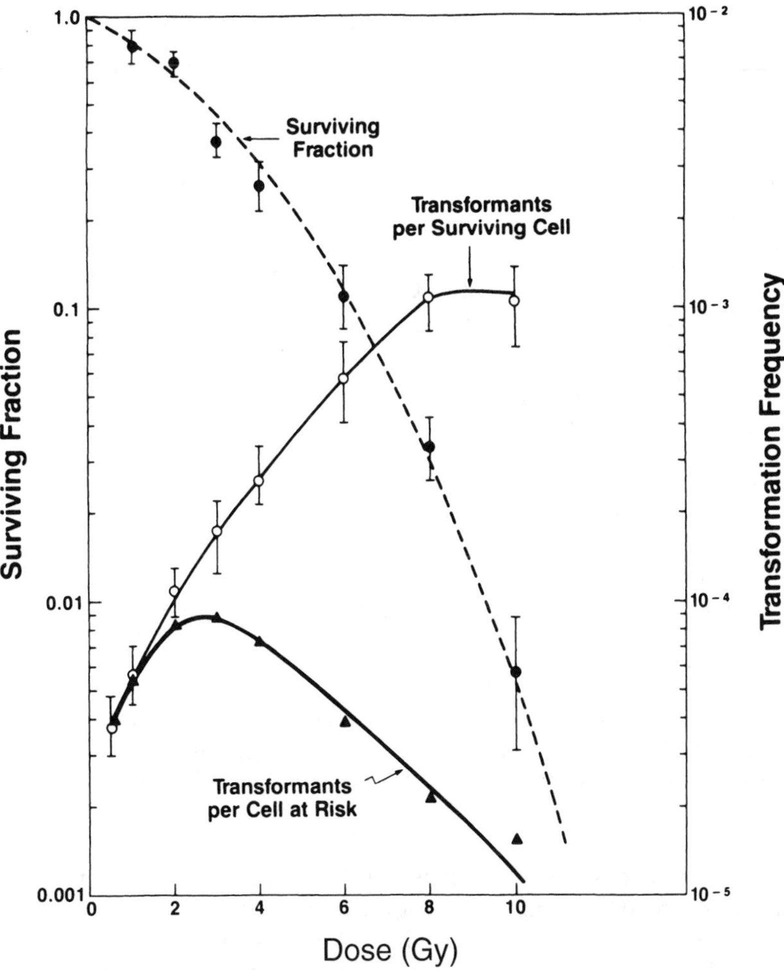

Fig. 7.6. Dose-response relationships for cell survival and for oncogenic transformation in mouse embryo (C3H10T½) cells. Transformation frequency per surviving cell rises with dose to a plateau by 6 to 8 Gy. When cell survival is factored in, transformation frequency per initial cell at risk rises rapidly with dose, reaches a maximum at about 2 to 3 Gy and then falls to parallel the cell survival curve (Hall and Hei, 1987).

A large collaborative study involving six European laboratories was recently published which concluded that the x-ray dose-response curve for oncogenic transformation in C3H10T½ cells "strongly supported" linearity (Mill *et al.*, 1998). However, the pooled data do not shed much light on the question of a threshold or linearity at low

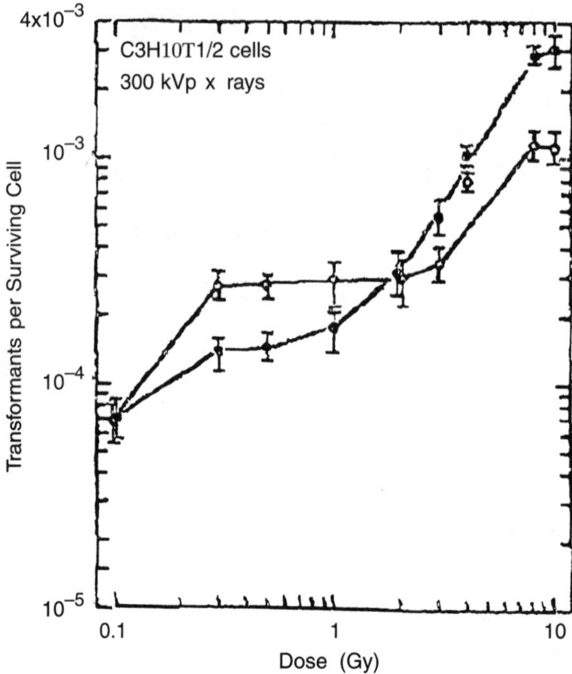

Fig. 7.7. Pooled data from many experiments for the oncogenic transformation incidence for C3H10T½ cells exposed to single (●) or split (○) doses of x rays. The interval between split doses was 5 h (Miller *et al.*, 1979).

doses since the lowest dose at which there was a significant excess of transformed foci over background was 1 Gy.

Miller *et al.* (1999) compared the transformation incidence on C3H10T½ cells induced by exactly one alpha particle through the nucleus (produced by a microbeam) with a Poisson distribution having a mean of one (produced by a broad beam irradiation in the track segment mode). It was found that exactly one particle was significantly less effective than a mean of one, and indeed the transformation incidence produced by the single particle was not significantly above the spontaneous level.

The authors interpreted the higher transformation incidence induced by the Poisson distribution of particles to be due to the minority of cells traversed by more than one particle. An alternative explanation suggested is that while cells irradiated with the microbeam received only a single particle through the nucleus, cells exposed under broad beam conditions may have received one or more particle traversals through the cytoplasm, as well as a mean of

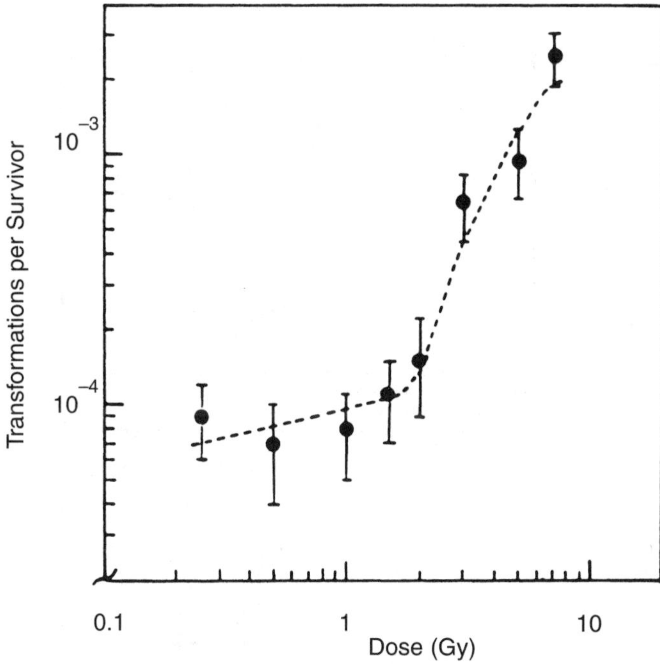

Fig. 7.8. Oncogenic transformation frequencies per surviving cells in C3H10T½ cells after exposure to different doses of 31 MeV protons (Bettega *et al.*, 1985).

one through the nucleus. Recent experiments with the same microbeam have shown that mutations can be induced by cytoplasmic irradiation, though this observation has not as yet been extended to transformation.

The data of Miller *et al.* (1999) are intriguing and while they do not cast much light on the question of linearity at low doses, they do cast doubt on the validity of a linear extrapolation from high to low doses. This has already been noted from other transformation experiments.

7.3 The Bystander Effect

Sigg *et al.* (1997) found that the oncogenic transformation incidence in C3H10T½ cells, induced by beta rays from ^{90}Y, was significantly enhanced in the presence of heavily damaged cells. They postulated that this was due to communication between cells *via*

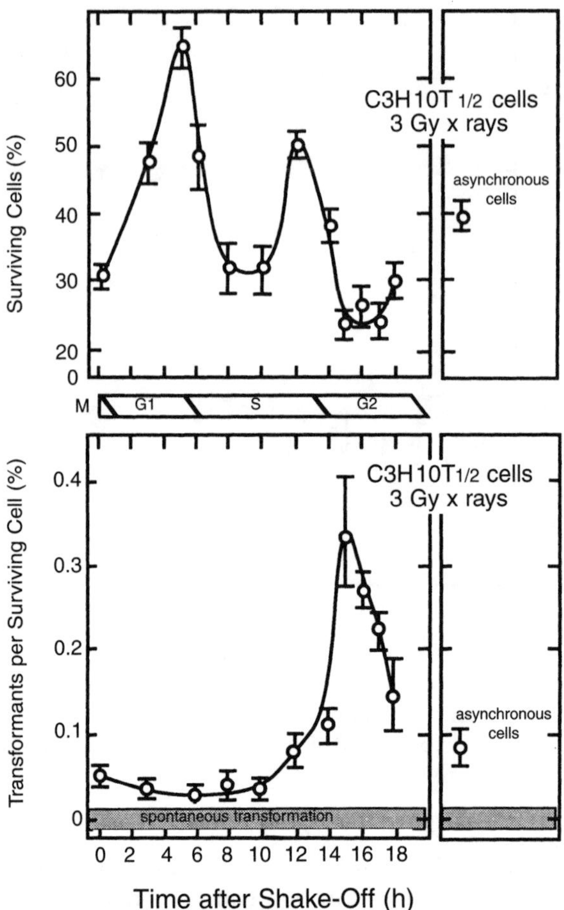

Fig. 7.9. (Top) Percentage of cells surviving a dose of 3 Gy as a function of time after mitotic shake-off. (Bottom) Transformants per surviving cell as a function of time after mitotic shake-off. In both cases the error bars are 1 SE from the mean (Miller *et al.*, 1992).

cell-to-cell contact, but were unsuccessful in isolating the signaling agent. This finding hints at a possible "bystander" effect in transformation. This observation is perhaps analogous to the report of Mothersill and Seymour (1998) of the killing of unirradiated cells by medium from cultures of irradiated cells, which implies the release of a cytotoxic substance by the irradiated cells. Similar effects were observed with late expressing chromosome aberrations as the endpoint using alpha particles and hemopoietic stem cells (Lorimore *et al.*, 1998).

7.4 Transformation by High Linear-Energy Transfer Radiations

There is an abundance of information concerning oncogenic transformation in C3H10T½ cells by a variety of high-LET radiations. Data for fission spectrum neutrons come from Hill *et al.* (1984; 1985a; 1985b) and for a range of essentially monoenergetic neutrons from Miller *et al.* (1989a). There is an equally large body of data for transformation incidence in C3H10T½ cells induced by charged particles covering a very wide range of LET values. Miller *et al.* (1995) used alpha particles of defined LET while Yang *et al.* (1985) used high-energy heavy ions, including carbon, neon, silicon, iron and uranium ions.

In all cases, whether neutrons or charged particles, the most biologically effective radiation had an LET of around 100 keV μm^{-1}. In this region, the dose-response curve for oncogenic transformation closely approximates a linear function of dose. These data will not be presented in detail since linearity is not in question for such densely ionizing radiations, and there is no evidence of a threshold since a single alpha particle has been shown to be capable of producing transformation in C3H10T½ cells (Miller *et al.*, 1999).

7.5 The Dose-Rate Effect

For high-LET radiations the effect of dose protraction on oncogenic transformation has been investigated in detail by Hill *et al.* (1984) (Figure 7.10). The slope of the curve becomes shallow for continuous low-dose rate or for a fractionated exposure, but in all cases there appears to be a linear relationship between dose and transformation incidence down to doses of around 0.2 Gy. Similar data have been obtained by Balcer-Kubiczek *et al.* (1987).

For high-LET radiations, such as neutrons and charged particles, there is evidence of an inverse dose-rate effect, *i.e.*, the same dose appears to be more effective at low-dose rate than at high-dose rate. This was first demonstrated for *in vitro* transformation by Hill *et al.* (1984; 1985a; 1985b) using modified fission spectrum neutrons (Figure 7.11). Using charged particles in the track segment mode, Miller *et al.* (1993) showed that the increase in oncogenic transformation (by a factor of about three) due to dose *protraction* for a given total dose depended on LET and was maximum for a LET of close to 100 keV μm^{-1}. These same investigators showed that dose protraction only led to an increase in transformation for a given dose of

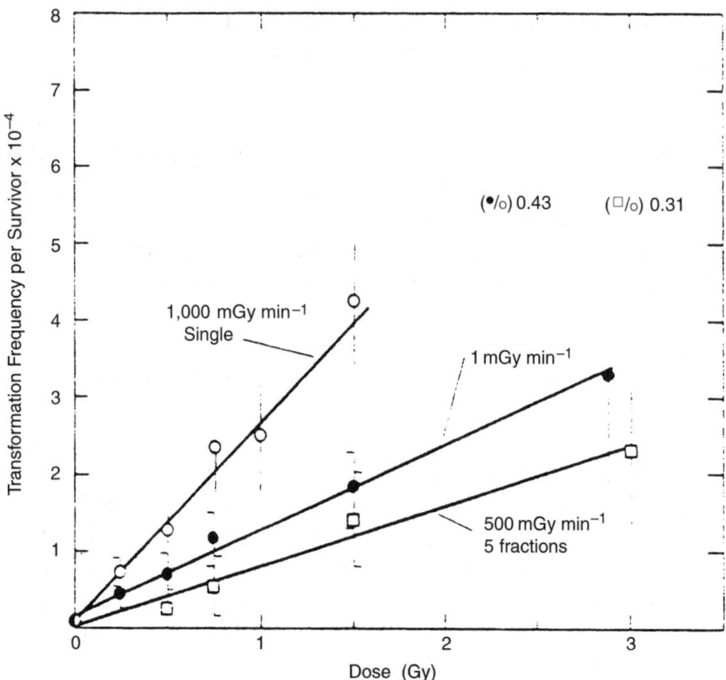

Fig. 7.10. The reduced induction of oncogenic transformation of C3H10T½ cells by fission spectrum neutrons due to a reduction in dose rate or the fractionation of high-dose-rate exposures. During irradiation at 1 mGy min^{-1}, cells were under active growth conditions (37 °C) and the five high-dose fractions were 24 h apart (Hill *et al.*, 1984).

high-LET radiation for cycling cells, and that the phenomenon disappeared for cells irradiated in plateau phase. This is consistent with the biophysical models that have been developed to account for the inverse dose-rate effect, all of which are based, in one way or another, on the variation in cellular response as cells move through the cell cycle (Brenner and Hall, 1990; Elkind, 1991; Rossi and Kellerer, 1986). Some investigators have failed to find an inverse dose-rate effect (Balcer-Kubiczek *et al.*, 1988; 1994) with the combination of dose, dose rate, and LET that they used.

7.6 Modulation

The frequency of radiation-induced oncogenic transformation can be modified by a variety of post-irradiation manipulations.

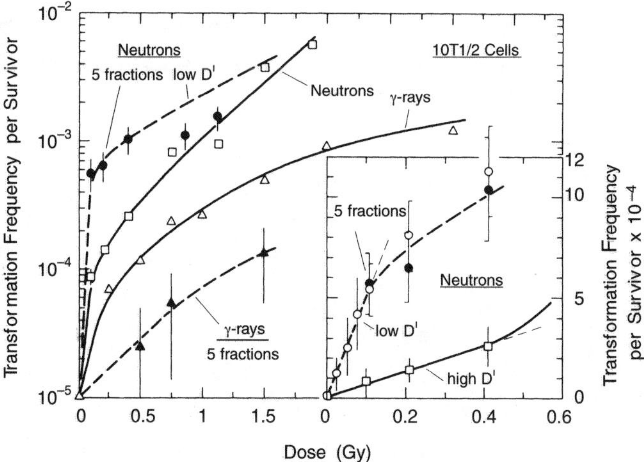

Fig. 7.11. Composite data for the effect of dose protraction on the induction of transformation due to fission-spectrum neutrons produced by the Janus Reactor. For orientation, the induction curves for high-dose rates of gamma rays (Δ) and neutrons (□) are shown. The dashed curve for neutrons, both in the inset and the main figure, is for low-dose rate exposures. The data on the dashed curve for fractionated gamma rays (Δ). On linear-linear coordinates, the inset shows the results at 0.83 mGy min^{-1} (○) and at a high-dose rate (□). Frequencies induced by five fractions of high-dose-rate neutrons (●) or gamma rays (Δ), 24 h apart, are shown in both parts of the figure. The slopes of the initial, linear parts of the two curves in the inset are in the ratio 8.1:1, low- to high-dose rates (21) (Hill *et al.*, 1985a).

First, the incidence can be significantly increased by the addition of 12-0-tetradicanoyl phorbol 13-acetate (TPA), a tumor promoting agent. The effectiveness of TPA is greatest for x rays and decreases progressively as the LET of the radiation increases. This is illustrated in Figure 7.12. In the case of x rays, a particularly interesting observation is that the addition of TPA "flattens" the age response function for transformation, *i.e.*, it has little or no effect on the most sensitive G_2/M cells, but it raises the incidence of transformation in the cells irradiated in the remainder of the cycle to equal that characteristic of the most sensitive cells. It has also been shown that the post-irradiation addition of TPA converts the linear-quadratic relationship between transformation frequency and dose into a purely linear one (Balcer-Kubiczek and Harrison, 1988). A variety of agents added post-irradiation can inhibit transformation, most notably the protease inhibitor antipain, but also such agents as DMSO and vitamin E. Balcer-Kubiczek *et al.* (1993) demonstrated the effectiveness of the aminothiols WR-1065 and WR-151326, each

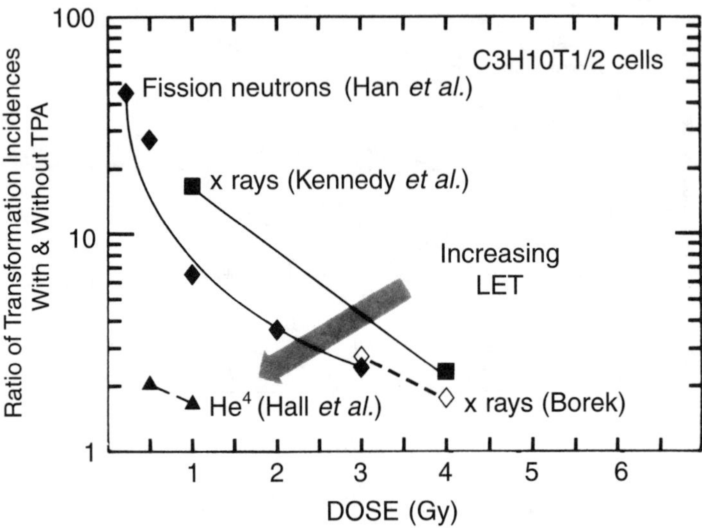

Fig. 7.12. Effects of TPA on the oncogenic transforming potentials of various high- and low-LET radiations (Hall *et al.*, 1989).

at a concentration of 1 mM, to protect C3H10T½ cells against the transforming effects of fission-spectrum neutrons.

7.7 Genomic Instability

Experiments with *in vitro* assays for oncogenic transformation have contributed significantly to our knowledge concerning genomic instability, which is one of the hallmarks of cancer.

Some years ago Kennedy (1985) showed that the yield of transformed foci from irradiated C3H10T½ cells could be increased substantially by successive dilutions which increased the number of cell divisions between irradiation and confluence. They interpreted these data to indicate that radiation induced a high-frequency initial event, which was followed by a second event of much lower frequency related to each successive cell division. The frequency of the initial event was much too high to be attributed to a mutation, which led to a great deal of speculation that possibly epigenetic events could be involved.

More recently, Hei *et al.* (1994; 1996) transformed immortalized human bronchial epithelial cells with a 0.3 Gy dose of alpha particles

(150 keV μm^{-1}); this corresponds to, on average, one alpha-particle traversal per nucleus. If injected into immunologically suppressed mice immediately after irradiation, the cells did not produce tumors, but became tumorigenic when subcultured for about 30 generations. During this period of subculture, a cascade of progressive events took place: increased saturation density (*i.e.*, number of cells growing per square centimeter), followed by loss of contact inhibition, decreased sensitivity to factors that induce terminal differentiation, and a gradual accumulation of a variety of chromosomal aberrations. Parallel unirradiated cultures exhibited no such changes, maintained a stable phenotype and did not produce tumors in animals. This is illustrated in Figure 7.13. This crescendo of events in irradiated cells would appear to be an excellent example of genomic instability: an accumulation of chromosomal aberrations and morphological changes, culminating in tumor formation in immunologically suppressed mice. It was further shown that this instability was not due to a deficiency in mismatch repair gene function (Xu *et al.*, 1997).

7.8 Adaptive Response

Azzam *et al.* (1994) showed that chronic exposure of plateau-phase C3H10T½ cells to ^{60}Co gamma rays at doses as low as 0.1 Gy

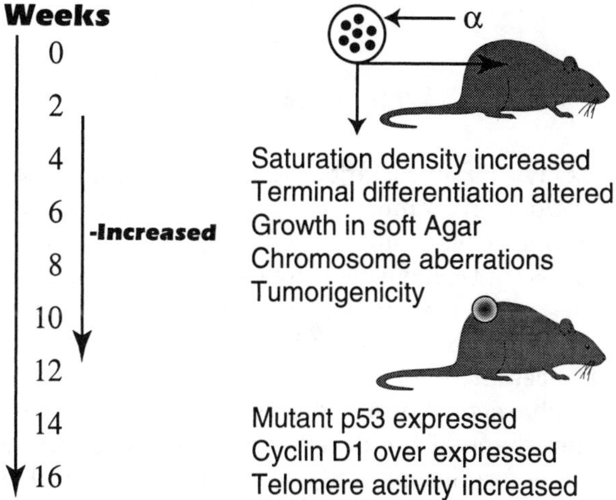

Fig. 7.13. Schematic diagram showing how immortalized human bronchial epithelial cells irradiated with alpha particles exhibit genomic instability. If injected into nude mice immediately after irradiation, no tumors are formed. When subcultured for about 30 generations, a cascade of events occurs leading to tumorigenicity.

protected the cells against subsequent oncogenic transformation by a subsequent large acute radiation exposure.

The same authors subsequently reported that a single exposure of quiescent C3H10T½ cells to gamma-ray doses as low as 1 mGy reduces the risk of oncogenic transformation, from the spontaneous level to a rate three to four-fold lower (Azzam *et al.*, 1996). Increasing the dose, up to 100 mGy at the same low-dose rate (2.4 mGy min^{-1}) did not further reduce the transformation frequency. This protective effect was seen only in cells with an abnormally high spontaneous transformation frequency. These results appear to show that low doses delivered at a low-dose rate can induce processes that protect the cell against "naturally" occurring as well as radiation-induced oncogenic transformation. As such, the effect would be an example of an adaptive response of particular relevance since the endpoint observed (oncogenic transformation) is closer to carcinogenesis than chromosomal aberrations or mutations. However, the results must be viewed with some caution since it is known that there is a very narrow window of sensitivity to radiation-induced transformation in G_2/M. The differential sensitivity between this window and the remainder of the cell cycle is striking. It is not unreasonable to assume that the "conditioning dose" depletes the population of the cells that are most likely to be transformed by a subsequent dose, or indeed transformed spontaneously.

7.9 Summary

In vitro assays for oncogenic transformation in cells of rodent origin have yielded an abundance of quantitative data. The incidence of transformation reaches about 10^{-3} at approximately 3 to 5 Gy. A significant excess of transformed clones or foci can be detected down to 100 mGy of gamma rays or 10 mGy of neutrons. The dose-response relationship tends to have a complex shape; the initial slope at low doses is not necessarily an extrapolation of the slope at high doses. An important characteristic of rodent systems, which may contribute to the complex shape of the dose-response relationship, is the dramatic variation of radiosensitivity with phase of the cell cycle. Cells are most sensitive to radiation-induced oncogenic transformation in G_2/M. For low-LET radiations, lowering the dose rate decreases the yield of transformants, but under some circumstances with high-LET radiations, an inverse dose-rate effect has been reported, *i.e.*, for the same total dose, a low-dose rate resulted in a higher yield of transformants.

Cells of human origin are relatively resistant to transformation by ionizing radiations; in fact there are no reports of radiation-induced transformation in human cells unless they have already been immortalized. Even then, the incidence of transformed foci (that cause tumors in nude mice) is of the order of only 10^{-7} at a dose of 0.3 Gy of alpha particles or high-energy iron ions. Because of this low incidence, dose-response relationships have not been produced, and so it is not possible to comment on their shape. However, it is pertinent to note that transformation of C3H10T½ cells can occur (albeit at low incidence) by the passage of an average of one alpha particle per nucleus.

The mechanism(s) of oncogenic transformation by radiation is not clear. In rodent cells, it appears to be a multi-step process, with an initial event that has too high a probability to be explained by mutation at any one genetic locus. In both rodent and human cells the endpoint of a morphologically identifiable focus, or a tumor in a nude mouse, is preceded by an accumulation of chromosomal aberrations and other changes that are characteristic of genomic instability.

The data for oncogenic transformation have uncertain implications for the question of the linear, nonthreshold model. For Syrian hamster embryo cells, the data are consistent with linearity from 0.01 to 1 Gy of gamma rays, though the error bars are large (Figure 7.4). For C3H10T½ cells, the data are linear down to 0.2 Gy for neutrons (Figure 7.10). If genomic instability induced by a mutation in a gene or genes responsible for genomic stability were the mechanism of radiation-induced transformation, then linearity at low doses would be a credible hypothesis; however this mechanism is by no means proven. Indeed, since genomic instability appears to be induced in a large fraction of the irradiated cell population (10 to 40 percent), the target would be too large for it to be a mutation in a gene or even a set of genes. While this is true of oncogenic transformation in rodent cells, it may not be true in human cells which are orders of magnitude less susceptible to transformation by radiation. For human cells, the target size is commensurate with a point mutation (Ward, 1991).

7.10 Research Needs

A principal need is to develop transformation assays based on human cells. Present systems involve cells immortalized by a virus which introduce artifacts. Also, there is need to define the molecular changes involved in the process of radiation-induced transformation in human cells, especially those involved in the high frequency of the initial transforming event. There is also a need to develop microbeam facilities to irradiate single cells with exactly one photon, to address the question of a threshold for a low-LET event.

8. Carcinogenic Effects in Laboratory Animals

8.1 Introduction

Laboratory animal experiments have helped to define: (1) the shapes of the dose-response curves over a wide range of doses; (2) the influence of the spatial and temporal distribution of the dose on the dose-response relationship; (3) the extent to which the dose-response relationship may vary with sex, age, the organ irradiated, and other variables; and (4) the molecular and cellular mechanisms of the effects in question (NAS/NRC, 1990; UNSCEAR, 1986; 1993; Upton *et al.*, 1986). The results of the various studies significantly extend and amplify the human data. Although a comprehensive review of the animal literature is beyond the scope of this Report, findings pertinent to evaluation of the linear-nonthreshold dose-response hypothesis will be discussed.

8.2 Characteristics and Multistage Nature of Carcinogenesis in Model Systems

The benign and malignant neoplasms of laboratory animals are similar, in general, to their human counterparts, comprising a diversity of histological types, which vary widely in their relative frequencies, rates of growth, and degrees of malignancy. Most, if not all, of the neoplasms appear to be clonal growths, arising through a succession of stages which for convenience have been termed "initiation," "promotion," and "progression" (UNSCEAR, 1993). Although the mechanisms underlying carcinogenesis are still incompletely known, genetic and epigenetic alterations of various types have been implicated.

Of particular interest in relation to the dose response for carcinogenic effects, especially following acute irradiation, is the frequency and nature of the initiation process, the first step in multistage carcinogenesis. Assuming that initiation has not already occurred *via* unknown causes, radiogenic initiation will be a significant factor in the dose-response relationship. In recent years, *in vivo/in vitro* experimental models have been developed with which aspects of radiation effects, cell interactions, and carcinogenesis can be investigated at the level of the cancer progenitor cell. These models include rat tracheal epithelial cells (Luebeck *et al.*, 1996; Terzaghi *et al.*, 1978; Terzaghi-Howe and Ford, 1994), mouse mammary epithelial cells (Adams *et al.*, 1984; 1987; Ethier and Ullrich, 1982), rat mammary epithelial cells (Clifton, 1990; Gould *et al.*, 1977; Kim *et al.*, 1993), and rat thyroid epithelial cells (Clifton *et al.*, 1978; Watanabe *et al.*, 1988). In these models, carcinogenic initiation per cancer-susceptible cell is frequent, as common as 1 cell in 10 for 5 Gy irradiated rat thyroid clonogens (Domann *et al.*, 1994; Mulcahy *et al.*, 1984) and 1 in 100 for 7 Gy irradiated rat mammary clonogens (Clifton *et al.*, 1986; Kamiya *et al.*, 1995). Induction of initiation in these models, like induction of neoplastic transformation of rodent cells in culture (see Section 7), is thus far more common than induction frequencies of known specific locus mutations (Clifton, 1996). Hence, more common alternative processes, some of which may persist for one to several cell generations, have been suggested as possible initiating mechanisms (Kennedy, 1985; 1991). These include increased transcription of specific genes (Boothman *et al.*, 1993), altered DNA methylation patterns (Feinberg *et al.*, 1988; Hardwick *et al.*, 1989; Holliday, 1991), and more recently, delayed genomic instability.

Such delayed instability is characterized by its development at very high frequencies in the progeny of irradiated susceptible cells, several population doublings after exposure, and results in high frequencies of both nonclonal chromosome aberrations and point mutations. It is influenced by the genetic background of the exposed cells, is dependent on radiation dose and type, and is reduced by dose fractionation. For a more detailed discussion of radiation-induced genomic instability see Section 6.3.5 of this Report.

Although these observations are of great interest, it remains to be shown whether any of the processes cited above, including delayed and persistent genomic instability, are common findings after irradiation of other cancer progenitor cell populations, how they may vary with radiation dose, and whether they in fact increase the probability of further changes relevant to malignancy. The alternative possibilities, that neoplastic initiation involves a mutational event in any one of a very large number of genes, or that there are one or more genes that are orders of magnitude more susceptible to radiogenic mutation than are currently recognized, also remain to be more thoroughly tested.

Finally, in considering dose-response relationships in experimental animals, it is necessary to keep in mind that the frequency of neoplasms of any particular type is strongly dependent on the genetic background, physiological condition, and environment of the animals at risk (UNSCEAR, 1986; Upton *et al.*, 1986). Dose-response data for a given neoplasm in animals of one strain or species are, therefore, of uncertain predictive significance for animals of other strains or species, including humans.

8.3 Dose-Response Relationships (Dose, Dose Rate, Linear-Energy Transfer) as Influenced by Homeostatic and Other Modifying Factors

8.3.1 *Background*

As has been emphasized elsewhere (UNSCEAR, 1977; 1986; 1993; Upton *et al.*, 1986), the dose-response relationships for neoplasms in irradiated laboratory animals vary markedly from one type of growth to another, for reasons yet to be determined. The diverse relationships include those with apparent thresholds, those without apparent thresholds, and those that actually bend downward with increasing dose (Figure 8.1).

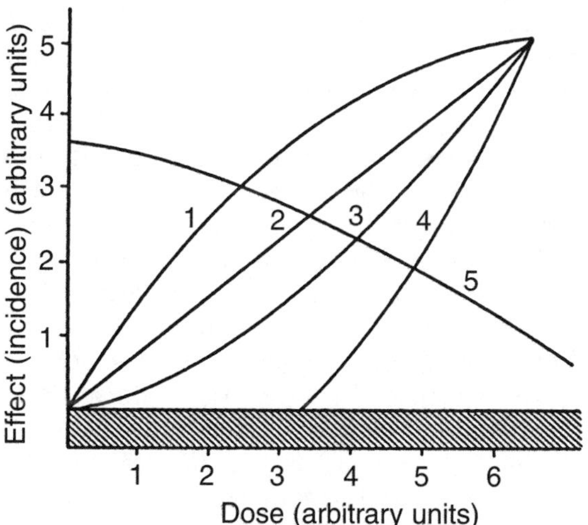

Fig. 8.1. Schematic representation of the variations in dose-response curves typically observed for different types of neoplasms in laboratory animals (UNSCEAR, 1986). (1) Concave downward (*e.g.*, Figures 8.10 and 8.12); (2) linear-nonthreshold type (*e.g.*, Figure 8.9); (3) concave upward (linear-quadratic type, *e.g.*, Figure 8.11); (4) threshold type (*e.g.*, Figure 8.6); and (5) effect declining with increasing dose (typical for reticulum cell sarcomas in mice, as noted in Section 8.3.2.3).

8.3.2 *Leukemia*

Leukemias and lymphomas of various types have been induced by irradiation in mice, rats, dogs, guinea pigs, swine, and other laboratory animals (UNSCEAR, 1986). The dose-response relationships for such neoplasms have varied widely from one type of growth to another, and none has been consistent with a simple linear-nonthreshold model (NAS/NRC, 1990; UNSCEAR, 1986), although the BEIR report states that the dose-response data for mouse myloid leukemia do not exclude a linear dose term in the low to intermediate dose range.

8.3.2.1 *Thymic Lymphoma.* The most thoroughly studied of radiation-induced hematologic growths is a T-cell neoplasm arising in the mouse thymus, the induction of which appears to involve indirect effects since it can be inhibited by shielding only a small portion of the hemopoietic bone marrow (Kaplan and Brown, 1952; UNSCEAR, 1986; 1993) or by restoring thymic lymphopoiesis post-irradiation through the grafting of intact marrow or the injection of tumor-necrosis factor (Humblet *et al.*, 1997). The dose-response curve for induction of thymic lymphomas by acute whole-body x irradiation alone is complex but typically of the threshold type (*e.g.*, Figure 8.2); however, the initiating effects of radiation appear to increase as a linear-nonthreshold function of the dose in mice injected with an appropriate promoting agent after exposure (Figure 8.3). Although low-LET radiation is typically several times less effective when the dose is highly fractionated or protracted than when it is delivered acutely, the effectiveness of fast neutrons is less dose-rate-dependent (NCRP, 1980; UNSCEAR, 1986) and has actually varied inversely with the dose rate in some experiments (*e.g.*, Figure 8.4). It should be noted that human T-cell leukemia-lymphoma does not appear to be increased in the Japanese atomic-bomb survivors.

The mechanism through which radiation induces thymic lymphomas remains to be defined precisely but appears to involve the depletion of thymic lymphocytes and T-cell precursors in the bone marrow (Sado, 1992; Sado *et al.*, 1991). Although recombinant, lymphomagenic, retroviruses are also characteristically detectable in the induced lymphomas (Janowski *et al.*, 1990), evidence suggests that the activation of such proviral agents is not essential for the initiation of the tumors (Okumoto *et al.*, 1990; Sado, 1992; Sado *et al.*, 1991). Susceptibility to the induction of the disease is increased, however, in knockout mice that lack *ATM*, a gene responsible for safeguarding genomic stability (Barlow *et al.*, 1996) and in mice with the severe combined immunodeficient (scid) trait (Lieberman *et al.*, 1992), in

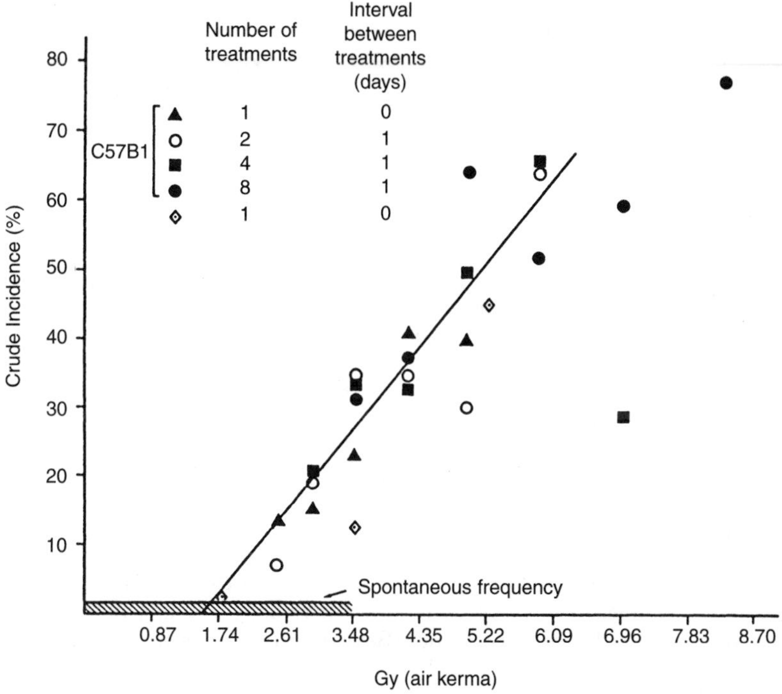

Fig. 8.2. Influence of single or fractionated whole-body irradiation on the incidence of thymic lymphoma in mice (UNSCEAR, 1986). Data for C57Bl mice (Kaplan and Brown, 1952); rhomboid symbols represent data for (C57Bl/6 JNrs × WHT/HtF1) mice (Saski and Kasuga, 1981).

which there is also a defect in DNA repair (Fulop and Phillips, 1990). Noteworthy, therefore, is the fact that a significant percentage of lymphomas exhibit mutations of K- and N-ras genes (Guerrero and Pellicer, 1987), marker chromosomal anomalies (Barlow *et al.*, 1996; McMorrow *et al.*, 1988), and/or hypermethylation of genes involved in the inhibition of cyclin/CDK complexes (Malumbres *et al.*, 1997).

8.3.2.2 *Myeloid Leukemia.* Myeloid leukemias, although induced experimentally by ionizing radiation in dogs, swine, and mice, have been investigated less extensively than thymic lymphomas (UNSCEAR, 1986). The dose-response relationship for induction of myeloid leukemia by whole-body x or gamma irradiation has been best documented in mice of the RF and CBA strains (*e.g.*, Figure 8.5), in which it is consistent with a linear-quadratic function combined with a competing function for cell-killing (DiMajo *et al.*, 1996; UNSCEAR, 1986). Gamma rays are several times less leukemogenic for such animals

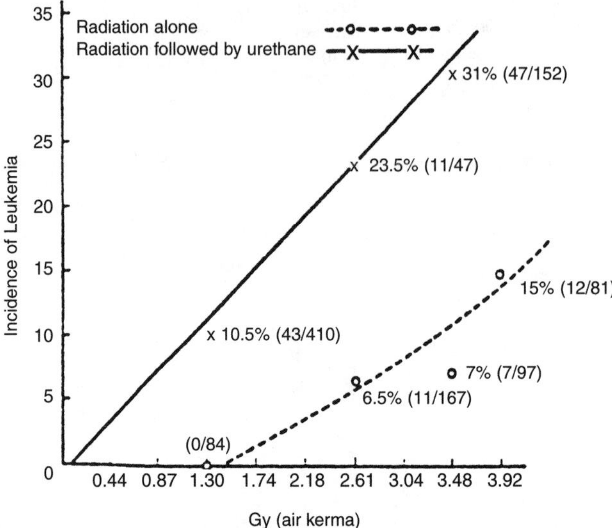

Fig. 8.3. Promoting effects of urethane on the induction of thymic lymphoma in C57Bl mice previously exposed to whole-body x radiation (no effects were observed with urethane alone) (Berenblum and Trainin, 1963).

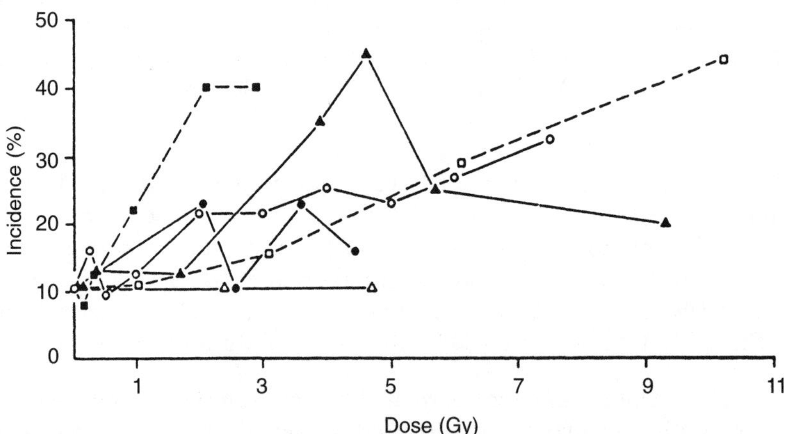

Fig. 8.4. The incidence of thymic lymphoma in whole-body irradiated RF female mice, in relation to the dose, dose rate, and quality of radiation (circles represent results with single exposures; squares, daily exposures; triangles, daily exposures for the duration of life; open symbols, x and gamma rays; shaded symbols, fast neutrons) (Upton *et al.*, 1970).

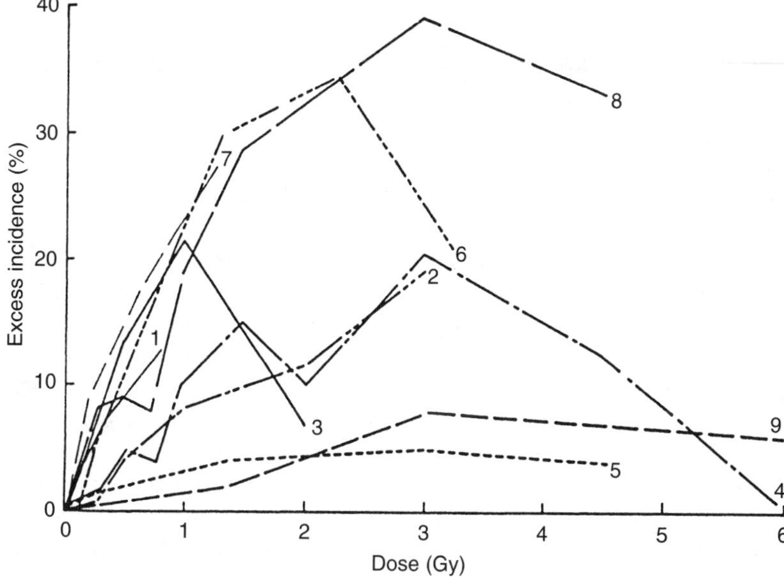

Fig. 8.5. Incidence of myeloid leukemia (in excess of control incidence) in whole-body irradiated male mice, in relation to the dose, dose rate, and quality of radiation. (1) acute fast neutron irradiation in RFM mice (Ullrich and Preston, 1987); (2) acute gamma irradiation in RFM mice (Ullrich and Preston, 1987); (3) acute fast neutron irradiation in CBA mice (Mole and Davids, 1982); (4) acute x irradiation in CBA mice (Mole *et al.*, 1983); (5) protracted gamma irradiation in CBA mice (Mole and Major, 1983); (6) acute fast neutron irradiation in RF/Up mice (Upton *et al.*, 1970); (7) protracted fast neutron irradiation in RF/Up mice (Upton *et al.*, 1970); (8) acute x irradiation in RF/Up mice (Upton *et al.*, 1970); (9) protracted gamma irradiation in RF/Up mice (Upton *et al.*, 1970) (NAS/NRC, 1990).

when administered at low-dose rates than when administered in a single brief exposure (Figure 8.5), but the leukemogenic effectiveness of fast neutrons appears to be marginally, if at all, dependent on the dose rate (*e.g.*, Figure 8.5).

Cytogenetic studies have revealed interstitial deletions in chromosome two as a consistent feature of the murine leukemia, suggesting that fragile sites on this chromosome, possibly influenced by genomic imprinting, may be involved in the initiation of the disease (Bouffler *et al.*, 1996; Breckon *et al.*, 1991; Clark *et al.*, 1996; Meijne *et al.*, 1996). Susceptibility to the leukemia is reduced in germfree mice (Walburg *et al.*, 1968) but is increased in mice injected with turpentine (Upton, 1959), paralleling changes in the rates of granulocytopoiesis in such animals.

8.3.3 *Osteosarcoma*

Of the various types of sarcomas that have been induced experimentally by ionizing radiation, osteogenic sarcoma has been investigated the most extensively. The induction of osteosarcomas by internal and/or external irradiation has been documented in mice, rats, guinea pigs, and dogs (Goldman, 1986; NAS/NRC, 1988; Reitmair *et al.*, 1995; UNSCEAR, 1986; 1993).

The frequency of osteosarcoma has been reported to increase as an apparently linear function of the dose in mice exposed to large doses (>3 Gy) of x rays (Finkel and Biskis, 1968). However, for mice exposed to bone-seeking radionuclides that emit low-LET radiation, there is a decidedly nonlinear dose response (*e.g.*, Figure 8.6). The available dose-response data for osteosarcoma come largely from studies of the tumorigenic effects of bone-seeking radionuclides.

Analysis of the dose-response relationship for osteosarcomas induced by locally deposited radionuclides is complicated by the nonuniformity of the distribution of the dose to bone (Stannard, 1988). Although the average dose to bone is often employed in such analyses, the

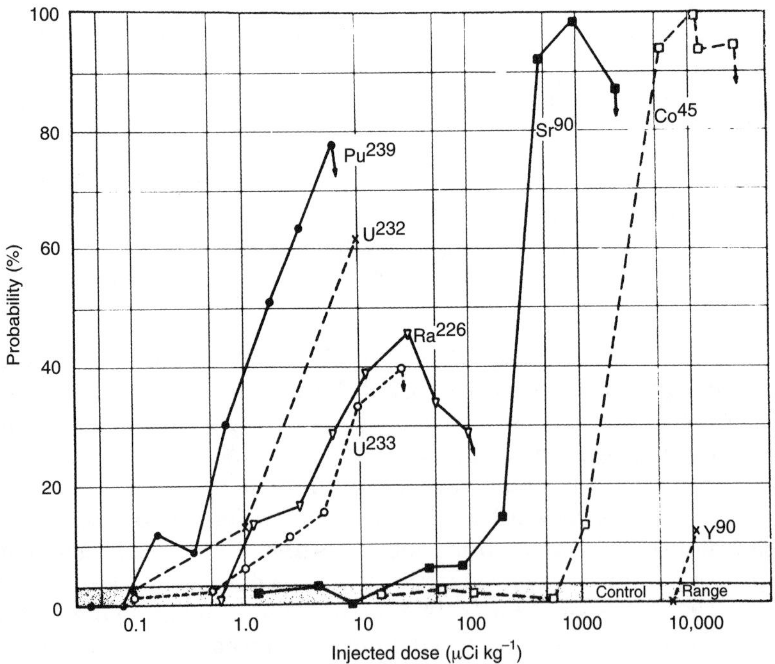

Fig. 8.6. Average probability of dying with a malignant bone tumor as a function of the injected dose of different radionuclides in mice (Finkel, 1959).

dose to osteoblasts (bone-forming cells) is more relevant, and this varies greatly throughout the skeleton because of inhomogeneities in radionuclide deposition. This source of uncertainty notwithstanding, the cumulative incidence of osteosarcomas in animals receiving low-to-intermediate doses of radionuclides typically increases as a power function of the average dose to bone from beta particles, although it has appeared in some instances to increase as an approximately linear function of the average dose to bone from alpha particles (Lloyd *et al.*, 1993; Muggenburg *et al.*, 1996; NAS/NRC, 1988; UNSCEAR, 1986) (*e.g.*, Figure 8.7). For both types of radionuclides, however, the dose-response relationships are complicated by the following: (1) the average latent period for tumor induction varies inversely with the dose rate (Figure 8.8), and (2) the tumorigenic response saturates at high-dose levels (*e.g.*, Figure 8.6). Therefore, to account for these features, more elaborate and nonlinear dose-response models have been formulated (*e.g.*, Gilbert *et al.*, 1998; Marshall and Groer, 1977; Whittemore and McMillan, 1982). The latter include the observation that if the dose rate is low enough the latent period for tumorigenesis can exceed the life span, resulting in an apparent threshold for the effect (Evans, 1974; Raabe *et al.*, 1983; 1990; Rowland, 1997).

Radiation-induced tumor cells have been observed to exhibit the activation of various oncogenes (Strauss *et al.*, 1992), the inactivation or loss of various tumor-suppressor genes (Strauss *et al.*, 1992), and the presence of oncogenic viruses (Finkel *et al.*, 1976; Janowski *et al.*, 1990). The precise roles that any of these changes may play in the induction of the neoplasms remain, however, to be determined,

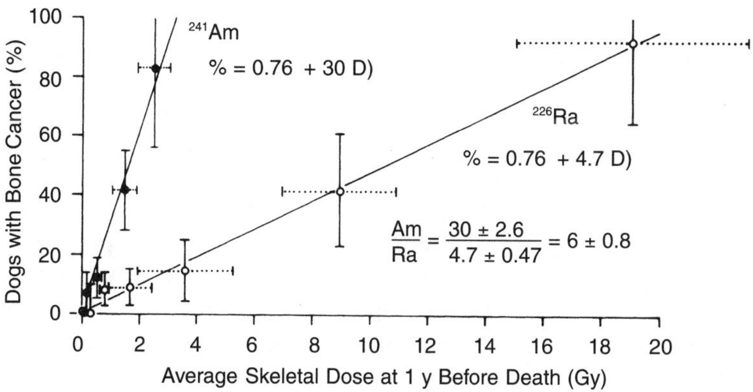

Fig. 8.7. Incidence of bone cancer in relation to the average accumulated skeletal dose in dogs injected with ^{241}Am or ^{226}Ra citrate (Lloyd *et al.*, 1994).

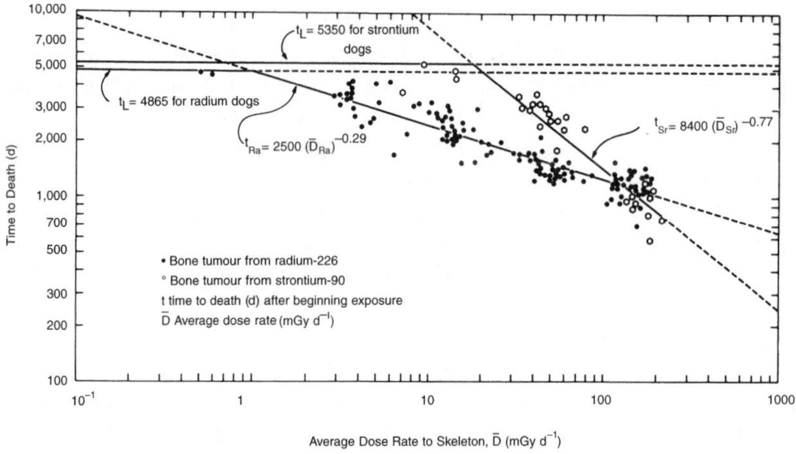

Fig. 8.8. Median time to death as a function of the average skeletal dose rate from ^{226}Ra or ^{90}Sr in dogs (Raabe *et al.*, 1981).

as does the mechanism through which the administration of steroids may enhance the tumorigenic effects of ^{90}Sr in mice (Nilsson and Broome-Karlsson, 1976).

8.3.4 *Mammary Gland Tumors*

Given the importance of human breast cancer (Henderson, 1993; Parker *et al.*, 1997) and the increased incidence among irradiated women (Land *et al.*, 1993a; 1994a; 1994b; Thompson *et al.*, 1994), experimental mammary cancer has received special attention (NRPB, 1995; UNSCEAR, 1986; 1993). Data from rats and mice, the most common experimental species, offer examples of the variations and complexities of the dose-response relationships.

After single, high-dose-rate x or gamma irradiation, mammary tumor dose-response relationships in susceptible Sprague-Dawley rats (Figures 8.9 and 8.10) and BALB/c mice are linear or bend upwards (Figures 8.11) (Bond *et al.*, 1960; Ullrich *et al.*, 1987). After a single acute exposure to neutron radiation, the dose-response curves for mammary neoplasias often rise steeply at low doses and bend over after higher doses (Figures 8.10 and 8.12). Downward-bending slopes of neutron dose-response relationships and linear or upward-bending gamma- and x-ray dose-response curves result in high neutron/gamma- or neutron/x-ray relative biological effectiveness (RBE) estimates for tumor induction, which may reach 100 at low doses and low-dose rates, and drop to ~10 at higher doses and

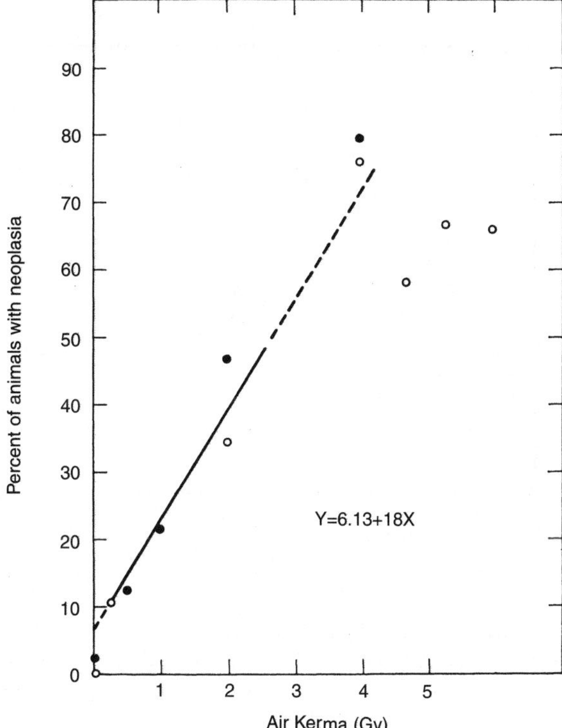

Fig. 8.9. Mammary tumor incidence in female Sprague-Dawley rats 11 months after acute whole-body 250 kVp x irradiation (two experiments, open and closed circles, respectively). Curve is least squares fit to data from rats exposed to air kermas[5] between 0 and 3.5 Gy (adapted from Bond *et al.*, 1960).

dose rates (Shellabarger *et al.*, 1980). However, comparable values of the dose dependence of the neutron/x-ray RBE have not been universally observed (Broerse *et al.*, 1986; Shellabarger *et al.*, 1982).

The effectiveness of dose fractionation depends on the size of each fraction, the interval between fractions, and the LET of the radiation. In BALB/c mice exposed to 0.1 to 0.25 Gy of low-LET radiation in multiple small doses per daily fraction or chronically at a low-dose rate, the dose response has been observed to be consistent with the linear term of the linear-quadratic curve that can be fitted to the response for single acute exposures (Figure 8.11); whereas the cancer

[5]The figure was originally published with the abscissa labeled "Dose (R)." The units were assumed to be roentgens and converted to air kerma.

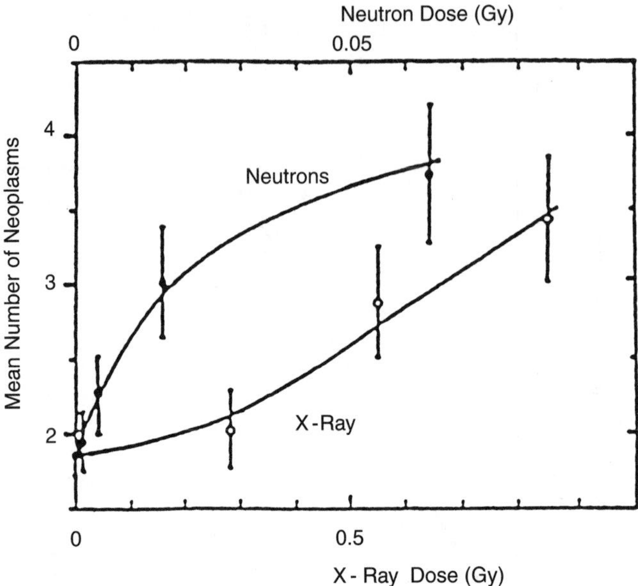

Fig. 8.10. Mean numbers of mammary tumors (fibroadenomas and carcinomas) per female Sprague-Dawley rat exposed acutely to 0.43 MeV neutrons or 250 kVp x rays. Note that neutron and x-ray dose scales differ by a factor of 10 (Shellabarger *et al.*, 1980).

incidence after the same total dose given in fewer, higher dose fractions approximated that expected after a single acute exposure (Figure 8.11). Similarly, the incidence of mammary carcinomas in Sprague-Dawley rats was observed to be four-fold greater after a 2.7 Gy exposure to ^{60}Co gamma rays delivered in a few minutes than after the same dose delivered chronically over several days, no such differences were observed after 0.9 Gy delivered similarly, and there was no effect of the dose rate at either dose on the incidence of fibroadenomas (Shellabarger and Brown, 1972). Also, in WAG/Rij rats exposed to 2 Gy x rays in a single exposure or cumulatively over 10 equal monthly fractions, no evidence of a dose-rate dependent change in the final mammary cancer incidence was observed (UNSCEAR, 1993).

In contrast to the dose-rate dependence exhibited with gamma rays, when BALB/c mice were exposed to low total doses of neutrons delivered at low-dose rates, the resulting increases in cancer showed a "reverse" dose-rate effect, *i.e.*, were about twice those seen after the same acute doses (Figure 8.12).

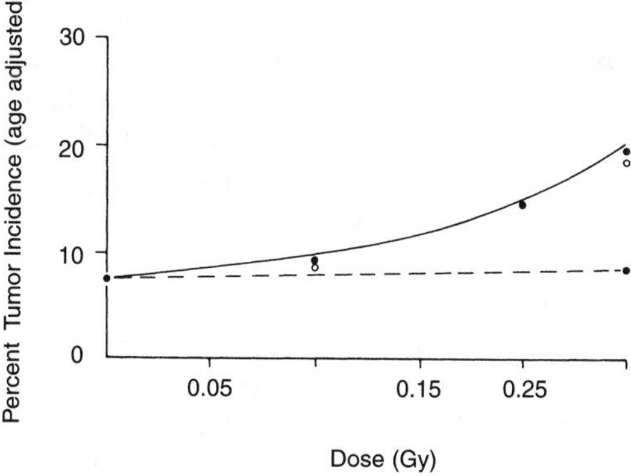

Fig. 8.11. Effect of dose rate and dose fractionation on the incidence of mammary carcinomas in female BALB/c mice exposed to ^{137}Cs gamma rays. Solid line: high-dose rate (0.4 Gy min^{-1}), linear-quadratic fit. Dashed line: low-dose rate (83 mGy d^{-1}), linear term of linear-quadratic equation. Solid circles near upper solid line: cancer incidence after single high-dose-rate exposures. Open circle at 0.25 Gy by solid curve is cancer incidence after five daily fractions of 50 mGy each. Solid circle at 0.25 Gy by dashed line is cancer incidence after 25 daily fractions of 10 mGy each (modified slightly from Ullrich *et al.*, 1987).

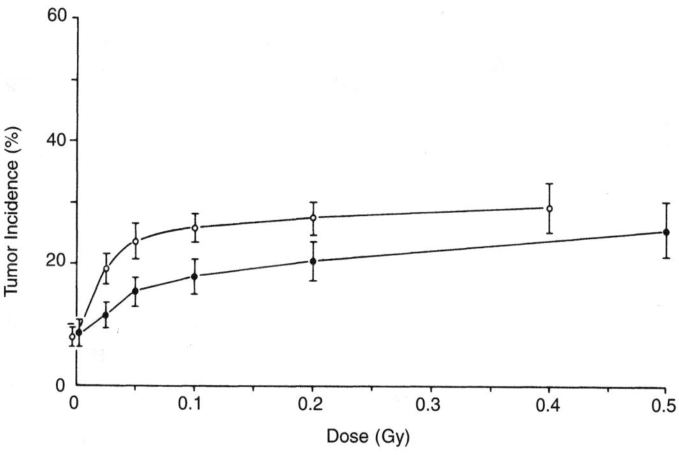

Fig. 8.12. Effect of dose rate on dose-response curves for induction of mammary carcinomas in female BALB/c mice by fission-spectrum neutrons administered in one or two fractions separated by 24 h (pooled) at 10 to 250 mGy min^{-1} (high-dose rate; ●) or administered at 10 or 100 mGy d^{-1} (pooled) (low-dose rate; ○) (Ullrich, 1984).

In groups of Sprague-Dawley rats given 5 Gy of ^{60}Co gamma rays in a single exposure or in 4 to 32 twice weekly fractions, the frequencies of mammary carcinomas progressively increased with fractionation such that after 32 exposures the cancer incidence per rat was more than twice that after a single exposure (Figure 8.13); the frequencies of fibroadenomas per rat tended to decrease with fractionation (Shellabarger *et al.*, 1966).

In some of the above studies on rats, the interpretation of the dose-response data is complicated by reported tumor frequencies that include both fibroadenomas and carcinomas; fibroadenomas are benign lesions, differing from mammary cancers in morphology, histogenesis and response both to hormones and to radiation. Other factors complicating interpretation of the dose-response relationships stem from complex hormonal and cellular interactions. For example, female mice are highly susceptible to radiation-induced

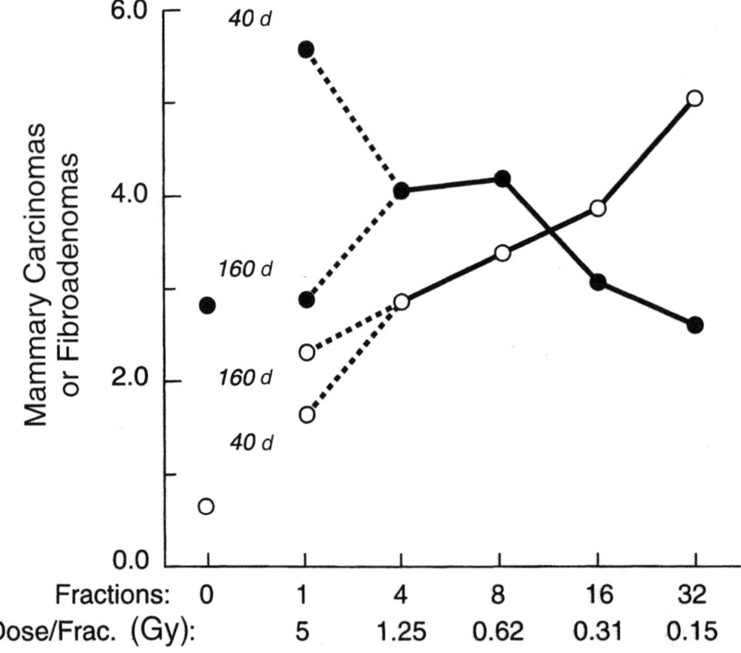

Fig. 8.13. Incidences per rat of mammary carcinomas (○) and mammary fibroadenomas (●) as a function of fraction number in Sprague-Dawley female rats exposed to a single dose of 5 Gy of ^{60}Co gamma rays at 40 or 160 d of age, or to the indicated numbers of twice-weekly fractions, beginning at 40 d of age, to yield total cumulative doses of 5 Gy (Shellabarger *et al.*, 1966).

ovarian tumors (Ullrich, 1983), which may or may not secrete progestins, estrogens or both (Clifton, 1959). Together with pituitary prolactin and adrenal glucocorticoids, these hormones control mammary gland growth and differentiation, and often determine whether radiogenically initiated cells will progress to neoplasia (Broerse *et al.*, 1987; Furth and Clifton, 1958; Kim and Furth, 1960; Russo and Russo, 1987; Shellabarger *et al.*, 1978; 1982; Yokoro *et al.*, 1977). For example, groups of Wistar-Furth female rats were exposed to 0.1, 0.25, 0.5 or 2 Gy 180 kVp x rays which alone would induce few or no mammary tumors. Three days after exposure, all but one group of 2 Gy irradiated rats were grafted with MtT (prolactin-secreting pituitary tumors) (Yokoro *et al.*, 1977). Mammary tumors developed in high incidence in the 2 Gy irradiated, MtT-grafted intact rats, and at progressively lower incidence and longer times after lower x-ray doses (Figure 8.14). Ovariectomy of 2 Gy exposed MtT-grafted rats reduced the frequency and increased the latency of mammary tumors. Rats that were exposed to 2 Gy and were otherwise untreated remained free of mammary tumors for several months after irradiation. However, mammary tumors developed in the latter rats after they were grafted with MtT seven months after exposure (Figure 8.14). Thus, radiation-initiated mammary cells remained dormant until stimulated to tumor development by the increased prolactin from the MtT grafts seven months after irradiation. Comparable effects were seen after neutron irradiation and MtT grafting (Yokoro *et al.*, 1977). By the same token, AXC rats developed mammary carcinomas after 0.43 MeV neutron irradiation only when also treated with the synthetic estrogen diethylstilbestrol; few fibroadenomas developed in any AXC rats. In contrast, Sprague-Dawley rats developed only fibroadenomas and only when neutron-irradiated without diethylstilbestrol (Stone *et al.*, 1979).

The frequency of radiogenic initiation of mammary cancer greatly exceeds the rate with which radiation is known to induce mutations at any one genetic locus (Clifton *et al.*, 1986; Kamiya *et al.*, 1995). The promotion-progression of the initiated cells is suppressed, however, by cell-number-dependent interactions (Figure 8.15). Despite the fact that various oncogenes and tumor-suppressor genes have been described in rodents (Bennett *et al.*, 1995; Chen *et al.*, 1996; Gould and Zhang, 1991), none has been clearly related to radiation-induced cancer initiation. In contrast, both radiation-induced transformation (Ullrich *et al.*, 1996), delayed chromatid instability and point mutations in the *p53* tumor suppressor gene (Ponnaiya *et al.*, 1997a) developed after many population doublings at very high frequency in the progeny of cancer susceptible BALB/c mouse mammary epithelial cells that had been exposed to radiation *in vivo*

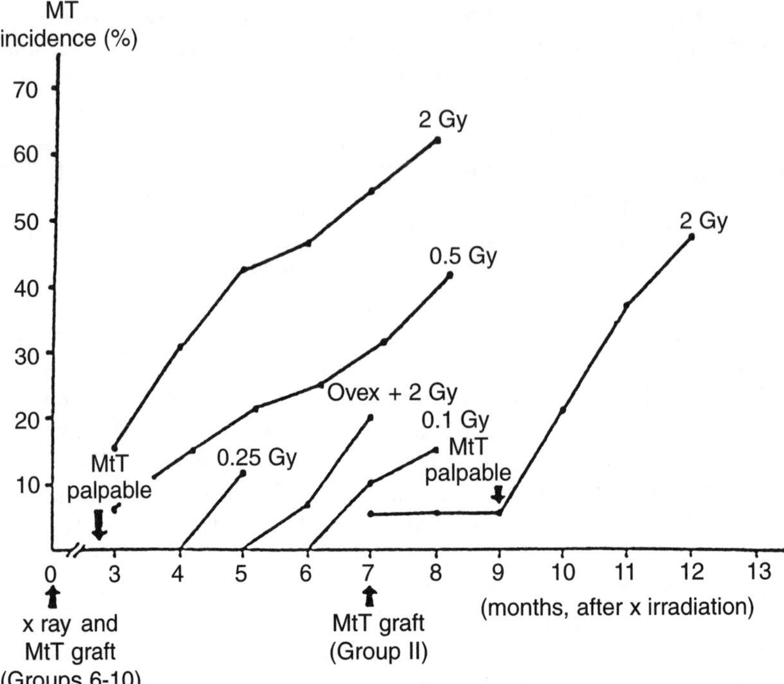

Fig. 8.14. Incidence of mammary tumors (MT) in Wistar-Furth female rats rendered hyperprolactinemic by grafts of prolactin-secreting pituitary tumors (MtT) 3 d or seven months after irradiation with the indicated doses of 180 kVp x rays. "MtT palpable": time when grafted MtT became palpable; "Ovex": ovariectomized shortly after irradiation. Radiation doses were used that produced few or no mammary neoplasms in the absence of hyperprolactinemia (Yokoro *et al.*, 1977).

or in culture. Such instability did not develop in the progeny of similarly treated mammary epithelial cells from cancer resistant C57BL/6 mice nor in F1 hybrids between these two strains (see Section 6.3.5). Genomic instability was suggested by the authors as a possible mechanism of high frequency initiation (Ponnaiya *et al.*, 1997a).

8.3.5 *Thyroid Neoplasia*

More than four decades ago it was established that a dose of 3 to 7 Gy to the rat thyroid gland would cause development of scattered hyperplastic nodules in the follicular epithelium, which would

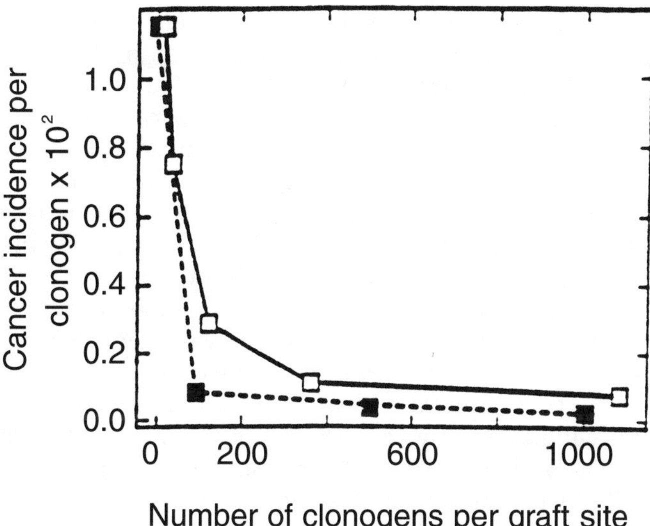

Fig. 8.15. Cancer incidence per grafted F344 rat mammary clonogen plotted against the total numbers of surviving mammary clonogens transplanted per graft site. (□) grafted clonogens from donor rats that had been irradiated 7 Gy of ^{137}Cs gamma rays; (■) clonogen donors had been treated with N-nitroso-N-methylurea. All graft recipients were endocrinologically manipulated to induce hyperprolactinemia and glucocorticoid deficiency, a highly effective mammary cancer-promoting hormonal combination (Kamiya *et al.*, 1995).

progress to form nodular adenomas and carcinomas (Doniach, 1963; 1970). Rat thyroid tumor induction by external x or gamma radiation or internally deposited ^{131}I was widely confirmed and has been utilized extensively in studies of the factors and conditions that may influence the neoplastic process (Dumont *et al.*, 1980; NAS/NRC, 1990; NCRP, 1985). Among the most important findings of these earlier studies are that (1) radiogenic thyroid cancer arises exclusively within the follicular epithelium, and (2) thyroid cancer induction is heavily dependent on the hypothalamic-hypophyseal-thyroid feed-back system. Any drug, dietary factor or condition, such as iodine deficiency, which elevates hypophyseal thyroid stimulating hormone levels will increase the probability of progression of initiated thyroid cells to neoplasia (Dumont *et al.*, 1980; NAS/NRC, 1990; NCRP, 1985), and some of these have been used to promote thyroid cancer after experimental initiation. Radiogenic initiation of thyroid clonogens has been shown to occur with high frequency after irradiation (Domann *et al.*, 1994; Mulcahy *et al.*, 1984; Watanabe *et al.*, 1988).

Most of the experiments on thyroid carcinogenesis have been inadequate for evaluation of dose-response relationships due to the small numbers and varying genetic backgrounds of the animals studied, questions concerning dosimetry, and other experimental limitations (Dumont *et al.*, 1980; NCRP, 1985). The most extensive dose-response data are from a study of thyroid neoplasia in Long-Evans rats exposed to the radiation from injected ^{131}I as compared with external x rays (Lee *et al.*, 1982). The radiation dosimetry in thyroid tissue from internally deposited ^{131}I was first thoroughly reexamined (Lee *et al.*, 1979), and then groups of rats were exposed to comparable radiation doses from one of the two sources. The resulting dose-response curves for the two radiation types were very similar. The slopes of the curves for carcinoma induction decreased moderately with increasing doses (Figure 8.16), while the slopes of the adenoma induction curves increased with increasing doses (Figure 8.17). The results differ from those of most earlier studies in not supporting the conclusion that radiation from internally deposited ^{131}I is appreciably less carcinogenic for the rat thyroid than is acute external x irradiation (Lee *et al.*, 1982; UNSCEAR, 1986).

8.3.6 *Lung Tumors*

The incidence of adenocarcinomas of the lung has been observed to increase linearly with the dose over the range of 0 to 2 Gy of ^{137}Cs gamma rays in acutely exposed BALB/c mice, and also linearly, but with a slope about one-third as steep, after chronic exposure at a rate of 83 mGy d^{-1} (Figure 8.18; Ullrich and Storer, 1979). Conversely, no comparable dose-rate dependent reduction in lung cancer incidence was seen with fractionation or protraction of neutron exposures of 0 to 0.5 Gy (Figure 8.19). In rats exposed to plutonium dioxide by inhalation, the dose response has been observed to vary in shape and slope, depending on the genetic background of the exposed animals (Sanders and Lundgren, 1995).

The induction of benign pulmonary adenomas by low-LET radiation has likewise been observed to be dose-rate dependent. Adenoma incidence increased as a linear-quadratic function of dose in RFM mice after acute x irradiation, but increased less steeply after split dose with a 1 or 30 d interval.

8.3.7 *Renal Neoplasms*

"Eker" rats, which are heterozygous for a mutated *Tsc2* gene, the rat homologue of the human tuberous sclerosis *Tsc2* gene (Kobayashi

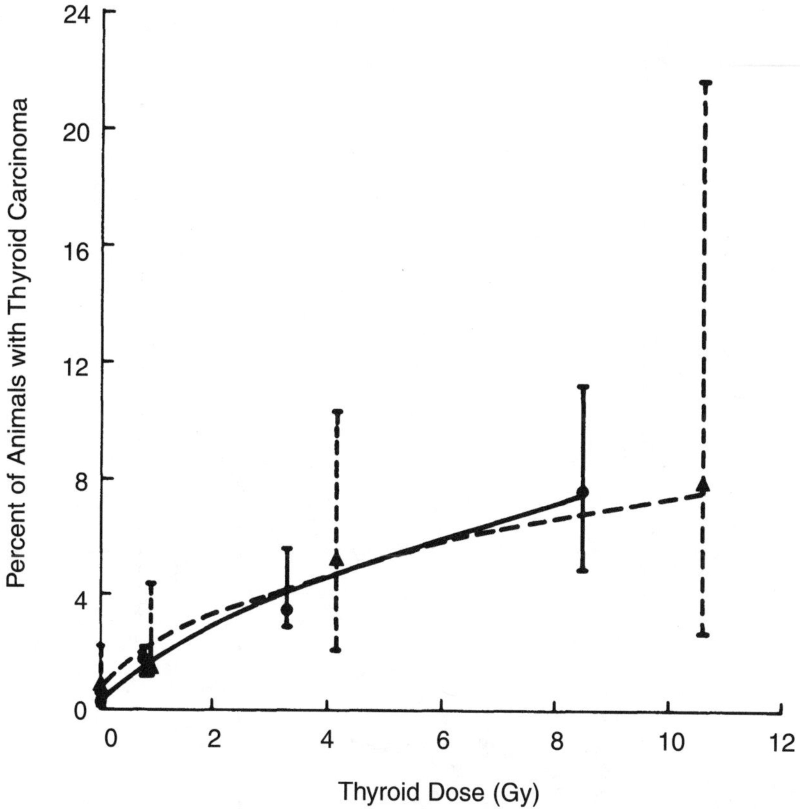

Fig. 8.16. Incidence of thyroid carcinoma in Long-Evans rats as a function of the dose from external x rays (▲, – – –) or internally incorporated ^{131}I (●, —). Curves are computer fits; vertical bars are 95 percent confidence limits (Lee *et al.*, 1982).

et al., 1997), are susceptible to radiogenic induction of multiple bilateral cystic or tubular renal adenocarcinomas (Eker and Mossige, 1961; Yeung *et al.*, 1993). These carcinomas increase linearly with gamma-ray dose over the range of 3 to 9 Gy (Figure 8.20). Carcinogenesis in this system has been interpreted as a two-step process, the first step being the inheritance of a mutant *Tsc2* gene, and the second step being the loss or mutation of the remaining wild-type allele (Hino *et al.*, 1993). The reason for the nearly two-fold difference in the renal cancer frequencies in males and in females remains to be determined, but strongly implies that other factors are also involved in the process.

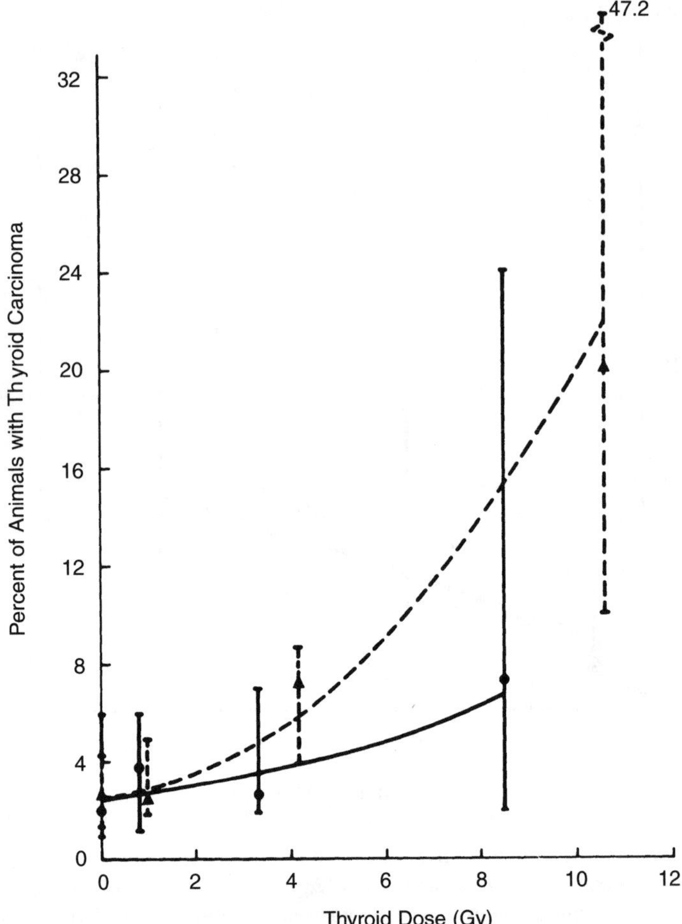

Fig. 8.17. The incidence of thyroid adenoma in Long-Evans rats as a function of the dose from external x rays (▲, – – –) or internally incorporated ^{131}I (●, —). Curves are computer fits; vertical bars are 95 percent confidence limits (Lee *et al.*, 1982).

8.3.8 *Skin Tumors*

Benign and malignant dermal and epidermal tumors of various types have been induced in laboratory animals, with dose-response relationships which vary depending on the type of tumor in question

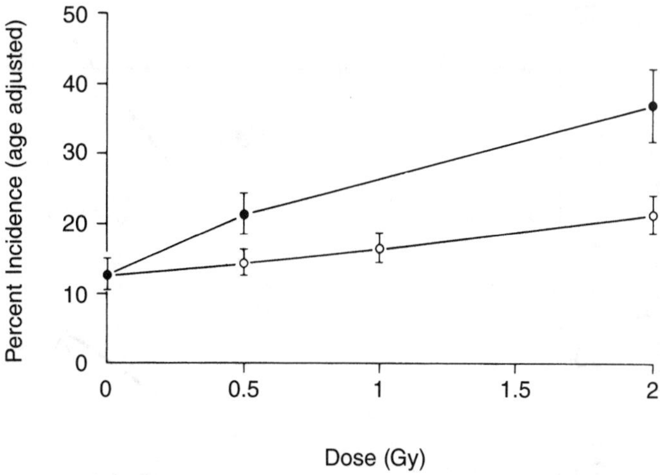

Fig. 8.18 Incidence (percent SE) of lung adenocarcinomas in BALB/c mice as a function of ^{137}Cs gamma-ray doses delivered at 400 mGy min^{-1} (●) or 83 mGy d^{-1} (○) (adapted from Ullrich and Storer, 1979).

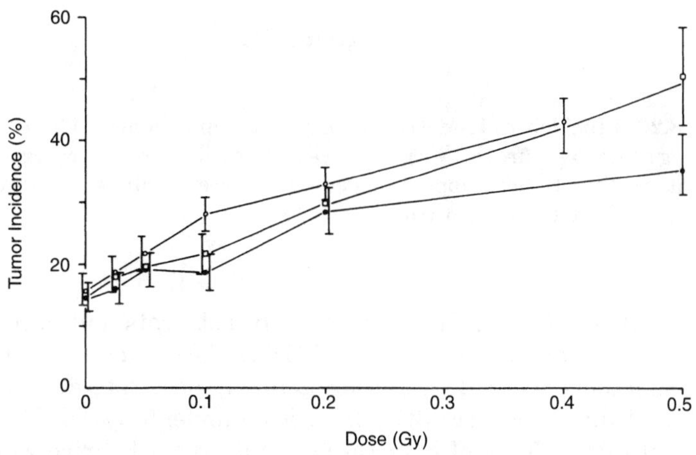

Fig. 8.19. The incidence of adenocarcinomas of the lung after neutron irradiation in BALB/c mice. Data from single acute high-dose-rate (10 to 250 mGy min^{-1}) neutrons pooled with 24 h split dose data (●, lower line), 30 d split high-dose-rate groups (□, middle line), pooled low-dose rate (10 and 100 mGy d^{-1}) groups (○, upper line). Vertical bars are SE (Ullrich, 1984).

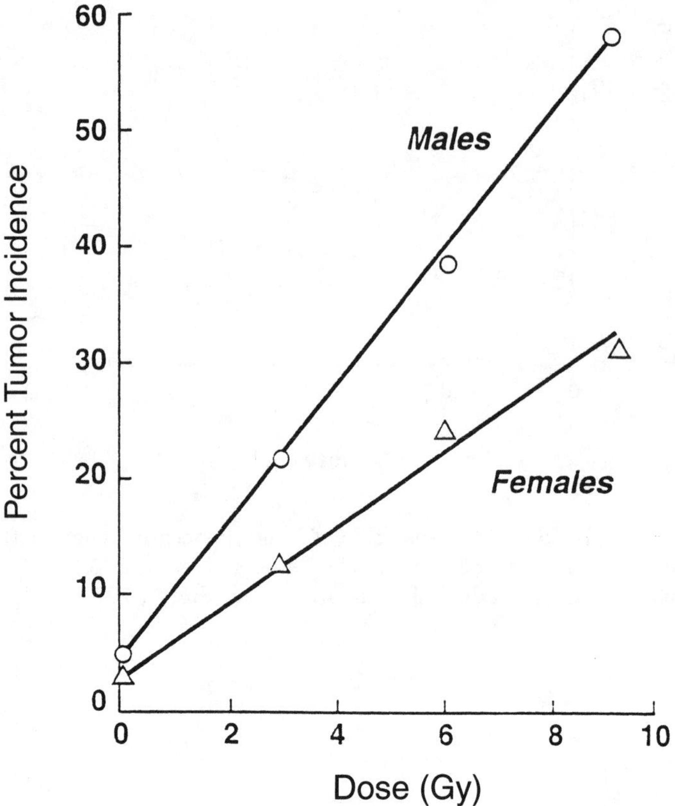

Fig. 8.20. Incidence of renal cancer in male (○) and female (Δ) "Eker" rats following acute exposure to ^{137}Cs gamma rays. The rats were heterozygous for the mutant *Tsc2* tumor suppressor gene and were irradiated at three to four weeks of age (data from Hino *et al.*, 1993).

(ICRP, 1991b; UNSCEAR, 1977; 1986). In both rats and mice, the dose-response relationships for low-LET radiation tend to vary as power functions of the dose and to be highly dose-rate dependent, with few, if any, tumors resulting from doses under 5 Gy (UNSCEAR, 1986). In fact, 0.5 Gy of beta radiation delivered thrice weekly throughout life failed to induce tumors in any of 50 exposed ICR/CRJ mice (Ootsuyama and Tanooka, 1993), whereas lifetime exposure to thrice weekly doses of 1.5 Gy or more induced tumors in 100 percent of such animals (Ootsuyama and Tanooka, 1988). In contrast to the highly curvilinear dose-response that is characteristically observed with low-LET radiation, the dose-response relationship with high-

LET radiation has often appeared consistent with a linear-nonthreshold function (*e.g.*, Figure 8.21), but could be consistent with other models as well.

Susceptibility to the induction of skin tumors varies markedly among species, being many times higher in the rat than in the mouse, for unknown reasons (Albert *et al.*, 1972). In rats exposed to beta radiation, the induction of skin tumors is inhibited by retinoids (Burns *et al.*, 1982) but enhanced by 4-nitroquinoline-1-oxide, the effects of which appear to be synergistic with those of radiation (Hoshino and Tanooka, 1975).

The mechanisms of the observed tumorigenic effects on the skin are still to be elucidated. It is of interest, however, that Burns and Albert (1986a; 1986b) reported that the kinetics for production and repair of DNA dsbs in rat skin resemble the corresponding dose-, dose-rate-, LET-, and age-dependent variations in dose-response relationships for tumorigenesis by irradiation. Also, K-ras and Myc oncogenes have both been observed to be activated in radiation-induced epidermal tumors of the rat (Sawey *et al.*, 1987), the latter appearing to be activated at a later stage in tumor development than the former (Garte *et al.*, 1990).

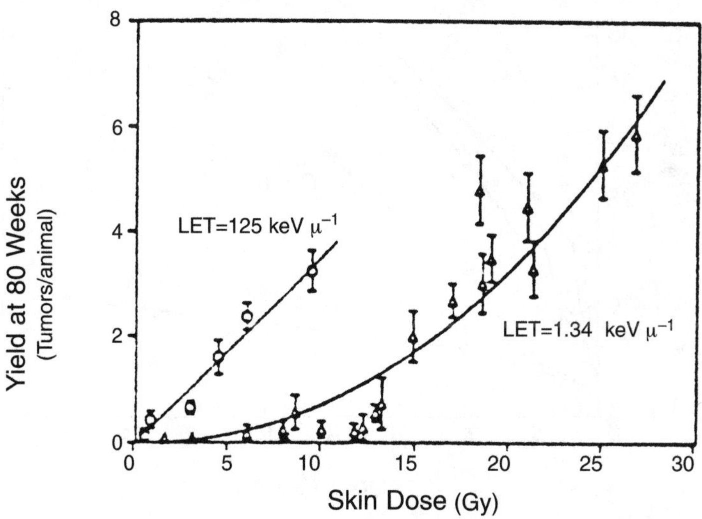

Fig. 8.21. Incidence of skin tumors as a function of the dose in rats irradiated acutely with electrons (Δ) (or argon ions (○) (Burns and Albert, 1986a).

8.3.9 *Mouse Harderian Gland Tumors*

The Harderian gland, an accessory lacrymal gland at the inner angle of the eye of the mouse and of some other mammals with nictitating membranes, has been used as an informative model in studies of the carcinogenic effects of accelerated heavy ions. In (C57BL/6J × BALBN/cJ)F1 female mice intrasplenically isografted with two pituitary glands each to ". . .enhance markedly the expression of radiation-induced Harderian gland tumors and advance the time of appearance" (Fry *et al.*, 1985) and then irradiated with ^{60}Co gamma rays or heavy charged particles from the BEVALAC accelerator, dose-dependent increases in the incidence of Harderian gland tumors were observed (Figure 8.22). RBE values for the heavy ions/^{60}Co gamma rays calculated from the initial linear slopes of the tumor prevalence curves increased with the masses and energies and hence with LET of the particles, from 5 for ^{4}He (228 MeV amu^{-1}) to 27 for both ^{40}Ar and ^{56}Fe (570 and 600 MeV amu^{-1}). The shapes of the dose-

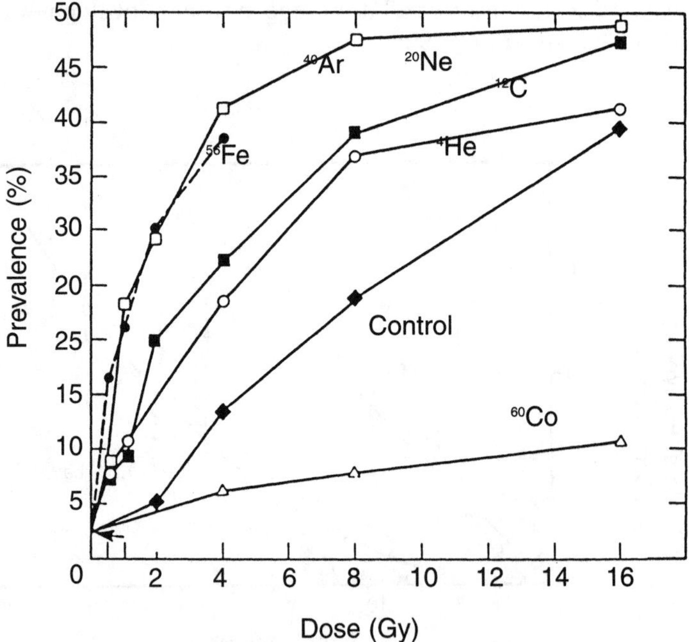

Fig. 8.22. Dose-response relationships for Harderian gland tumor induction in (C57BL/6J × BALB/c)F1 female mice after exposure to various heavy ions from the BEVALAC accelerator, or to ^{60}Co gamma rays. The mice were intrasplenically isografted with two pituitary glands before irradiation to promote radiogenically initiated cells to neoplasia (Fry *et al.*, 1985).

response curves and magnitudes of the RBE values for ^{56}Fe and ^{40}Ar are similar to those observed for fission-spectrum neutrons, suggesting the possibility that tumor induction, as a function of LET, reaches a point of maximum effectiveness (Fry *et al.*, 1985). Although in previous experiments isografted pituitary glands had markedly increased the incidence of Harderian gland tumors following exposure of such mice to either neutrons or gamma rays, Harderian gland tumor frequencies were not increased by pituitary grafts in mice exposed to BEVALAC-accelerated ^{56}Fe ions (Alpen *et al.*, 1993). Lack of tumor promotion by pituitary grafts suggests that the high-LET ^{56}Fe radiation elicits a maximum tumorigenic response at low fluences.

8.4 Life Shortening

In laboratory animals, shortening of the life span by low-to-intermediate doses of whole-body radiation has been observed to result primarily from increased or accelerated rates of neoplasia in different organs of the body (UNSCEAR, 1982; Walburg, 1975). In this respect, such radiation-induced life shortening may be considered to represent an integrated measure of the tumorigenic effects of whole-body radiation at low doses and low-dose rates (Fry, 1994). Thus, the overall death rate from all causes, which typically rises exponentially with age throughout adult life, has been observed to be displaced upward with increasing dose of whole-body radiation in animals exposed early in life (*e.g.*, Figure 8.23). Furthermore, the magnitude of the dose-dependent displacement appears to be similar in mice, rats, guinea pigs, and dogs, in spite of marked differences among these species in the natural life span (Figure 8.24).

The most extensive dose-response data come from experiments with laboratory mice, in which the extent of life shortening generally appears to: (1) increase linearly with the dose after acute x or gamma irradiation (Figure 8.25); (2) increase less steeply, but still linearly, with the dose after highly fractionated or protracted x- or gamma-irradiation (Figure 8.26), even at doses of 10 to 250 mGy d^{-1} (Figure 8.27); (3) increase more steeply after acute exposure to fast neutrons (Figure 8.26) than after acute exposure to gamma rays; and (4) increase no less steeply with chronic neutron irradiation than with acute neutron irradiation in the dose range below 0.5 Gy, but to increase less steeply, or vary inversely, with the dose fractionation at doses above this range (Figure 8.26). In interpreting these data, it is noteworthy that the extent of life shortening in mice exposed

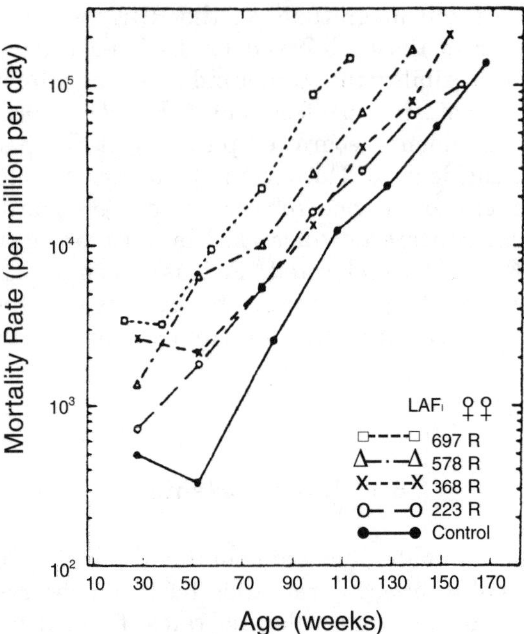

Fig. 8.23. Effects of acute whole-body radiation on the daily death rate of aging LAF1 mice. The symbols represent experimentally determined Gompertzian values, plotted at the midpoints of the intervals over which they were calculated [Sacher (1966) based on data from Upton *et al.* (1960)].

acutely to gamma radiation averages only about five percent per gray, with little, if any, effect on longevity detectable in such animals at doses below 1 Gy (Figure 8.25).

Ostensibly at variance with the above dose-response relationships for life shortening are the results of experiments in which irradiation at low doses and/or low-dose rates has failed to cause any detectable life shortening in exposed animals or has even appeared to enhance their survival, at least in males (*e.g.*, Boche, 1967; Caratero *et al.*, 1998; Carlson *et al.*, 1957; French *et al.*, 1974; Friedberg *et al.*, 1976; Gowan and Stadler, 1964; Grahn *et al.*, 1972; Langendorff, 1963; Lorenz *et al.*, 1954; Luning, 1960; Maisin *et al.*, 1988; 1996; Nishio, 1969; Spalding *et al.*, 1982; Storer *et al.*, 1979; Upton *et al.*, 1967). With few exceptions, however, such results do not depart significantly from the dose-response relationships illustrated above because of the small numbers of animals involved; hence, they may represent no more than the random variations to be expected at low doses and low-dose rates (UNSCEAR, 1982; 1993). Furthermore, in those cases where the mean survival time of lightly irradiated

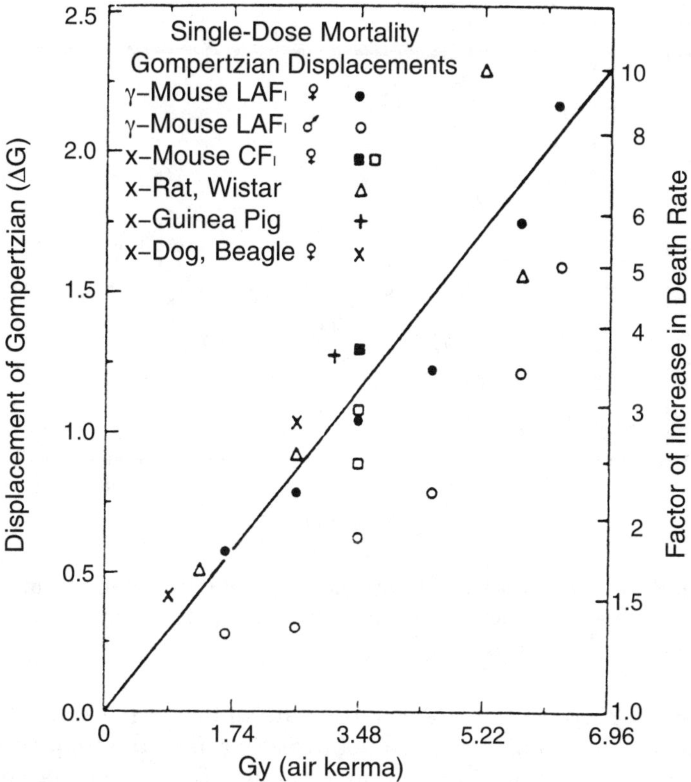

Fig. 8.24. Upward displacement of the Gompertz slope for daily mortality in aging laboratory mice, rats, guinea pigs, and dogs as a function of the dose of acute whole-body radiation received earlier in life (adapted from Sacher, 1966).

animals has significantly exceeded that of the controls, the differences have generally been attributable to radiation-induced reduction in the rate of mortality from intercurrent infectious diseases early in adult life (*e.g.*, Figure 8.28), as opposed to protection against tumor development and prolongation of the ultimate life span (Congdon, 1987; UNSCEAR, 1994). In animals maintained under conditions conducive to long-term survival, low-level irradiation has not been observed consistently to confer protection against the neoplastic growths associated with normal aging.

8.5 Summary

The data available from radiation carcinogenesis experiments in animals permit few generalizations concerning the shapes of the

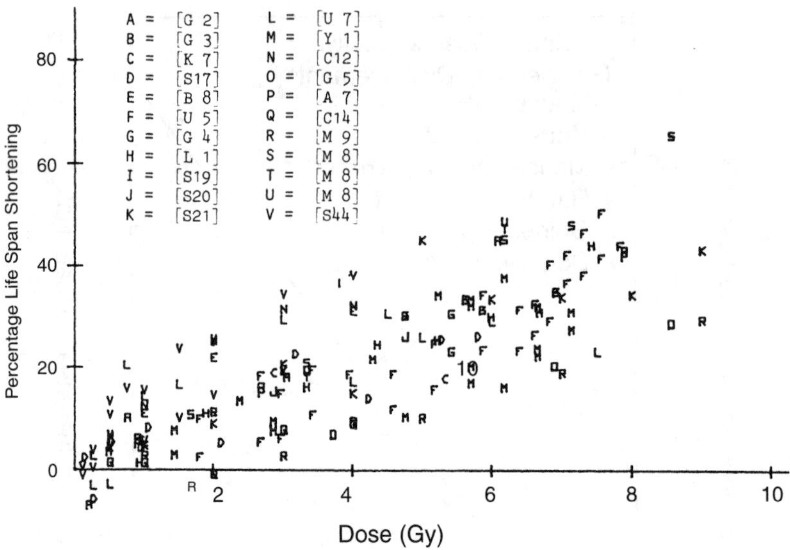

Fig. 8.25. Dose-response relationship for life shortening in male and female mice exposed acutely to whole-body x or gamma rays from various experiments (UNSCEAR, 1982).

dose-responsive curves. Indeed, the data indicate that the observed frequencies of neoplasms need not directly reflect the frequencies of radiogenic initiation. In addition, animals of the same species but of different genetic strains may differ markedly in susceptibility to development of a given type of neoplasm. For example, some strains of mice and rats do not show an increase in the frequency, or an acceleration in the appearance, of hemopoietic neoplasms or mammary cancers following irradiation; whereas such malignancies are major causes of mortality following exposure in animals of other strains. Furthermore, among radiogenic neoplasia-susceptible animals, several types of dose-response curves have been observed after acute radiation exposure. These include linear-nonthreshold curves, curves that are linear or curvilinear with a threshold, linear-quadratic curves, curves that bend upward or downward as power functions of dose, and finally, curves that appear to change slope abruptly at a given dose. Despite these variations, radiation-induced life shortening in the low-to-intermediate dose range, which has usually been predominately attributable to the cumulative effects of radiogenic malignancies has generally tended to increase linearly with dose.

In otherwise untreated, susceptible animals, the effects of dose protraction and dose fractionation depend on the dose rate, the dose

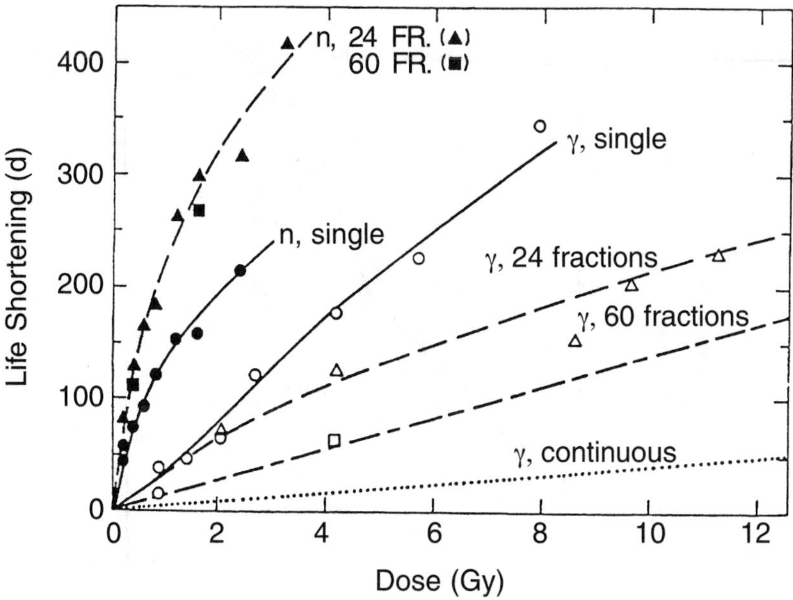

Fig. 8.26. Influence of fractionation or protraction on the dose-response relationship for life shortening in B6CF1 mice exposed to whole-body gamma or fission neutron radiation; (○) one single, brief high-dose-rate exposure; (Δ) one high-dose-rate exposure each week for 24 weeks; (□) one high-dose-rate exposure each week for 60 weeks; (· · ·), daily, 23 h exposure for the direction of life (gamma rays) [Thomson *et al.* (1982) based on data from Thomson *et al.* 1981a; 1981b)].

fraction sizes, the interval between fractions, and the radiation type. The incidence of neoplasms following fractionated doses or chronic exposures to low-LET gamma, x or beta rays is frequently significantly reduced, as compared to that following acute exposures. In the extreme case of skin tumors, very large total beta-ray doses have been administered to mice in thrice weekly fractions of 0.5 Gy each throughout life without induction of dermal or epidermal neoplasms. However, mice are more resistant in general to skin carcinogenesis than are rats. In cases in which neoplasms have appeared to be induced as linear-quadratic functions of the dose following acute exposure, doses accumulated as small fractions or at very low dose rates have induced such neoplasms at the frequency predicted by the linear portion of the linear-quadratic relationship. Values of the dose-rate effectiveness factor thus may vary from 2 to 10.

In contrast, dose-response relationships for high-LET neutrons often tend to rise steeply at doses in the 5 to 50 mGy range, above

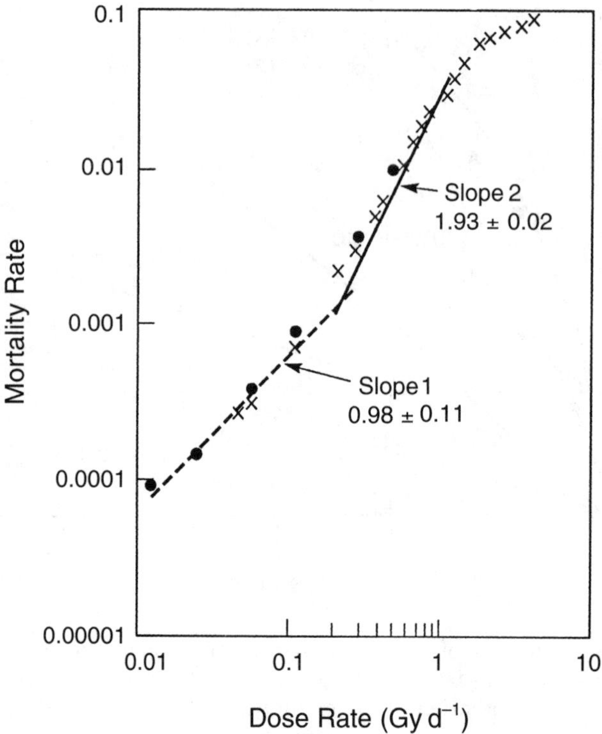

Fig. 8.27. Influence of the dose rate on the rate of mortality in mice exposed daily throughout life to whole-body ^{60}Co gamma radiation [based on data from Grahn (1970) and Sacher (1973)].

which they tend to become less steep. Furthermore, fractionated or protracted doses of neutrons have in some cases been found to be more carcinogenic than single, acute neutron doses, in keeping with the "inverse dose-rate effects" of fractionation or protraction of neutron irradiation that have been observed in cell transformation and mutation studies. Hence, in some cases, the RBE values of high-LET neutrons tends to increase with decreasing dose and dose rate.

In evaluating the dose-response relationships for tumor induction by internally deposited radionuclides—*e.g.*, osteogenic sarcomas from bone-seeking nuclides; thyroid cancer from radioiodides—estimation of doses is rendered difficult by uneven distribution of the radionuclides and in the variety of biological half-lives.

Whether or when a radiation-initiated cell will give rise to an overt neoplasm often depends on the influence of stimulatory and/or inhibitory agents (*e.g.*, hormones, promoting agents) and initiated cells may lie dormant for prolonged periods before being stimulated

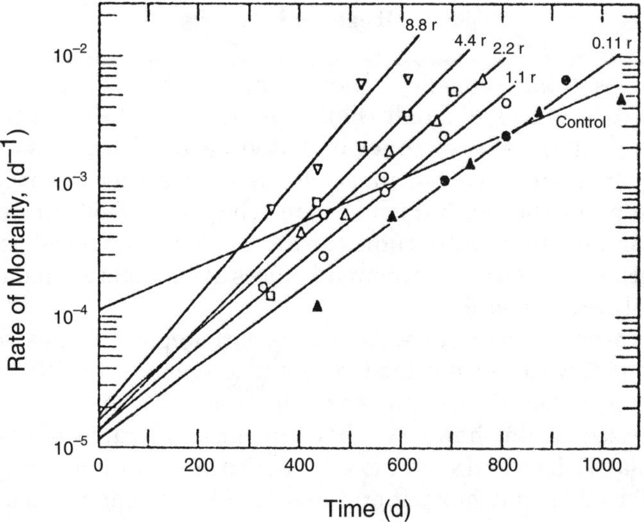

Fig. 8.28. Death rate in relation to time after the start of irradiation in mice exposed daily to 1.1 to 8.8 mGy d^{-1} of whole-body gamma radiation [UNSCEAR (1982) based on data of Lorenz *et al.;* (1954)]. (In all irradiated mice, the death rate was lower throughout most of the life span than that in the nonirradiated controls, owing primarily to the higher rate of intercurrent mortality from infectious diseases in the latter.)

to progress to neoplasia by such agents or conditions (*e.g.*, prolactin and mammary tumors). Even in the putative "two-step" induction of renal neoplasia that is dependent on mutational inactivation of both alleles of a cancer suppressor gene in the Eker rats, the probability of tumor formation is also dependent on sex-related susceptability.

Development of radiation-induced neoplasms is also strongly influenced by cancer-susceptibility and cancer resistance genes, and molecular analyses of several types of radiation-induced malignancies have revealed the presence of a variety of activated oncogenes or inactivated suppressor genes. However, with few exceptions, it has not been possible to attribute radiogenic mutation at any specific genetic locus. Furthermore, experiments with quantitative transplantation systems from which conservative estimates of cancer risks per susceptible cell could be made have shown radiogenic neoplastic initiation typically to be orders of magnitude more frequent than any known mutational event at any one genetic locus. It has thus been suggested that radiogenic tumor initiation may involve a radiation-induced event or process that is more common on a cellular basis as, for example, genomic instability.

8.6 Research Needs

Changes in the frequency of neoplasms in animals exposed to low-level radiation are so small that assessment of the relevant dose-response relationships cannot be based on empirical observations of such changes alone but must be based on models incorporating knowledge of the mechanisms underlying the effects in question. Therefore, in addition to studies at the whole-animal level, research approaches exploiting pertinent advances at the cellular and molecular levels are called for.

The various research needs and research opportunities that are pertinent have been outlined in detail elsewhere (NCRP, 1993c). Hence, no attempt to summarize them again is made here. It is emphasized again, however, that ongoing and rapid advances in molecular biology and somatic cell genetics afford promising opportunities for advancing our understanding of the mechanisms of radiation carcinogenesis and of adaptive responses influencing the process. It is important, therefore, that such opportunities are exploited in suitable laboratory animal model systems, through which the full spectrum and sequence of relevant molecular and cellular changes can be elucidated.

9. Carcinogenic Effects in Human Populations—Epidemiological Data

9.1 Considerations in Using Epidemiologic Data for Low-Dose Risk Assessment

Epidemiologic data are highly relevant in the sense that risks can be directly determined for human populations, rather than having the uncertainties associated with extrapolation from animal or *in vitro* model systems.

On the other hand, epidemiologic data have limited capability to evaluate the shape of the dose-response curve at low doses. This is because at low doses, the signal to noise ratio is poor, *i.e.*, the number of cancers caused by the radiation as compared to the number developing due to other or unknown causes is very small so that one cannot reliably detect the elevation in incidence. This is exacerbated by the fact that with a heterogeneous human population, with individuals having widely varied life styles, one is not dealing with a radiation exposure to organisms that exist in an otherwise controlled, uniform environment. Instead, humans have many other exposures to variable degrees, some of which could be carcinogenic cofactors or protectors. The limited data on carcinogenic cofactors suggest for instance, that they can change not only the absolute probability of developing cancer but also the shape of the dose-response curve (Fry and Ullrich, 1986), which has important implications with regard to low-dose linearity. Also, human studies often have a fair degree of uncertainty in the dose estimates. In high-dose studies the dose uncertainties have a smaller effect on the risk estimates (*e.g.*, Pierce *et al.*, 1991) than in low-dose studies, where the ratio of the magnitude of dose uncertainties to the imputed doses may be relatively large and hence may obscure or distort associations.

Another source of imprecision is the heterogeneity in susceptibility to radiogenic cancer in exposed human populations. Genetic heterogeneity, which may be a significant source of variability in susceptibility to radiation effects, but about which we currently have little information, is discussed in Section 9.2.7.2. There is also heterogeneity in radiation cancer induction associated with gender [*e.g.*, for thyroid cancer, breast cancer (Ron *et al.*, 1995; Shore, 1992)], smoking and other life-style factors (Lubin *et al.*, 1994; NAS/NRC, 1999; NCRP, 1997), age-at-irradiation [*e.g.*, for thyroid cancer, breast cancer and leukemia (Ron *et al.*, 1995; Tokunaga *et al.*, 1994; UNSCEAR, 1994)], and age, or years, at risk (Kellerer and Barclay, 1992; Thompson *et al.*, 1994).

A final category of imprecision relates to variations in length and completeness of follow-up, accuracy of tumor diagnoses, and other methodological variations in epidemiologic studies.

There is thus substantial uncertainty about the magnitude of risks resulting from low doses and low-dose rates. It is intrinsically difficult to assess risk accurately and precisely when doses are below, say, 50 mSv or when they are delivered at rates of a few millisievert per year. The limitations occur for at least two reasons. First, observational epidemiologic studies do not have the benefits of randomization to exposed or control groups. Randomization "guarantees" with a defined probability that the exposed and control subjects will not differ on other factors which might bias or "confound" the results. With a nonrandomized, low-dose, observational study the magnitude of confounding effects may be as large as, or larger than, the exposure-caused effects and hence may give "false positive" or "false negative" results. An epidemiologic study may have subtle biases present because of uncontrolled, and often uncontrollable, methodologic inadequacies of the study. Commonly, most of the potential confounding variables are either unknown or data on them are not available, so the confounding and biases cannot be adequately controlled. These biases, being typically of a small magnitude, have relatively little impact on a high-dose study where the dose-related effects are apt to be large, but in a low-dose study such biases could easily be similar in size to the size of expected effects and hence could seriously mask or inflate the apparent risk estimates.

In addition, any study at low doses will have inadequate statistical power unless the sample size is extremely large (Land, 1980; Shore, 1995), due to the intrinsic problem of detecting a small effect because the standard deviation will be relatively much larger than the size of the effect to be detected, and in epidemiologic studies the variance may be further inflated by added uncertainties, as detailed above.

Epidemiologic data therefore have limited ability to define the shape of the dose-response curve at low doses. Nevertheless, since extrapolations from animal data and mechanistic understanding are currently insufficient to define the shape of cancer dose-response curves in humans unequivocally, the epidemiologic data provide a risk estimate that integrates the results of the various biological processes leading to cancer induction in humans.

In summary, with low-dose data, unless the study size is exceedingly large, whether it be people, laboratory animals, or cells, a study will have a low statistical power to detect an effect. This implies it will also have little ability to discriminate between different shapes of the dose-response curve (*e.g.*, linear versus threshold). It is also important to note a corollary, that when the statistical power is low, the study probably will have little ability to rule out high risk estimates per unit dose.

One important principle, motivated by the preceding discussion of the imprecision of low-dose epidemiologic studies and their potential for invalidity, is that one should not rely on a single low-dose study, especially not a small one, as the primary basis for inferring there is no effect or for inferring a large effect. It is even less defensible to single out a particular *post hoc* subgroup in a study as demonstrating no effect (or a large effect), for this capitalizes on chance in such a way as to invalidate the statistical conclusion.

The best one can do with low-dose epidemiologic data is use a weight-of-evidence approach (Ashby *et al.*, 1990) in which all pertinent, high-quality data are incorporated into the assessment of risk. If the raw data of the relevant studies are available, this assessment can be done formally by means of a pooled analysis (Cardis *et al.*, 1995; Lubin *et al.*, 1994; Ron *et al.*, 1995), or second-best, a meta-analysis of the available dose-response tabulations can be undertaken (Lubin and Boice, 1997). However, for the purpose of an overall qualitative comparison and evaluation of quantitative results, tables summarizing the dose-response data from various studies are themselves very useful. An UNSCEAR report used this approach to good advantage (UNSCEAR, 1994). When properly carried out, a weight-of-evidence approach affords a balanced, broad-based assessment of the strength of an association of interest. A full weight-of-evidence approach would also incorporate experimental and mechanistic data to provide a comprehensive assessment of the association.

9.2 Types of Epidemiological Studies and Their Strengths and Weaknesses

9.2.1 *Cluster Studies*

An apparent cluster of cases of some disease(s) is sometimes noticed in a geographic area, organizational unit, or subgroup of treated persons, and a study is undertaken to determine if the cluster is "real" by comparing the number of cases found with an expected number based on published general-population data to determine if there is a "significant" excess. The human mind is agile in seeing patterns in some geographic unit or subgroup, so as to create or highlight an apparently elevated group. The results of such a study are frequently invalid, for the comparison of a group pre-selected as having an elevated rate to the average population rate is apt to be seriously biased. Any estimates of risk deriving from a cluster study are also likely to be spuriously high and would not be a sound basis for evaluating low-dose effects, although there have been attempts to draw such inferences from cluster studies (Najarian and Colton, 1978). For further discussion of this problem, the reader is referred to Boice (1991), Rothman (1990), and Shore (1995).

9.2.2 *Ecologic Studies (Studies of Aggregated Epidemiologic Data)*

In epidemiology, studies based on aggregated data are often called "ecologic" studies. Examples are studies of cancer rates in various geographic areas as a function of levels of background exposure (Frigerio and Stowe, 1983; Nambi and Soman, 1987; Wei *et al.*, 1990; Weinberg and Brown, 1987) or proximity to nuclear plants (Jablon *et al.*, 1991; Johnson, 1987), of lung cancer rates in relation to estimated residential radon levels (Cohen, 1995), and of birth defect rates in regions subjected to Chernobyl fallout (Harjulehto *et al.*, 1991; Lie *et al.*, 1992; Luning *et al.*, 1989). Ecologic studies are common because they are easy to perform since they often can use pre-existing data, and they appear to give precise answers since they are based on large numbers of deaths or incident cases.

Unfortunately, ecologic studies have the potential for several serious statistical problems that are difficult to detect but that can invalidate them. With the potential invalidity, such studies may be useful for hypothesis generation but seldom for hypothesis testing. Detailed technical discussions of the potential intrinsic weaknesses of ecologic studies can be found in Greenland and Morgenstern (1989), Greenland and Robins (1994), Piantadosi (1994), and Lubin (1998). Some of the main points can be summarized as follows:

1. In an individual-level analysis (*e.g.*, cohort or case-control study), standardization of the outcome variable (*e.g.*, using an age-adjusted cancer rate) is a common and valuable procedure. However, if one works with a standardized outcome variable in an ecologic analysis, the covariates and exposure variable also need to be standardized, or else bias may even be worsened by the standardization of the outcome variable.

2. If confounder variables are to be adequately controlled in an ecologic study (*e.g.*, control for cigarette smoking in a geographic study of radon and lung cancer), one needs the *detailed distribution* of the confounder variables *and* the exposure variable across individuals within each region. Summary variables (*e.g.*, percent smoking or mean smoking levels) are likely to be seriously inadequate to control confounding. This is especially true if there is nonlinearity or nonadditivity of effects at the individual level.
3. Unlike individual-level studies where nondifferential (*i.e.*, random) exposure misclassification or measurement error usually causes bias toward the null, in ecologic studies nondifferential misclassification can produce either serious overestimation or underestimation of exposure-disease associations.
4. While uncontrolled, nonlinear effects (*e.g.*, cancer rates that vary as the fifth or sixth power of age) can confound any type of study, they can be detected and controlled in an individual-level study, whereas in an ecological study this is frequently difficult or impossible to do with the data that are available. In addition, an ecologic study may show no confounding with respect to the ecologic association and yet still be biased for individual-level effects, which are the effects of interest.
5. Extraneous risk factors responsible for ecological bias may not be confounders or effect modifiers at the individual level, and they may not even appear to be confounders or modifiers at the ecological level. Yet the amount of ecological bias they produce "may be substantial (even reversing the direction of an observed association), especially when the observed range of average exposure level across groups is small or the exposure under study is not a strong risk factor" (Greenland and Morgenstern, 1989, pages 272 to 273). For example, a reversal of effects occurred in a study of radon and leukemia that depended on the size of the geographic areas that were studied (Muirhead *et al.*, 1991).
6. Use of a large number of geographic regions will not do away with biases. More observations will generally increase precision but may have little or no impact on validity, *i.e.*, the degree of bias.
7. Unless risk factors for the disease under study and the exposure of interest are uncorrelated *within states* (where "states" is a shorthand for whatever geographic unit is used in the ecological analysis) or unless risks from the exposure of interest and the other risk factors are strictly additive, the ecological exposure-response relationship will be biased (*i.e.*, not valid). Nor will

the addition of state-wide summary variables correct the bias (Lubin, 1998).

Given the intrinsic problems with analyses of ecologic data described above, such data cannot be regarded as trustworthy and should not be relied upon to assess either the presence or absence of excess radiation-induced cancer at low doses.

9.2.3 *Case-Control Studies*

Case-control studies, in which a series of cases of a disease from a defined population are compared to a matched or stratified random sample of people from the same population with respect to exposures of interest, can be effective ways to investigate relatively rare diseases and common exposures. However, when it comes to investigating the effects of low exposures and putatively small risks, modest biases can produce misleading results because the magnitude of the biases is likely to rival or exceed the magnitude of potential risks. Case-control studies are particularly subject to two kinds of bias. The first is selection bias, caused when the cases and controls that can be enlisted differ in the degree or ways to which they are unrepresentative of the population at risk. The second is the potential for recall bias when exposure information is based on self-reports. For example, diseased persons may have more motivation to report exposures than unaffected controls, so that the cases have more over reporting and/or less under reporting of exposure. An instructive example of this kind of bias was shown in Portsmouth Naval Shipyard studies in which an initial study based on self-report alleged an association between leukemia and radiation exposure at the Shipyard (Najarian and Colton, 1978), but subsequent studies based on recorded exposures found no association (Rinsky *et al.*, 1981; Stern *et al.*, 1986). [A subsequent methodological study showed the nature of the initial study's bias (Greenbert *et al.*, 1985).]

As another example, several case-control studies have shown apparent associations between self-reported exposures to medical or dental diagnostic radiation and various cancers (see, *e.g.*, Gibson *et al.*, 1972; Preston-Martin *et al.*, 1985a; 1989; Stewart *et al.*, 1962), but the two studies that have assessed the adequacy of "from memory" reporting of diagnostic radiation procedures found that 25 to 50 percent of such reports were inaccurate (Graham *et al.*, 1963; Preston-Martin *et al.*, 1985b). Case-control studies of diagnostic irradiation that rely on comprehensive medical records (*e.g.*, Boice *et al.*, 1991a; Inskip *et al.*, 1995) are therefore generally more credible than those based on self-reported exposures because the medical-record study is not likely to be biased.

However, case-control studies that are nested within a cohort study and that seek to obtain documented information on exposures or other risk factors for disease from objective records, are not subject to the biases mentioned above. The nested case-control study permits the investigator to obtain more detailed information on the persons of most interest (namely, the disease cases and a matched comparison group) without the cost of obtaining such information on the entire cohort.

9.2.4 *Cohort Studies*

In general, well-conducted cohort studies are the strongest type of epidemiologic study, short of randomized trials, but there are several weaknesses to look out for in evaluating and using their data in estimating low-dose risks. (1) Probably the single greatest weakness is that there may be too few cases of the disease in question to have much precision in assessing the low-dose region of the dose-response curve. (2) Biases can be introduced if a high rate of follow-up for cancer outcomes is not achieved for a cohort, especially if the rate of follow-up varies across the dose range. Self-selection factors in the study groups can produce artificially high or low risks, especially if comparisons are with the general population as the "baseline" group. Thus, for instance, studies of breast irradiation for preexisting diseases of the breast could have a built-in high risk for breast cancer. (3) Most especially, the "healthy worker effect" is ubiquitous in occupational studies: workers tend to be selected for initial good health at time of employment and those who continue working for longer times (which may well be correlated with cumulative dose attained) are further self-selected for good health. The "healthy worker effect" is especially prominent for cardiovascular and infectious diseases, but is often operative to some degree with regard to cancer endpoints as well. Hence, small elevations in risk in radiation worker studies may be partially or wholly masked by the "healthy worker effect" if standardized mortality ratios (SMR) or standardized incidence ratios (SIR) are used, because SMR and SIR are based on a comparison with the general population. Studies that use internal comparisons, *e.g.*, dose-response analyses, largely avoid the "healthy worker" bias.

9.3 Examination of Linearity of Dose Responses and Low-Dose Risks in Epidemiologic Data

An ideal type of tumor for examining the issue of low-dose linearity is one that has a high radiation sensitivity for induction and a relatively low rate of "spontaneous" occurrence. This combination affords the maximum ability to detect an elevation in cancer risk if one exists. However, caution should be exercised in extrapolating the dose-response relationship observed for one tumor type to another tumor type. The two types of tumor that best fit these criteria are leukemia (either mortality or incidence data) and thyroid cancer (incidence data only since the case fatality rate is low). Another tumor with a relatively high ratio of radiation-induced to spontaneous occurrence is female breast cancer (NAS/NRC, 1990). To cover other major radiogenic tumor endpoints, lung cancer and total solid tumors also should be examined. In view of the marked variations in dose-response relationships among the different types of cancer, no attempt is made herein to consider those types of cancer for which the available dose-response data are more limited or those types, such as osteosarcoma (see, *e.g.*, Rowland, 1997) for which the existing data suggest the existence of effective thresholds.

Studies to be examined should have reasonably accurate dosimetry, substantial numbers of persons in the dose range of interest, a long follow-up period, and a high rate of follow-up. The major studies with average doses less than 1 Sv and those with substantial fractionation or protraction of radiation exposures will be examined in the following. The Japanese atomic-bomb study has an appreciable number of subjects in the low-to-moderate dose range, as do several medical irradiation studies. An additional group of studies of interest includes those with appreciable dose fractionation or dose protraction. This includes medical irradiation cohorts with fractionated or low-dose-rate exposures and radiation-worker studies, of which a few of the largest will be examined. When possible, information on the dose-response curves will be presented.

9.3.1 *Total Solid Cancers*

Table 9.1 provides summary estimates of the risk of total solid tumors (or total cancers if only that information was available) from pertinent radiation studies, and Table 9.2 provides dose-response tabulations for the lowest dose groups available in various publications of such studies.

In the Japanese atomic-bomb life-span study cohort, between 1950 and 1990 there were 4,565 deaths from solid cancers (*i.e.*, all cancers

TABLE 9.1—*Risk estimates for radiation-induced total solid tumors (or total cancers), from epidemiologic studies where the mean bone-marrow dose was less than 1 Sv or the doses were fractionated or protracted.*

Study Group (reference)	Mean Dose (mSv)	Observed/ Expected Cancers[a]	Percent ERR Sv^{-1} (95% CI)	Absolute Risk 10^{-4} PY Sv
Japanese atomic-bomb survivors: mortality (Pierce *et al.*, 1996)[b]	240	4,565/4,231	40 (31, 51)[c,d]	10.6[c]
Japanese atomic-bomb survivors: incidence (Thompson *et al.*, 1994)	264	8,613/7,385[e]	63 (52, 74)[c]	53.9[c]
^{226}Ra implant or teletherapy for uterine bleeding (Inskip *et al.*, 1993)	~500	1,457/1,096	66 (52, 80)	82
Radiotherapy for skin hemangiomas (Furst *et al.*, 1988; 1990)	~70	224/190	257 (44, 490)	15.4
Multiple fluoroscopic exams for TB pneumothorax (Davis *et al.*, 1987)	120	173/169	17.7 (−104, 154)	4.67
^{131}I for hyperthyroidism (Hoffman *et al.*, 1982)	~60[f]	93/93	0 (−500, 500)	0
^{131}I for hyperthyroidism (Holm *et al.*, 1991)	~60[f]	1,543/1,456	100 (13, 190)	148
^{131}I for thyroid cancer (Hall *et al.*, 1991)	~280	99/69	154 (60, 262)	158
China background radiation (Luxin, 1980)	~47	45/47	−78 (−736, 890)	−0.14
U.S.S.R. Techa River (Kossenko, and Degteva, 1994)	133	774/589	235 (161, 316)	32.9
Early U.K. radiologists (Smith and Doll, 1981)[b]	~3,000[g]	136/136[c]	0 (−5.5, 6)	−0.03

TABLE 9.1—*Risk estimates for radiation-induced total solid tumors (or total cancers), from epidemiologic studies where the mean bone-marrow dose was less than 1 Sv or the doses were fractionated or protracted.* (continued)

Study Group (reference)	Mean Dose (mSv)	Observed/ Expected Cancers[a]	Percent ERR Sv^{-1} (95% CI)	Absolute Risk 10^{-4} PY Sv
Early U.S. radiologists (Matanoski *et al.*, 1987)[b]	~3,000 (~2,400–6,000)	345/263	10.3 (5.8, 15)	2.2
Chinese medical x-ray workers (Wang *et al.*, 1990b)	≈700	332/274	30 (12, 50)	2.6
Japanese medical x-ray workers (Aoyama *et al.*, 1998)	~470[f]	437/538	−40 (−56, −23)	−7.9
U.S. Army radiation technologists (Jablon and Miller, 1978)[b]	~80[h]	145/152	−63 (−300, 250)	−6.4
Danish radiotherapy workers (Ennow *et al.*, 1989)	18	148/148	0 (−860, 970)	−1.3
U.K. radium dial painters (Baverstock and Papworth, 1989)	≈400 (386 γ + small α dose)	95/86	26 (−25, 86)	5.6
Combined study: U.S., U.K., Canada (Cardis *et al.*, 1995)[i,j]	40.2	1,596/1,602	−0.7 (−39, 30)	—
Chelyabinsk Nuclear Plant (Koshurnikova *et al.*, 1994)	1,472	372/367	0.9 (−5.9, 8.2)	0.23
U.S. nuclear shipyard workers (Matanoski, 1991)[b]	~50[f]	603/632	−91.8 (−240, 65)	−23.7
U.S. Mound DOE Facility (Wiggs *et al.*, 1991)[b]	29.7	66/75[k]	−404 (−1,060, 380)	−63.4

U.S. Savannah River (Cragle *et al.*, 1998)[b]	46.2	413/505	−394 (−561, −215)	—
U.S. Oak Ridge Y-12 uranium fabrication workers (Checkoway *et al.*, 1988; Frome *et al.*, 1990)[b]	9.6[l]	196/193	140 (−1,290, 1,730)	25.5

≈ Rough estimate given by the authors.

[a]Observed and expected values for the exposed groups, as defined by the particular study. For the Japanese atomic-bomb survivors, it was defined as all those with estimated doses ≥5 mSv.

[b]Mortality data only.

[c]Risk estimate based on the dose-response relationship.

[d]90 percent CI rather than 95 percent CI.

[e]Solid cancers only; lymphopoietic cancers not given.

[f]Estimated for this tabulation from the indirect information available.

[g]3,000 mGy assumed; authors indicated 1,000 to 5,000 mGy for the 1920 to 1945 cohort and even higher doses prior to 1920.

[h]80 mGy assumed; authors indicated "probably <100 mGy."

[i]The combined study included worker cohorts at Hanford, Oak Ridge, Los Alamos, and Rocky Flats in the United States; Selafield, the Atomic Energy Authority, and the Atomic Weapons Establishment in the United Kingdom; and the Atomic Energy of Canada, Limited. Hence, results for these cohorts are not listed separately.

[j]All cancers except leukemia.

[k]Dose-response slope positive ($p = 0.11$). Some exposures to ^{210}Po, ^{238}Pu, ^{3}H.

[l]Also alpha dose to lung.

Table 9.2—*Total cancer or total solid tumors among groups with low-dose or fractionated irradiation, according to cumulative radiation dose.*[a]

Japanese atomic-bomb survivors: solid tumor mortality (Pierce *et al.*, 1996)[b]	5–100 1.05 (2,795)	100–200 1.05 (504)	200–500 1.15 (632)	500–1,000 1.30 (336)	>1,000 1.73 (298)
Japanese atomic-bomb survivors: solid tumor incidence (UNSCEAR, 1994)[b]	10–100 1.02 (2,223)	100–200 1.08 (599)	200–500 1.19 (759)	500–1,000 1.44 (418)	1,000–2,000 1.87 (273)
U.K., U.S. & Canada combined occupational cohorts (Cardis *et al.*, 1995)	10–19 0.99 (462)	20–49 0.93 (445)	50–99 1.09 (276)	100–199 1.03 (196)	200+ 1.01 (217)
U.S. Mound DOE Facility (Wiggs *et al.*, 1991)	<10 1.00[c] (49)	10–49 0.62 (10)	50+ 1.49 (7)		
U.S. DOE Los Alamos (Wiggs *et al.*, 1994)	0–10 1.00[c] (230)	10–50 0.76 (52)	50–100 1.30 (22)	100+ 1.02 (18)	
Japanese radiologic technicians (Aoyama *et al.*, 1998)	<500 0.87 (24)	500–749 0.96 (14)	750–999 1.02 (6)	≥1,000 1.56 (11)	

[a]Dose ranges (mSv) are underlined; second line gives RR plus observed number of cancers in parentheses.
[b]Analyses performed in such a way that the RR in the lowest dose, baseline group (not shown) was one.
[c]Reference group, set to one by definition.

except those of the lymphatic or hematopoietic systems) in the exposed group (dose to the colon ≥5 mSv) and 3,013 in the unexposed group (dose <5 mSv). A detailed analysis of the dose-response relationship for total solid cancers recently was reported by Pierce *et al.* (1996), with consideration for other factors such as age, gender and the temporal pattern of risk. Figure 9.1 shows the overall dose-response curve. They estimate an overall slope for the excess relative risk (ERR) of 0.40 Sv^{-1} (90 percent CI = 0.31, 0.51) (Table 9.2). Table 9.3 shows the age-gender adjusted ERR estimates for the five lowest non-zero dose categories that were reported in each study. While the variances are relatively large, so that most of the

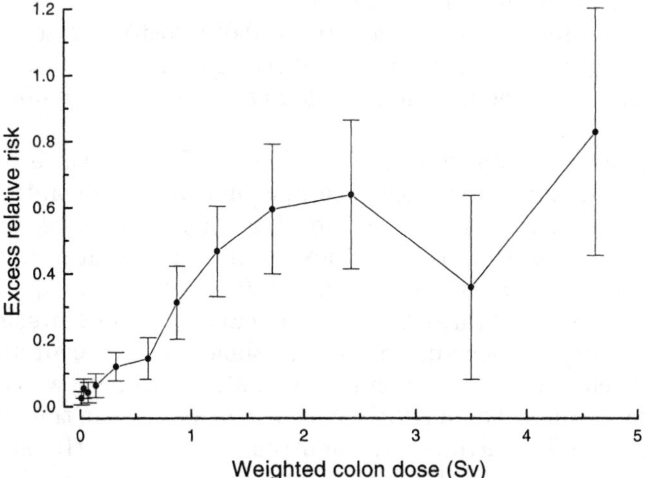

Fig. 9.1. Shape of the dose-response curve for total solid cancer mortality in the Japanese atomic-bomb survivors. There is no statistically significant nonlinearity in the range 0 to 3 Sv, however, the data do not suffice to define the shape of the curve below 5 mSv (Pierce *et al.*, 1996).

TABLE 9.3—*Life-span mortality study: ERR at low doses for total solid cancers (Pierce et al., 1996).*

Dose category (mSv)	ERR Sv^{-1} ± SE
5 – 20	2.6 ± 2.1
20 – 50	1.6 ± 0.90
50 – 100	0.60 ± 0.40
100 – 200	0.43 ± 0.25
200 – 500	0.38 ± 0.13

individual dose groups are not statistically significant by themselves, it is noteworthy that the data suggest that the risk per unit dose in the low-dose range may be as high or higher than that at higher doses. The data do not indicate a threshold below which there is no excess risk. Pierce *et al.* (1996) noted that an examination of the incidence (rather than mortality) data showed a similar pattern, except that there was less suggestion of supralinearity. One caveat to these findings is the possibility of ascertainment bias such that the diagnosis of cancer was more complete in the irradiated group than in the unexposed group.

Also of interest was an analysis of the lowest dose at which a statistically significant result could be seen for total solid cancers in the atomic-bomb study (Pierce *et al.*, 1996). The dose range 0 to 50 mSv produced a significant ($p = 0.02$) positive dose-response relationship once gender and age at irradiation were controlled for appropriately, although the data do not exclude the possibility of a threshold below 5 mSv.

Hoel and Li (1998) recently modeled the dose-response relationship for various cancer incidence and mortality endpoints for the atomic-bomb survivor data, comparing linear and dose-threshold curves. For total solid tumors they found that the fit of the linear curve and curves with a threshold of 100 mSv fit about equally well for the incidence data but the linear curve provided a somewhat better fit for the mortality data. Thresholds above about 100 mSv fit more poorly than a linear curve. A similar conclusion was reached by Kellerer and Nekolla (1997), based on evaluation of the influence of dose-dependent errors in the neutron dosimetry at Hiroshima. In a similar comparison of models, Little and Muirhead (1998) found the dose response for solid tumor mortality in the atomic-bomb study provided no evidence for a threshold, with the best estimate of a threshold value being <0. For solid tumor incidence the data were consistent with there being no threshold and the best estimate of a possible threshold was about 0.04 Sv (Little and Muirhead, 1996). Likewise, Vaeth *et al.* (1992) indicated that only low-dose extrapolation factor values near to linearity were consistent with the solid tumor incidence data. The results suggest that the low-dose data are too imprecise to definitively rule out one type of curve versus another.

The studies of therapeutic medical irradiation are generally not informative about total cancer since the radiation exposures were to localized areas of the body only. One exception is the studies of ^{131}I therapy, which also provide information on protracted exposure. A study of ^{131}I treatment for thyroid cancer reported a subsequent excess of total cancer (less thyroid): SIR = 1.51 (95 percent CI = 0.95, 2.30) based on 20 cases in 258 patients (Edmonds and Smith,

1986). The whole-body doses appear to have been in the range 1 to 4 Gy. Another study of 834 patients treated for thyroid cancer (Hall *et al.*, 1991), with a whole-body dose of about 0.3 Gy, also reported a total-cancer excess (SIR = 1.44, 95 percent CI = 1.05 to 1.92, 46 cancers). A large study of ^{131}I for hyperthyroidism (Holm *et al.*, 1991) reported an excess of total cancers after a whole-body dose of about 60 mGy (SIR = 1.06, 95 percent CI = 1.01 to 1.11; 1,543 cancers), although there is no dose-response relationship and the association was limited to those with toxic nodular goiter but not Graves disease, suggesting the association might be due to the condition rather than the radiation. Another large thyrotoxicosis study also reported mixed results with regard to irradiation (Ron *et al.*, 1998). On the other hand, a smaller study of hyperthyroidism treatment (Hoffman, 1984) showed no excess (SIR = 0.81, 95 percent CI = 0.7, 1; 105 cancers).

Several environmental or occupational studies have information on total cancer risks following fractionated and protracted exposures. Most of the studies of earlier cohorts of radiologists (Matanoski *et al.*, 1984; Smith and Doll, 1981) or radiological technicians (Wang *et al.*, 1990b) have reported excesses of total cancers, but one did not (Aoyama *et al.*, 1998). The population living along the Techa River in Russia, who were exposed to 0.07 to 1.4 Sv over the course of several years, had a total of 774 cancers and showed a statistically significant exposure-response relationship (Kossenko and Degteva, 1994). These studies indicate that highly fractionated and protracted radiation exposures produce cancer when the cumulative doses are large enough.

The large combined study of 95,673 United States, United Kingdom, and Canadian radiation workers (Cardis *et al.*, 1995), with a mean dose of 40 mSv, had 3,976 cancer deaths. The dose-response analysis showed no clear excess risk, with an ERR coefficient of $-0.07\ Sv^{-1}$ (90 percent CI = -0.4, 0.3), although this result was statistically compatible with risks obtained by linear-extrapolation from high-dose data. A large occupational study of nuclear shipyard workers likewise found no excess of total cancer mortality in the group with $\geq$5 mSv (SMR = 0.95, 95 percent CI = 0.88, 1.03; 603 cancer deaths) (Matanoski, 1991). A study of 143,000 United States radiologic technologists showed a deficit of cancers (observed/expected = 1,589/1,929 = 0.82 among those certified before 1960). The doses are unknown in this study (Doody *et al.*, 1998).

The dose-response data in Table 9.2 show apparent risk at low doses for the atomic-bomb survivor data, but the results are mixed in the various worker studies. Two of the occupational studies show apparent positive trends at doses below 100 or 200 mSv (Kendall *et al.*, 1992; Wing *et al.*, 1991), but the remaining studies in the table do not. Certain investigators have published analyses that purport to show that cancer risk from protracted occupational radiation exposures is much higher than one would estimate by linear extrapolation from the atomic-bomb data if the age and latent periods are defined in certain ways (Frome *et al.*, 1997; Kneale and Stewart, 1993; Wing *et al.*, 1991), but these analyses do not appear to be supported by the bulk of available occupational data.

In summary, the Japanese atomic-bomb study provides suggestive evidence that acute exposures, even at low doses, increase the risk of solid malignant tumors. Several studies of medical exposures also provide substantial evidence that fractionated and protracted radiation exposures increase total-cancer risk. However, when the total doses are low, occupational data do not provide clear evidence of risk. This is not surprising in view of their limited precision and the potential of confounding of small radiation effects by the "healthy worker effect" commonly found in occupational studies.

9.3.2 *Leukemia*

Among atomic-bomb survivors, analyses of both leukemia mortality (Pierce *et al.*, 1996) and leukemia incidence (Preston *et al.*, 1994) have shown that a linear fit of the dose-response data is inadequate but that a linear-quadratic model with concave upward curvature provides a good fit to the data (see Figure 9.2). When Preston *et al.*

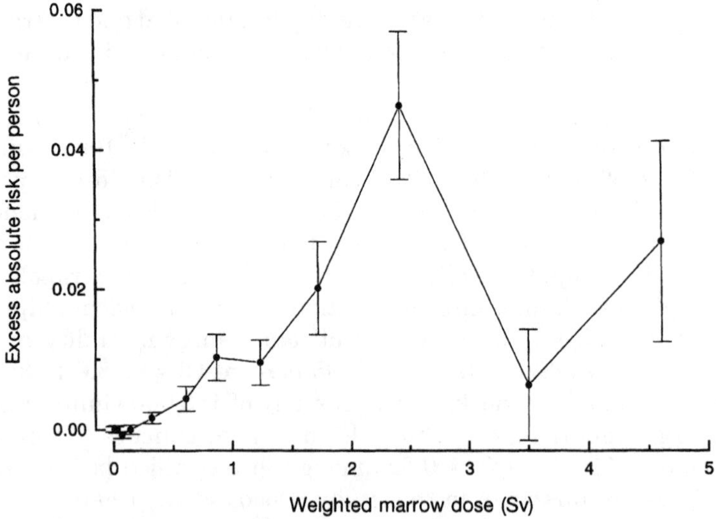

Fig. 9.2. Shape of the leukemia mortality dose response in the Japanese atomic-bomb survivors (Pierce *et al.*, 1996). There is statistically significant upward curvature in the dose range 0 to 3 Sv. The negative risk for the second dose category is not statistically significant.

(1994) tried to fit a threshold curve with a threshold at 0.5 Sv, they found they could statistically reject that model. However, it was not clear whether they could reject a threshold at a lower dose level (*e.g.*, 0.1 or 0.2 Sv). More recently, Little and Muirhead (1996; 1998) and Hoel and Li (1998) reported that curves with thresholds at 100 to 200 mSv do provide a better fit to the leukemia data than does a linear curve. Pierce *et al.* (1996) found that the degree of curvilinearity in the fitted linear-quadratic curve was such that the risk per unit dose was approximately three times as great at 1 Sv as it was at 0.1 Sv.

A number of radiation studies with low-to-moderate bone-marrow doses are informative, as shown in Table 9.4, including several studies of medical irradiation. A Swedish study of 20,024 patients given x-ray treatment for arthrosis or spondylosis has evaluated the risk of non-chronic lymphatic leukemia (CLL) (Damber *et al.*, 1995). Overall the mean bone-marrow dose was 0.39 Gy and the SIR was 1.27 (CI = 1, 1.6; 66 cases) for non-CLL, with a suggestion of a dose-response trend (see Table 9.5). A study of 4,483 women given radiotherapy (primarily intracavitary ^{226}Ra) for uterine bleeding, with a mean bone-marrow dose of about 0.65 Gy, showed a statistically significant excess of leukemia deaths (RR = 2.5, CI = 1.4, 5.2; 62 cases), although there was no dose-response relationship (Inskip *et al.*, 1993).

Among 14,624 infants treated in Sweden for skin hemangiomas, of whom about 83 percent were treated with ^{226}Ra applicators and most of the remainder with x rays (Lundell and Holm, 1996), and in whom the mean marrow dose was 130 mGy, there were 14 leukemias in those with average bone-marrow doses >10 mGy. For those in the marrow dose range 11 to 100 mGy, there was no excess (RR = 0.9, N = 6), while for those with >100 mGy there was a nonsignificant excess (RR = 1.7, 95 percent CI = 0.7 to 3.4). Another study of 11,807 infants given ^{226}Ra treatments for skin hemangiomas found a nonsignificant excess of leukemias (SIR = 1.54, 95 percent CI = 0.8 to 2.6; 13 leukemias) (Lindberg *et al.*, 1995). Given the somewhat lower organ doses in this study as compared to the parallel study by Lundell, the mean marrow dose was probably about 100 mGy.

There are various sources of data on leukemia risk following fractionated or protracted exposures, including several medical irradiation studies. The Massachusetts multiple fluoroscopy series received a mean marrow dose of about 120 mGy. The leukemia risk was not elevated (SMR = 1.1) based on six leukemia cases (Davis *et al.*, 1987). Persons administered ^{131}I for thyroid cancer or hyperthyroidism have received protracted irradiation. Three studies with a total of about 1,800 thyroid cancer patients given ^{131}I therapy have reported collectively five leukemia cases when two to three were expected (Dottorini

TABLE 9.4—*Risk estimates for radiation-induced leukemia, from epidemiologic studies where the mean bone-marrow dose was* <*1 Sv or the doses were fractionated or protracted*[a]

Study Group (reference)[b]	Mean Dose (mSv)	Observed/ Expected[c,d] Cancers	Percent ERR Sv^{-1} (95% CI)	Absolute Risk 10^{-4} PY Sv
Japanese atomic-bomb survivors (Pierce *et al.*, 1996)	240	176/98	462 (328, 640)[e,f]	2.6[e]
^{226}Ra treatment for skin hemangiomas (Lindberg *et al.*, 1995)	~120	13/8.4	450 (−120, 1,310)	1.1
^{226}Ra for uterine bleeding (Inskip *et al.*, 1990)	~650	34/17.3	190 (80, 320)	2.6
X-ray treatment for arthrosis or spondylosis (Damber *et al.*, 1995)	390	66/51.9 (NCLL)	70 (−2, 156)	1
^{131}I for hyperthyroidism (Saenger *et al.*, 1968)	~100	11/10.4 (ML)	58 (−444, 838)	0.6
^{131}I for hyperthyroidism (Holm *et al.*, 1991)	60	34/36.17	−100 (−560, 500)	−2.8
Diagnostic/therapeutic ^{131}I (Holm, 1991)	14	195/178.9	643 (−410, 1,820)	12.2
Diagnostic ^{131}I (Holm *et al.*, 1989)	<10 (~8 assumed)	119/88.8	4,250 (1,380, 7,550)	56.6
Utah fallout study (Stevens *et al.*, 1990)[g]	19	17/9.9	3,800 (−320, 11,200)	—
U.S., Project Smoky (Caldwell *et al.*, 1980; 1983)	4.7	10/4	32,200 (5,800, 74,100)	219

U.S.S.R. Techa River releases (Kossenko, 1996)	370	27/12.3[e]	324 (48, 600)[c]	0.9
U.S. radium dial painters (Stebbings *et al.*, 1983; 1984)	~2,000[h]	8/7.29	4.9 (−25, 54)	0.04
Early U.K. radiologists (Smith and Doll, 1981)[i]	~3,000[j]	8/3.24	49 (4.9, 123)	0.3
Early U.S. radiologists (Matanoski *et al.*, 1987)	~3,000 (~2,400-6,000)	33/19.8	22 (5.6, 44)	0.4
Chinese medical x-ray workers (Wang *et al.*, 1990b)	≈700	34/14.1	202 (43, 440)	0.9
Japanese medical x-ray workers (Aoyama *et al.*, 1998)	470[k]	20/15.3	66 (−38, 210)	0.4
U.S. Army radiation technologists (Jablon and Miller, 1978)	~80[l]	11/6.73 (NCLL)	800 (−430, 3,900)	3.6
Chelyabinsk Nuclear Plant (Koshurnikova *et al.*, 1994)	2,450	20/4.4	145 (76, 241)[e]	8[e]
U.S. nuclear submariners (Charpentier *et al.*, 1993)	~5[k]	21/23.08	−178 (−840, 730)	−1.3
Combined study: U.S., U.K., Canada (Cardis *et al.*, 1995)	40.2	119/109.4	218 (13, 570)	1.2
U.S. Savannah River workers (Cragle *et al.*, 1998)	46.2[k,m]	25/19.6	591[e,n]	—
U.S. nuclear shipyard workers (Matanoski, 1991)	~50	21/24.1	−260 (−920, 660)	−2.6

TABLE 9.4—*Risk estimates for radiation-induced leukemia, from epidemiologic studies where the mean bone-marrow dose was <1 Sv or the doses were fractionated or protracted*[a] (continued)

≈ Rough estimate given by the authors.

[a]Abbreviations used: (ML) myeloid-monocytic leukemia; (NCLL) all leukemias except chronic lymphocytic leukemia.

[b]Studies were included if they had at least 10 observed leukemias or 10 expected leukemias (where the expected value was estimated by summing the baseline expectation and the expected excess based on the linear leukemia risk estimate from the Japanese atomic-bomb study (Shimizu *et al.*, 1990). Over 15 studies failed to meet either of these criteria and were excluded. Reports of individual worker studies were excluded if their data were included in the combined study reported here (Cardis *et al.*, 1995).

[c]Observed and expected values for the exposed groups, as defined by the particular study. For the Japanese atomic-bomb survivors, the exposed group was defined as those with estimated doses ≥5 mSv.

[d]All types of leukemia unless specified otherwise.

[e]Risk estimate based on the dose-response relationship.

[f]90 percent CI rather than 95 percent CI.

[g]Case-control study, all ages. Dose based on geographic locations. Similar RR for CLL suggest possible biases.

[h]Average dose of ~1,800 mSv alpha and 200 mSv external gamma.

[i]Began radiology practice 1897 to 1954.

[j]3,000 mGy assumed; authors indicated 1,000 to 5,000 mGy for the 1920 to 1945 cohort and even higher doses prior to 1920.

[k]Mean dose estimated for this tabulation from the indirect information available.

[l]80 mGy assumed; authors indicated "probably <100 mGy."

[m]Approximately 15 percent of dose from internal depositions. 5,000 individuals with ^3H exposure. Leukemia excess reported among a subset of early employees.

[n]Dose-response statistically significant.

TABLE 9.5—*Leukemia among groups with low-dose or fractionated irradiation, according to cumulative radiation dose.*[a]

Japanese atomic-bomb survivors: incidence (Preston *et al.*, 1994; UNSCEAR, 1994)	<10 1.11 (90)	10–100 0.90 (38)	100–200 0.73 (8)	200–500 2.25 (27)	>500 6.80 (68)
Japanese atomic-bomb survivors: mortality (Shimizu *et al.*, 1990; UNSCEAR, 1994)[b]	5–55 0.99 [0.7–1.4][c]	55–95 0.61 [0.3–1.2][c]	95–195 1.08 [0.6–1.8][c]	195–495 1.79 [1.2–2.7][c]	495–995 4.15 [2.8–6.2][c]
^{226}Ra treatment for uterine bleeding (Inskip *et al.*, 1990)	10–250 0 (1.2 expected)	251–500 1.94 (11)	501–750 2.09 (13)	751–1,000 3.27 (7)	1,001–2,800 2.67 (2)
^{226}Ra treatment for uterine bleeding (Inskip *et al.*, 1993)	10–250 0 (0.78 expected)	251–500 2.60 (10)	501–750 1.41 (9)	751–3,300 2.46 (7)	
X-ray treatment for arthroses (Damber *et al.*, 1995)	1–199 1.07 (25)	200–500 1.27 (18)	>500 1.61 (23)		
^{131}I exposure: diagnostic or treatment of hyperthyroidism (Hall *et al.*, 1992a)	0–0.01 1.04 (12)	0.02–0.1 1.06 (48)	0.11–10 1.23 (92)	11–100 0.77 (32)	>100 1.82 (11)
Combined study: U.S., U.K., Canada (Cardis *et al.*, 1995)[d,e]	10–19 1.06 (19)	20–49 0.78 (14)	50–99 0.86 (8)	100–199 1.21 (8)	200+ 1.38 (10)

TABLE 9.5—*Leukemia among groups with low-dose or fractionated irradiation, according to cumulative radiation dose.*[a] (continued)

U.S. nuclear shipyards (Matanoski, 1991)[f]	<u><5</u> 0.44 (4)	<u>5–10</u> 0.41 (2)	<u>10–49</u> 1.25 (13)	<u>50–99</u> 0.85 (3)	<u>100+</u> 1.06 (3)
U.S. DOE Los Alamos (Wiggs *et al.*, 1994)	<u>0–10[e]</u> 1.0 (21)	<u>10–50</u> 0.34 (2)	<u>50–100</u> 0.67 (1)	<u>100+</u> 1.28 (2)	

[a] Dose ranges (mSv) are underlined; second line gives RR (or SMR) plus observed number of leukemias in parentheses.
[b] Analyses performed in such a way that the RR in the lowest dose, baseline group (not shown) was one.
[c] Number of cases not reported; 90 percent CI given instead.
[d] Excluding CLL.
[e] Trend test, $p < 0.05$ with a positive slope.
[f] Analysis incorporates a 2 y lag in assigning exposures.

et al., 1995; Edmonds and Smith, 1986; Hall *et al.*, 1992a); the mean bone-marrow doses in the studies ranged from 0.25 Gy to several gray. A large study of 10,552 hyperthyroidism patients in Sweden found that, after a mean marrow dose from ^{131}I of 60 mGy, there was no elevation in risk (SIR = 0.94, N = 34) (Holm *et al.*, 1991). A United States study of 18,379 hyperthyroid patients given ^{131}I found no excess leukemia (RR = 0.8, N = 17) as compared to a group treated surgically (Ron *et al.*, 1998; Saenger *et al.*, 1968).

Other potential sources of information on fractionated exposure and leukemia are studies of diagnostic irradiation history and of radioactive fallout. There are several studies of diagnostic irradiation but all except one rely upon anamnestic reports of medical irradiation procedures (Gibson *et al.*, 1972; Gunz and Atkinson, 1964; Linos *et al.*, 1980; Preston-Martin *et al.*, 1989; Stewart *et al.*, 1962), which have been shown to be quite unreliable. The one based on medical records found a nonsignificant RR of 1.42 for non-CLL when exposures were lagged by 2 y and no dose-response relationship (Boice *et al.*, 1991a).

Studies of fallout and leukemia have been reported from several geographic areas. A study of leukemia was conducted in Utah, downwind from the Nevada atomic-bomb test site, where the bone-marrow doses from fallout were estimated to range from about 1.5 to 19 mGy. A suggestive, but not statistically significant ($p = 0.08$), dose-response trend in non-CLL was seen (Stevens *et al.*, 1990). Preliminary reports on leukemia in areas with the greatest fallout from the Chernobyl accident generally have been negative (Auvinen *et al.*, 1994; Ivanov *et al.*, 1996; Parkin *et al.*, 1996).

Leukemia mortality has also been studied in the population along the Techa River in Russia that was exposed over the course of several years to bone-marrow doses ranging from 0.17 to 1.6 Gy from releases of fission products into the river. The exposed population of about 28,000 had a significantly elevated RR for leukemia of 2.2 based on 27 cases (Kossenko, 1996). The authors found an adequate fit of a linear dose-response curve which showed an excess absolute risk (EAR) of 0.85 (90 percent CI = 0.24 to 1.45) 10^{-4} person-year (PY) Gy, which is about a third as large as the coefficient for the Japanese atomic-bomb survivors. It is unclear whether this reflects a sparing effect of dose protraction or is simply due to uncertainties in the risk estimates.

The last category of information on leukemia risk after fractionated or protracted exposures consists of occupational studies. Early radiologists, or medical radiation technologists, had highly fractionated doses, but their cumulative doses are believed to have been large (little or no dosimetry data exist for these cohorts). These studies all showed an excess of leukemia. Among early United States radiologists the leukemia SMR = 1.67 (CI = 1.2 to 2.3, N = 33), and the average cumulative dose has been crudely estimated as 3 Gy (Matanoski *et al.*, 1987). A smaller study of early United Kingdom radiologists, with similar doses, had a leukemia SMR = 2.5 (CI = 1.2 to 4.7, N = 8) (Smith and Doll, 1981). A large study of United States radiologic technologists did not find an excess of leukemia [observed/expected = 60/66 = 0.81 among those certified before 1960 (Doody *et al.*, 1998)]. A study of Japanese medical x-ray workers found a SMR = 1.3 (CI = 0.8 to 2, N = 20) with an estimated average dose of about 0.5 Gy (Aoyama, 1989). Finally, Wang *et al.* (1990b) reported that Chinese medical x-ray workers, with an average dose estimated as perhaps 0.7 Gy, had a leukemia SMR = 2.4 (CI = 1.3 to 4.1, N = 34).

Three large studies of occupational exposures at lower dose levels are pertinent. One study followed up 3,145 men who worked at Department of Energy (DOE) facilities or in the Navy Nuclear Reactor Propulsion Program and who in at least one calendar year had received ≥50 mSv. No excess of leukemia was seen (observed/expected = 2/4.3) (Fry *et al.*, 1996). A United States study of nuclear shipyard workers by Matanoski (1991) is of interest because over 9,000 workers had exposures ≥50 mSv, and another 18,000 had between 5 and 50 mSv. No excess of leukemia was observed (SMR = 0.87, CI = 0.5 to 1.3, N = 21). Unfortunately, a dose-response breakdown was not reported. The most comprehensive study of leukemia in workers exposed to relatively low doses is that by Cardis *et al.* (1995). This study of >95,000 workers in the United States, United Kingdom, and Canada who were monitored for external exposure to ionizing radiation contained 119 cases of leukemia (excluding CLL). The doses ranged from less than 10 mSv to greater than 400 mSv, with about 20 percent having doses greater than 50 mSv. The dose-response trend was statistically significant ($p < 0.05$) and yielded an estimate of risk (ERR = 2.2, 90 percent CI = 0.1 to 5.7) that was about 60 percent as high as, but statistically compatible with, the linear risk coefficient among the Japanese atomic-bomb survivors, again suggesting that at low-dose rates there is a reduced efficiency of the neoplastic effects of ionizing radiation.

Available dose-response summary data are shown in Table 9.5. A couple of studies suggest some leukemia risk at doses less than about 200 mSv (Damber *et al.*, 1995; Hall *et al.*, 1992a) and two large occupational radiation studies have mixed results in that range (Cardis *et al.*, 1995; Matanoski, 1991), while several others are negative.

In summary, the Japanese atomic-bomb survivor study found a curvilinear dose-response relationship (upward concave) for leukemia with no apparent risk below about 0.2 Sv. Several studies of fractionated or protracted irradiation with relatively high cumulative doses showed leukemia excesses; most of those with low cumulative doses did not, although a notable exception was the large study of combined radiation-worker cohorts. The bulk of the data indicate that the risk per millisievert at low doses is probably less than at higher doses. However, one can find both low-dose studies that suggest there is no risk and others that suggest there is risk. It is probably beyond the capability of existing epidemiologic data to resolve the issue of low-dose leukemia risk.

9.3.3 *Thyroid Cancer*

In the Japanese atomic-bomb survivor study of cancer incidence (Thompson *et al.*, 1994), 132 thyroid cancers were observed among those with thyroid doses greater than 10 mSv. Forty thyroid cancers were observed in survivors who were exposed prior to age 20 (Table 9.6). There was a linear dose-response relationship ($p < 0.001$) with no significant nonlinearity ($p = 0.17$) (see Table 9.7). A strong effect of age at exposure was also seen, such that the ERR per sievert were 9.5, 3, 0.3 and −0.2 at ages 0 to 9, 10 to 19, 20 to 39 and 40+, respectively.

An Israeli study of 10,834 children x irradiated for ringworm of the scalp found 43 thyroid cancers (RR = 4, CI = 2.3, 7.9) (Ron *et al.*, 1989). A dosimetric study showed that the average dose was about 90 mGy (Werner *et al.*, 1968); two other studies also have essentially confirmed this dose estimate (Lee and Youmans, 1970; Schulz and Albert, 1968). This is one of the strongest pieces of evidence for an effect at a relatively low dose. A similar, but four times smaller, study of patients irradiated for scalp ringworm did not find a substantial excess of thyroid cancer (two observed, 1.3 expected), but the two studies are statistically compatible (Shore, 1992).

A study of 2,634 patients x irradiated for enlarged tonsils in Chicago showed a statistically significant excess of thyroid cancer after a mean dose of about 600 mGy (RR = 2.5, CI = 1.5, 16; 309 cancers) (Ron *et al.*, 1995; Schneider *et al.*, 1993). A study of x irradiation for lymphoid hyperplasia (Pottern *et al.*, 1990) also showed an excess after a dose of about 240 mGy.

A cohort of 2,657 infants x irradiated for enlargement of the thymus gland have been followed up (Shore *et al.*, 1993), of whom 56 percent had doses <0.5 Gy. There was a strong dose-response

TABLE 9.6—*Thyroid cancer ERR and EAR estimates for cohort studies with acute, external irradiation before age 20.*

Study (reference)	Mean Dose (mSv)	Observed/ Expected Cancers[a]	Percent ERR Sv^{-1} (95% CI)[b]	Absolute Risk 10^{-4} PY Sv[b]
Japanese atomic-bomb (<15 y at exposure) (Ron *et al.*, 1995; Thompson *et al.*, 1994)	230	40/19.2	4.7 (1.7, 11)	2.7
Enlarged thymus (Shore *et al.*, 1993)	1,360	37/2.8	9.1 (3.6, 29)	2.6
Tinea capitis (Ron *et al.*, 1989)	90	44/11.2	32.5 (14, 57)[c]	7.6
Enlarged tonsils (Schneider *et al.*, 1993)	590	309/125	2.5 (0.6, 26)	3
Skin hemangioma (Lundell *et al.*, 1994)	260	17/7.5	4.9 (1.3, 10)	0.9
Skin hemangioma (Lindberg *et al.*, 1995)	120	15/8	7.5 (0.4, 18)	1.6
Lymphoid hyperplasia (Pottern *et al.*, 1990)	240	13/~2.2	~20 (9.5, 37)	15.1

[a]Observed and expected values for the exposed groups, as defined by the particular study. For the Japanese atomic-bomb survivors, it was defined as all those with estimated doses ≥5 mSv.
[b]Estimates were based on dose-response analyses.
[c]When an indicator variable for irradiated versus control group was included, the ERR slope dropped to 6.6 Gy^{-1} (Ron *et al.*, 1995).

TABLE 9.7—*Thyroid cancer among groups with low-dose or fractionated irradiation, according to cumulative radiation dose.*[a]

Japanese atomic-bomb survivors (Thompson *et al.*, 1994)	<10 1 (93)[b]	10–1,000 (170) 1.38 (115)	>1,000 (1,830) 3.44 (17)		
Scalp ringworm study (Ron *et al.*, 1989)	47–80 3.3 (15)	80–150 4.2 (24)	150–500 6.1 (4)		
Thymus irradiation (Shore *et al.*, 1993)	10–250 3.85 (2)	250–500 13.6 (3)	500–2,000 7.1 (1)	2,000–4,000 42.3 (11)	4,000–5,990 78.6 (11)
^{226}Ra for skin hemangioma (Lundell *et al.*, 1994)	20–1,000 1.14 (9)	>1,000 10.1 (4)			
Diagnostic ^{131}I (Hall *et al.*, 1996)[c]	1–250 0.55 (5)	260–500 0.68 (4)	510–1,000 0.47 (5)	>1,000 1.04 (11)	
Cervical cancer treatment (Boice *et al.*, 1988)	<50 1.0 (3)[b]	50–100 1.86 (38)	100–150 2.39 (10)	150+ 3.42 (21)	
U.K. National Dose Registry: mortality (Kendall *et al.*, 1992)	<10 1.21 (4)	10–49 0.80 (2)	50–199 0.83 (2)	200–399 0 (E = 0.5)	400+ 3.6 (1)
U.S. DOE Hanford: mortality (Gilbert *et al.*, 1993)	<10 1.2 (2)	10–49 1.0 (1)	50–99 0 (0.2 expected)	100–199 0 (0.1 expected)	200+ 0 (0.05 expected)

[a]Dose ranges (mSv) are underlined (with mean doses parenthesized); second line gives RR plus observed number of cancers in parentheses.
[b]Reference group set to one by definition.
[c]Excluding those referred for suspicion of a thyroid tumor.

relationship over the entire dose range. A statistically significant dose-response relationship was found even when the dose range was limited to ≤0.3 Gy, but not when the range was limited to ≤0.2 Gy. The small number of thyroid cancers at low doses limited the statistical power of these analyses.

Among a Swedish cohort of 14,351 infants who were treated, primarily with ^{226}Ra, for skin hemangiomas, the thyroid cancer SIR was 2.28 (CI = 1.3 to 3.7) based on 17 cancers (Lundell *et al.*, 1994). The mean dose was estimated as 260 mGy. A similar study of infants with a mean thyroid dose of 155 mGy from ^{226}Ra treatment of skin hemangiomas also found an excess of thyroid cancer (SIR = 1.88, CI = 1.05 to 3.1; 15 cancers) (Lindberg *et al.*, 1995).

A summary of available dose-response data for thyroid cancer doses under 1 Sv is shown in Table 9.7. Several studies suggest that there is increased risk at fairly low doses, although the dose groupings were often wider than would be desirable for an examination of the low-dose region. Nevertheless, the thyroid cancer incidence data for external irradiation are reasonably consistent in showing apparent risk in the lowest dose groups in which it was tabulated.

A pooled analysis of five of the largest cohort studies of thyroid cancer risk from external irradiation during childhood showed that a linear dose-response model fit the data well (Ron *et al.*, 1995). Figure 9.3 shows that the risk per unit dose at low doses was as large as, or larger than, that at higher doses.

Several case-control studies have been performed to determine the effects of medical diagnostic irradiation on thyroid cancer rates (Hallquist *et al.*, 1994; McTiernan *et al.*, 1984; Ron *et al.*, 1987). Of these, only one has used objective information (Inskip *et al.*, 1995) rather than anamnestic reports of diagnostic irradiation, with their potential for recall bias. This study did not find an association between thyroid cancer and number of x-ray examinations of the head-neck-upper spine (trend p = 0.54) or examinations of the chest-shoulders-upper gastrointestinal tract (p = 0.50). Nor was there an association for diagnostic x-ray examinations before 1960, when doses were probably much higher.

Several studies of ^{131}I exposure to juveniles shown in Table 9.8 have involved high doses and will not be discussed, but several others with lower doses are of note. Hall *et al.* (1996) have followed up 34,104 patients who were administered ^{131}I for diagnostic purposes, of whom about 10,800 were being examined for suspicion of thyroid tumor. Among the latter, there was a subsequent excess of thyroid cancer, but none was found among the remainder (SIR = 0.75), for whom the thyroid dose was about 0.9 Gy. Of particular interest was

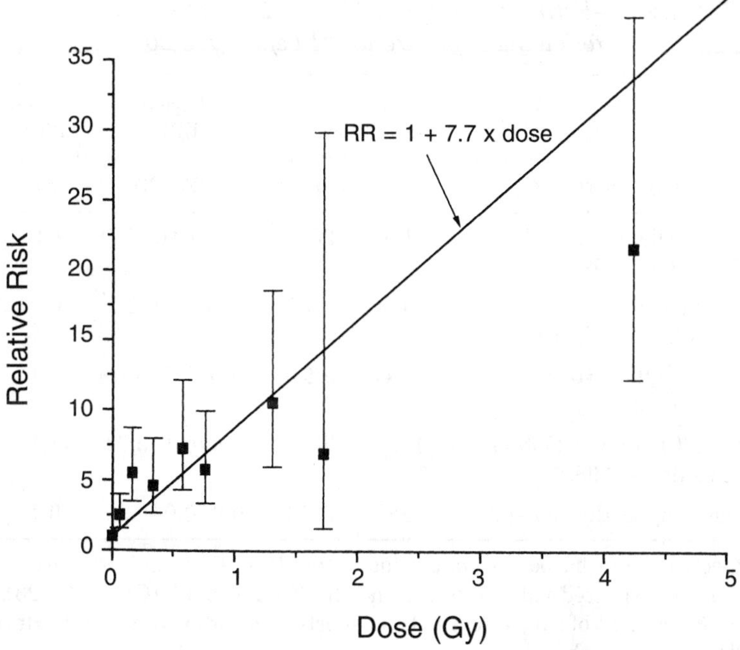

Fig. 9.3. Dose-response data points and curve for thyroid cancer from five pooled studies of external radiation exposure before 15 y of age (Ron *et al.*, 1995).

the subset of 1,764 patients who were exposed before 20 y of age and who were not being evaluated for suspicion of tumor; among these there were two thyroid cancers (SIR = 1.38, CI = 0.2 to 5). Results for the total number of juvenile patients from several ^{131}I studies are shown in Table 9.8.

A study conducted by the U.S. Food and Drug Administration (FDA) (Hamilton *et al.*, 1989) reported a small excess of thyroid cancer among 3,503 juveniles given diagnostic ^{131}I, where the thyroid doses ranged from <0.1 to >10 Gy, with a median dose of about 0.35 Gy and a mean dose of about 0.8 Gy. When compared to their unirradiated control group, there appeared to be a small excess of thyroid cancer (observed/expected = 4/1.4), but when compared to general population rates there was no excess (4/3.7).

In Utah there have been two rounds of thyroid examinations (about 1965 and 1984) of a cohort of school children exposed to ^{131}I and gamma rays from fallout from the atomic-bomb testing in Nevada. The mean thyroid dose was estimated as 170 mGy. The composite

TABLE 9.8—*Estimates of the ERR and EAR of thyroid cancer following exposure to ^{131}I before age 20.*

Study (reference)	Mean Dose (Gy)	Observed/ Expected Cancers	Percent ERR Sv^{-1} (95% CI)	Absolute Risk 10^{-4} PY Sv
Swedish Diagnostic ^{131}I (Hall *et al.*, 1996)	1.5	3/1.8	0.3 (<0–2.7)	0.2
FDA diagnostic ^{131}I (Hamilton *et al.*, 1989)	~0.8	4/3.7[a]	0.1 (<0–2)	0.05
Utah ^{131}I fallout (Kerber *et al.*, 1993)	0.17	8/5.4	7.9 (<0–16)	3.1
Marshall Islands (Robbins and Adams, 1989)	12.4[b]	6/1.2	0.3 (0.1–0.7)	1.1
Juvenile hyperthyroidism[c]	~88	2/0.1	0.3 (0.0–0.9)	0.1

[a]Expected value based on population rates. Based on their small control group, the expected value was 1.4 and the ERR was 3.1 (CI = <0–23).
[b]Over 80 percent of this dose was from short-lived radioiodines and external radiation rather than ^{131}I.
[c]Composite of nine studies: Crile and Schumacher (1965), Dobyns *et al.* (1974), Goldsmith (1972), Hayek *et al.* (1970), Holm *et al.* (1991), Kogut *et al.* (1965), Safa *et al.* (1975), Sheline *et al.* (1962), Starr *et al.* (1969).

of those results showed eight thyroid cancers in the irradiated group when about 5.4 would have been expected (Kerber *et al.*, 1993). The dose-response analysis was not statistically significant but the risk estimate was similar to that seen for higher doses of external radiation. A couple of potential methodological biases (higher screening referral rates in the higher-exposure areas and ascertainment of milk-drinking habits after the thyroid screening results were known) may have also contributed to the putative radiation effect.

Reports from Ukraine, Belarus and, to a lesser extent, Russia indicate a substantial excess of thyroid cancer among children exposed to ^{131}I fallout from Chernobyl. To date the studies have been mainly aggregate or "ecologic" studies of thyroid cancer rates by geographic region with the limitations attendant on this type of study. Nevertheless, the childhood thyroid cancer excesses are convincing; the excess of histologically confirmed thyroid cancer has been so large that it cannot be attributed only to increased surveillance (Astakhova *et al.*, 1998; Nikiforov and Fagin, 1998). For example, during 1990 to 1994 a total of 315 thyroid cancers in children were observed in Belarus, which was a 30-fold excess over the numbers observed there in the previous 10 y (Demidchik *et al.*, 1996). The excess rates appear to conform reasonably well to the magnitude of dose by region and to the fact that one would expect a larger effect when irradiation occurred early, rather than later, in childhood. It is particularly notable that insofar as thyroid dosimetry has been undertaken to date for individual cancer cases, most of the cancer cases had only moderate doses (*i.e.*, under 1 Gy).

Recently, Jacob *et al.* (1998) have made the most systematic attempt to date at evaluating the dose response for childhood thyroid cancer in these regions. They reported a reasonably linear dose-response relationship and estimated that the slope of the curve was about half as great as for external irradiation, albeit it was statistically compatible with the latter. The results should be treated cautiously in view of uncertainties in the dose estimation, possible cancer surveillance biases, and the ecologic nature of the study design.

Several questions about the increase in thyroid cancer from Chernobyl fallout have yet to be fully addressed. The first question is whether the thyroid doses may have been systematically underestimated or have large inaccuracies. The thyroid dosimetry in the best case, when a thyroid measurement was made, assumes a model of the timing of ^{131}I exposures and the duration of the ^{131}I retention curves. To the degree that these assumptions are not met, the projected doses may be in error. Furthermore, for most of the thyroid cancer cases, no thyroid ^{131}I measurements are available, so the doses are estimated indirectly by modeling from area ^{137}Cs deposition measurements or from thyroid measurements of other children in the region. A second question is whether the goitrogenic effects of low iodine levels in drinking water and foodstuffs in some of these regions may have synergized the induction of thyroid cancer. A third question is whether synergy between radiation and other environmental exposures in some of the regions that are highly industrialized may occur resulting in a greater dose response to ^{131}I than might be seen under other circumstances. However, in spite of these and other questions that may be raised, the fact remains that regions with the protracted but high exposures clearly have an excess of thyroid cancer, and a recent case-control study confirmed the association between thyroid cancer risk and ^{131}I exposure (Astakhova *et al.*, 1998).

In summary, a number of studies indicate a risk of thyroid cancer among children exposed to relatively low doses of acute radiation. The limited data available indicate that there is little thyroid cancer risk following irradiation in adulthood. Several studies of ^{131}I exposure suggest that the risk may be lower than for external, high-dose-rate irradiation, but the studies have small numbers of exposed children. The most extensive data on childhood ^{131}I exposure, that from Chernobyl, although based on an ecologic study, suggest that the risk may be appreciable in the moderate-dose as well as the high-dose range.

9.3.4 *Breast Cancer*

Female atomic-bomb survivors have been followed for both breast cancer mortality and breast cancer incidence. The latest report on this population included 398 incident cases with ≥1 mGy (Tokunaga *et al.*, 1994) and 193 with <1 mGy (controls, excluding the not-in-city controls). The observed dose response was highly consistent with a linear model, although a threshold at 50 mSv cannot be excluded. Adding a dose-squared term did not improve the fit (p = 0.92). The investigators were able to observe a statistically significant dose-response relationship even over the dose range 0 to 250 mSv. The ERR per sievert decreased at older ages at exposure; the ERR of 3.2 for those irradiated at ages 0 to 9 y decreased to where there was no evidence of an excess for irradiation at age 55 or over. Table 9.9 presents a summary of these and other breast cancer data, and Table 9.10 summarizes the dose-response data.

In a study of infants irradiated for purported enlarged thymus glands, in which about 1,200 girls were followed up to examine breast cancer risk (Hildreth *et al.*, 1989), there was a significant dose-response relationship that was consistent with linearity. There was also a significant elevation of risk in the subgroup that received less than 0.5 Gy (RR = 2.7, 95 percent CI ≠ 1.1 to 6.7).

The incidence of breast cancer has been studied among 2,573 women who received multiple chest fluoroscopic examinations during pneumothorax therapy for TB (Boice *et al.*, 1991b). The exposure was highly fractionated; on average the women were fluoroscopically examined nearly 100 times. The authors estimated the average breast dose as 0.79 Gy (range 0.01 to 6.4 Gy) (Boice *et al.*, 1978), although there may be uncertainty in the doses. The data were fit well by a simple linear model. The risk estimate for this highly fractionated exposure protocol was somewhat lower than, but compatible with, those from studies with unfractionated exposures. A tabulation of the breast cancer risk for the 1,675 women with total breast doses less than 1 Gy showed a modest elevation in risk (RR = 1.21, N = 75, mean dose 0.36 Gy). The absolute risk coefficient among those with less than 1 Gy was as large or larger than those

TABLE 9.9—*Estimates of breast cancer risk from irradiation before age 25.*

Study (reference)	Observed/ Expected Breast Cancer Cases[a]	Mean Breast Dose (Sv)	Percent ERR Sv^{-1} (90% CI)	Absolute Risk 10^{-4} PY Sv
Japanese atomic bomb (Tokunaga *et al.*, 1994)	298[b]	0.28	241 (163–344)	≈12
Swedish skin hemangioma (Lundell *et al.*, 1996)	60/36.4	0.2	325 (160–525)	2.3
Hodgkins disease (Hancock *et al.*, 1993)	6/0.18	40	81 (31–171)	0.6
New York thymus (Hildreth *et al.*, 1989)	22/6.7	0.69	177 (70–387)	6.4
Benign breast disease (Mattsson *et al.*, 1993)	23/6.8	≈4.2	100 (—)[c]	—
U.S. scoliosis (Hoffman *et al.*, 1989)	11/6	0.13	637 (55–1,490)	18
Massachusetts fluoroscopy (Boice *et al.*, 1991b)	92/52.7	0.79	80 (44–122)	13.5
Canada fluoroscopy—non-Nova Scotia (Howe and McLaughlin, 1996)[d]	269/204	0.79	40 (13, 77)	—
Canada fluoroscopy—Nova Scotia (Howe and McLaughlin, 1996)[d]	80/9.3	2.13	356 (185, 682)	—

[a]Observed and expected values for the exposed groups, as defined by the particular study. For the Japanese atomic-bomb survivors, it was defined as all those with estimated doses ≥5 mSv.
[b]All observed cases, including those with <0.01 Gy.
[c]Excess risk at 1 Gy for irradiation at age 20, at 25 y of follow-up, based on their model.
[d]Numbers given are for all ages, but risk estimate was based on a model containing age at exposure and estimated for age 15 y at exposure.

TABLE 9.10—*Breast cancer incidence (or mortality where noted) among groups of women with low-dose or fractionated irradiation, according to cumulative radiation dose.*[a]

Japanese atomic-bomb survivors (UNSCEAR, 1994)	1–9 0.96 (60)	10–99 0.96 (132)	100–249 1.15 (55)	250–499 1.48 (46)	500–999 2.35 (51)
Multiple fluoroscopic examinations: mortality (Howe and McLaughlin, 1996)	10–499 1.09 (120)	500–999 1.11 (73)	1,000–1,999 1.38 (75)	2,000–2,999 1.69 (27)	3,000–3,999 2.36 (20)
Multiple fluoroscopic examinations (Boice *et al.*, 1991b)	10–999 1.21 (75)	1,000–1,999 1.76 (44)	2,000–2,999 2.46 (14)	3,000+ 3.60 (9)	
Swedish benign breast disease patients (Mattsson *et al.*, 1993)	0 1 (95)[b]	1–200 1.19 (12)	200–500 1.50 (16)	500–2,000 1.92 (7)	
U.S. DOE Hanford: mortality (Gilbert *et al.*, 1993)	<10 1 (45)[b]	10–49 1.3 (7)	50–99 1.1 (1)	100–199 1 (1)	200+ 0 (0.3 expected)

[a]Dose ranges (mSv) are underlined; second line gives RR plus observed number of cancers in parentheses.
[b]Reference group; RR of one by definition.
[c]The expected value is for mortality rather than incidence.

at higher total doses, which suggests there is additivity of doses even at the low end of the range.

Howe and McLaughlin (1996) have ascertained breast cancer mortality among about 13,000 Canadian women who received ≥10 mSv from fluoroscopic examinations in treating pulmonary TB, among whom there were 349 breast cancer deaths, and another 18,800 who received 0 to 9 mSv from such procedures. The doses were highly fractionated; single doses were on the order of 10 mSv each. The cumulative doses ranged up to 10 Sv, although the majority were less than 1 Sv. For the breast cancer incidence data, the results strongly supported a linear dose-response curve, and the results indicate that the additive effects of small dose fractions can cause breast cancer.

Girls who received multiple diagnostic x rays for scoliosis have been followed up to determine breast cancer risk (Hoffman *et al.*, 1989). The 970 patients received an average of 41 examinations, and the total dose averaged 130 mGy. Eleven breast cancers were observed versus six expected (RR = 1.8, 95 percent CI = 0.9 to 3.2), and the dose-response trend was nearly significant, but the relatively small size of this study limits its statistical power. The tendency toward decreased parity in this cohort may be a potential methodological bias contributing to this putative radiation effect as well.

In a group of Chinese diagnostic x-ray workers, 20 breast cancers were observed when 13.2 were expected (RR = 1.5, 95 percent CI = 0.9 to 2.3) (Wang *et al.*, 1990b). The excess tended to be larger among those employed in the earlier years when the exposures were thought to be larger. Among 79,000 radiological technologists, a nested case-control study of the 528 breast cancer cases and 2,500 controls found no association between amount of occupational radiation exposure and breast cancer risk, although the larger mortality cohort had some suggestion of an effect among those with the highest exposure (Boice *et al.*, 1995; Doody *et al.*, 1998). A study of radium dial painters in the United Kingdom showed some suggestion of elevated breast cancer risk among those with doses of less than 0.2 Gy from gamma rays (SMR = 1.86, CI = 0.9 to 3.3) (UNSCEAR, 1994); however, lack of a clear excess for those with higher doses, plus the fact that the reproductive risk factors for this working population may have placed them at higher-than-average risk, call for caution in interpreting this finding.

Breast cancer risk has also been evaluated among women who were treated with ^{131}I for hyperthyroidism (Goldman *et al.*, 1988). The study population included 1,406 with ^{131}I therapy, with or without thyroid drugs or surgery, and 356 with only thyroid drugs or surgery. The administered dose ranged from 0.1 mCi (3.7 MBq) to greater than 10 mCi (370 MBq), but the doses to the breast were not calculated. Follow-up averaged 17 y. The RR for breast cancer (N = 51) in the ^{131}I-treated group was 1.2 (95 percent CI = 0.9 to 1.6). Another study of breast cancer in 1,000 women receiving similar doses of ^{131}I for hyperthyroidism was negative (23 cancers, RR = 0.8, 95 percent CI = 0.5 to 1.4) (Hoffman and McConahey, 1983). Breast doses were not reported but were probably about 0.1 Gy. In a study of ^{131}I treatment for thyroid cancer by Edmonds and Smith (1986), 258 patients received a mean breast dose of 1 Gy and were followed for an average of 11 y. Six breast cancers were observed (RR = 2.4, 95 percent CI = 1 to 4.9).

The dose-response data for breast cancer, shown in Table 9.10, are generally indicative of a risk at low doses. Even in the atomic-bomb survivor study in which the two lowest dose groups showed no clear evidence of increased risk, the overall dose-response relationship was compatible with linearity, with no statistical evidence of quadratic curvature. Nor was there evidence of strong supra-linearity of the dose-response curve, such as would produce large estimates of risk from low doses (*e.g.*, mammography) corresponding to those projected by Gofman (1996).

In summary, a number of studies of breast irradiation have involved fractionated exposures, protracted exposures and/or subgroups that received relatively low doses. Most of the studies either have shown significant breast cancer effects or have at least been consistent with such effects. The data are supportive of an interpretation that low doses or fractionated doses are largely additive (*i.e.*, linear) in their effects upon breast cancer risk.

9.3.5 *Lung Cancer*

9.3.5.1 *Low Linear-Energy Transfer Irradiation.* Several studies have shown an excess of lung cancer following relatively high doses of low-LET radiation (*e.g.*, Griem *et al.*, 1994; Inskip *et al.*, 1994; Weiss *et al.*, 1994), and a number of studies have examined lung cancer risk at low to moderate doses (Table 9.11). The Japanese atomic-bomb survivor study showed an excess of lung cancer (Pierce *et al.*, 1996), and the dose-response curve was approximately linear (Thompson *et al.*, 1994). Table 9.12 shows the lower end of the dose-response curve for lung cancer mortality in this study; there is an indication of risk at low doses (UNSCEAR, 1994). Land *et al.* (1993b) recently showed that both low- and high-LET radiations preferentially increase the small-cell histologic type of lung cancer.

TABLE 9.11—*Risk estimates for radiation-induced lung cancer.*

Lung Cancer Study Group (reference)	Mean Dose (mSv)	Observed/ Expected Cancers[a]	Percent ERR Sv^{-1} (95% CI)	Absolute Risk 10^{-4} PY Sv
Japanese atomic-bomb survivor: mortality (Pierce *et al.*, 1996)	240	939/872	53 (28–84)[b,c]	1.7[b,c]
Japanese atomic-bomb survivor: incidence (Thompson *et al.*, 1994)	250	456/368.8	95 (60, 136)[b]	4.4[b]
Ankylosing spondylitis (Weiss *et al.*, 1994)	2,540	563/469 = 1.20	5 (0.2, 9)[b]	0.9[b]
Breast cancer radiotherapy (Inskip *et al.*, 1994)	9,800	61/33.9 = 1.8	20 (−62, 103)	—
Radiotherapy for peptic ulcer (Griem *et al.*, 1994)	1,170	99/58.24 = 1.70	66 (47, 93)	2.3
Multiple fluoroscopic exams for TB pneumothorax (Howe, 1995)	1,020	455/455 = 1.00	0 (−6, 7)	0
Multiple fluoroscopic exams for TB pneumothorax (Davis *et al.*, 1987)	910	26/30.55 = 0.85	−16 (−47, 25)	−0.3
^{131}I for hyperthyroidism (Holm *et al.*, 1991)	70	105/79.5 = 1.32	458 (115, 856)	37
U.S.S.R. Techa River release (Kossenko and Degteva, 1994)	133	119/114.3 = 1.04	31 (−50, 111)[b]	0.6[b]
U.S. radium dial painters (Stebbings *et al.*, 1984)	~690[d]	19/13.65 = 1.39	57 (−20, 164)	0.8
Early U.K. radiologists (Smith and Doll, 1981)	~3,000[e]	31/27.49 = 1.13	4.3 (−7.3, 19)	0.2
Chinese medical x-ray workers (Wang *et al.*, 1990b)	≈700	45/49.9 = 0.90	−14 (−48, 28)	−0.3
Japanese medical radiation workers (Aoyama *et al.*, 1998)	470[f]	57/92.4 = 0.62	−81 (−112, −44)	−2.8

66426$T12B—Manuscript table 9.11—Operator 20—Proof/Edit —05-22-01 22:49:58

TABLE 9.11—*Risk estimates for radiation-induced lung cancer.* (continued)

Lung Cancer Study Group (reference)	Mean Dose (mSv)	Observed/ Expected Cancers[a]	Percent ERR Sv^{-1} (95% CI)	Absolute Risk 10^{-4} PY Sr
U.S. nuclear shipyard workers (Matanoski, 1991)	~50.5[f]	237/221.5 = 1.07	139 (−123, 426)	13
Savannah River Nuclear Plant (Cragle *et al.*, 1998)	46.2	175/185	−117 (−409, 210)	—
Combined study: U.S., U.K., Canada (Cardis *et al.*, 1995)	40.2	562/548.6 = 1.02[g]	61 (−146, 280)	3.9
U.S.S.R. Mayak Nuclear Reactors (Koshurnikova *et al.*, 1996)	870	48/46.9 = 1.02	2.7 (−27, 40)	0.1

~Estimate made for the present tabulation from the available information.
≈Rough estimate given by the authors.
[a]Observed and expected values for the exposed groups, as defined by the particular study. For the Japanese atomic-bomb survivors, it was defined as all those with estimated doses ≥5 mSv.
[b]Risk estimate based on the dose-response relationship.
[c]90 percent CI given.
[d]Average dose ~120 mSv alpha from radium + ~370 mSv from radon + ~200 mGy external gamma.
[e]3,000 mGy assumed; authors indicated 1,000 to 5,000 mGy for the 1920 to 1945 cohort and even higher doses prior to 1920.
[f]Estimated for this tabulation from the indirect information available.
[g]Observed and expected values for workers with cumulative dose ≥10 mSv (as compared to workers with <10 mSv). The dose-response slope was in the negative direction.

TABLE 9.12—*RR (and CI) for lung cancer among the Japanese atomic-bomb survivors according to lung dose, especially at low-dose levels (UNSCEAR, 1994).*

Dose Range (mGy)	RR (90% CI)
5 – 55	1.30 (1.11–1.54)
55 – 95	1.21 (0.89–1.62)
95 – 195	1.02 (0.75–1.36)
195 – 495	1.54 (1.22–1.93)
495 – 995	1.63 (1.19–2.19)
995 – 1,995	2.45 (1.73–3.38)
≥1,995	2.14 (1.16–3.59)

Of the studies with protracted or fractionated low-LET radiation exposure, a study of patients administered ^{131}I for hyperthyroidism was positive (SIR = 1.32, CI = 1.08 to 1.60; N = 105) after a lung dose of about 70 mGy; however, the excess occurred within the first 5 y after ^{131}I treatment in a particular diagnostic subgroup, so it may represent an anomaly (Holm *et al.*, 1991). Two studies following highly fractionated exposures from multiple fluoroscopic examinations of TB patients showed no excess of lung cancer after substantial cumulative doses. The results of the larger study are shown in Table 9.13 in parallel to the results from the atomic-bomb survivors. This study included 25,007 irradiated persons with an estimated mean dose of 1.02 Gy (Howe, 1995). The authors observed 455 lung cancers among those with lung doses >10 mGy, but the RR was not elevated (RR = 1) (Howe, 1995). Similar results were found in a second, smaller multiple fluoroscopy study (Davis *et al.*, 1987) (see Tables 9.11 and 9.13). A caveat should be added that the results of these two studies might be confounded by cigarette smoking history which has not been fully documented for these cohorts, nor is it known whether severe pulmonary TB may affect subsequent lung cancer risk. Hence, these data suggest, but fall short of definitively demonstrating, that *fractionated* low-LET lung exposures carry little or no lung cancer risk.

Several studies have examined lung cancer risk following environmental or occupational exposure to low-LET irradiation. A study of people exposed to the Techa River releases (Kossenko and Degteva, 1994) and four studies of medical radiation workers (Aoyama *et al.*, 1998; Jablon and Miller, 1978; Smith and Doll, 1981; Wang *et al.*, 1990b) showed no significant excesses of lung cancer, even though several of these studies consisted of workers in an earlier era when doses were much higher than today. Likewise, the study of United

TABLE 9.13—*Lung cancer risk in relation to radiation dose for the Canadian multiple fluoroscopy study and the Japanese atomic-bomb survivor study (Howe, 1995).*

Lung Dose (mSv)	Multiple Fluoroscopy		Atomic Bomb	
	Number of Cancers	RR (95% CI)	Number of Cancers	RR (95% CI)
<10	723	1.00	248	1.00
10–499	180	0.87 (0.7–1.0)	290	1.26 (1.1–1.5)
500–999	92	0.82 (0.7–1.0)	38	1.45 (1.0–2.1)
1,000–1,999	114	0.94 (0.8–1.2)	30	1.93 (1.3–2.9)
2,002–2,999	41	1.09 (0.8–1.5)	10	2.65 (1.5–4.7)[a]
≥3,000	28	1.04 (0.7–1.5)	3	

[a]This estimate is for all with doses >2,000 mSv.

States nuclear shipyard workers (Matanoski, 1991) and the large combined study of 95,000 nuclear workers (Cardis *et al.*, 1995) showed no excesses of lung cancer. Among additional studies of radiation workers, a few with relatively low exposures to low-LET radiation showed suggestive associations (Checkoway *et al.*, 1988; Frome *et al.*, 1997), although most have not shown statistically significant elevations in lung cancer risk (Charpentier *et al.*, 1993; Cookfair *et al.*, 1983; Cragle *et al.*, 1988; Dupree *et al.*, 1987; Wiggs *et al.*, 1991). However, only the Hanford study has collected data on smoking so as to be able to control for its potential confounding effects (Petersen *et al.*, 1990).

Available data on lung cancer dose-response curves for (primarily) low-LET irradiation are shown in Table 9.14. None of the studies provide clear-cut evidence of lung cancer risk below about 200 mSv for low-LET radiation.

9.3.5.2 *High Linear-Energy Transfer Irradiation.* A joint analysis of 11 studies of miners that included some 2,700 lung cancer deaths has clearly shown that exposure to high-LET alpha particles from radon progeny is associated with an increased risk of lung cancer (Lubin *et al.*, 1994; 1995a; NAS/NRC, 1999). The studies show a linear dose-response relationship if an inverse protraction effect is factored in (otherwise it is convex upward). The miner data have been examined to determine whether the inverse dose-rate effect extends down to the levels characteristic of residential radon levels, and they appear to indicate that the inverse dose-rate effect disappears, or at least diminishes, at low-dose rates and low total doses (Lubin *et al.*, 1995b; 1997). The animal data are generally concordant with that observation as well (Cross, 1994). Mechanistically, one would not expect a dose-rate effect at low doses, when the probability of more than one alpha-particle traversal of a cell is low (Brenner, 1994).

Of particular note is the fact that the combined studies, although complicated by the confounding effects of smoking, even showed significant elevations in RR for the 50 to 74 and 75 to 99 working level month (WLM) categories, which are comparable to some cumulative residential exposures, *i.e.*, lifetime residence in a home at the EPA action level of 4 pCi L^{-1} (150 Bq m^{-3}) would result in 40 to 80 WLM (Lubin, 1994). A Russian study of workers at a plutonium facility, where about 30 percent of the lung dose was from alpha particles and the remainder from gamma rays, showed an excess of lung cancer down to levels of about 1 Sv (Hohryakov and Romanov, 1994; Koshurnikova *et al.*, 1997; Tokarskaya *et al.*, 1997).

TABLE 9.14—*Lung cancer mortality among groups with low-dose or fractionated irradiation, according to cumulative radiation dose.*[a]

Japanese atomic-bomb survivors (UNSCEAR, 1994)	10–60 1.30 [1.1–1.5][b]	60–100 1.21 [0.9–1.6][b]	100–200 1.02 [0.8–1.4][b]	200–500 1.54 [1.2–1.9][b]	500–1,000 1.63 [1.2–2.2][b]
Multiple fluoroscopic examinations (Howe, 1995)	10–500 0.87 (180)	500–1,000 0.82 (92)	1,000–2,000 0.94 (114)	2,000–3,000 1.09 (41)	3,000+ 1.04 (28)
Multiple fluoroscopic examinations (Davis *et al.*, 1989)	<500 0.81 (29)	500–999 0.59 (9)	1,000+ 0.60 (17)		
U.K. National Dose Registry (Kendall *et al.*, 1992)	<10 0.94 (200)	10–49 0.92 (115)	50–199 1.19 (119)	200–399 1.35 (45)	400+ 0.62 (12)
DOE Hanford (Gilbert *et al.*, 1993)[c]	<10 0.97 (256)	10–49 1.03 (107)	50–99 1.25 (24)	100–199 1.13 (17)	200+ 0.96 (14)
DOE Oak Ridge National Laboratory (Wing *et al.*, 1991)[d]	0 0.95 (13)	1–39 0.93 (50)	40–79 1.2 (8)	80–119 1.1 (3)	120+ 1.8 (9)
DOE Oak Ridge Y-12: Gamma radiation (Checkoway *et al.*, 1988)	0–9.9 1.0 (49)[e]	10–49 0.9 (9)	50–99 1.3 (1)		
DOE Oak Ridge Y-12: Alpha radiation (Checkoway *et al.*, 1988)	0–9.9 1.0 (18)[e]	10–49 0.9 (11)	50–99 0.7 (5)	100+ 1.1 (11)	

DOE Mound (Wiggs *et al.*, 1991)	<10 1.0 (15)[e]	10–49 0.9 (4)	50+ 1.4 (3)		
DOE Los Alamos (Wiggs *et al.*, 1994)	0–10[e] 1.0 (57)	10–50 0.50 (9)	50–100 0.88 (4)	100+ 1.06 (5)	
Portsmouth Naval Shipyard (Rinsky *et al.*, 1988)[f]	0 1.2 (21)	1–9.9 1.1 (62)	10–49 1.8 (25)	50+ 1.4 (13)	
U.S. nuclear shipyards (Matanoski, 1991)[f,g]	<5 1.19 (91)	5–9 0.95 (41)	10–49 1.00 (93)	50–99 1.21 (37)	100+ 1.40 (34)
Chinese medical x-ray workers (Wang *et al.*, 1990b)	<5 y[h] 1.1 (3)	5–9 y[h] 0.7 (4)	10–14 y[h] 0.6 (4)	15–19 y[h] 1.5 (12)	20+ y[h] 0.8 (22)

[a] Dose ranges (mSv) are underlined; second line gives RR plus observed number of cancers in parentheses.
[b] Number of cases not reported; 90 percent CI given instead.
[c] Analysis with 10 y lag in doses.
[d] Non-monthly pay status workers only. Total worker tabulation appears to be confounded by smoking.
[e] Reference category set to one by definition.
[f] Asbestos exposure was a likely confounder in this study.
[g] Analysis with 5 y lag in doses.
[h] Duration of radiation work used in lieu of dose.

The miner data indicate that the joint effects of smoking and radon upon lung cancer risk are more than additive (*i.e.*, additivity is statistically rejected) and are compatible with, but somewhat less than, a multiplicative association (Lubin, 1994). This implies that larger absolute risks for radon-induced lung cancer will be seen among smokers than nonsmokers although the RR is larger among nonsmokers.

At least eight major case-control studies of lung cancer in relation to long-term measurements of exposure to residential radon have been conducted (Alavanja *et al.*, 1994; Auvinen *et al.*, 1996; Blot *et al.*, 1990; Darby *et al.*, 1999; Letourneau *et al.*, 1994; NAS/NRC, 1999; Pershagen *et al.*, 1992; 1994; Ruosteenoja, 1991; Schoenberg *et al.*, 1990). These studies included 4,263 lung cancer cases and 6,612 controls. The results are shown in Figure 9.4. Of the eight studies, five showed positive associations within some subgroup(s), but for only three was the principal overall analysis significant in the positive direction. A major limitation of the individual studies,

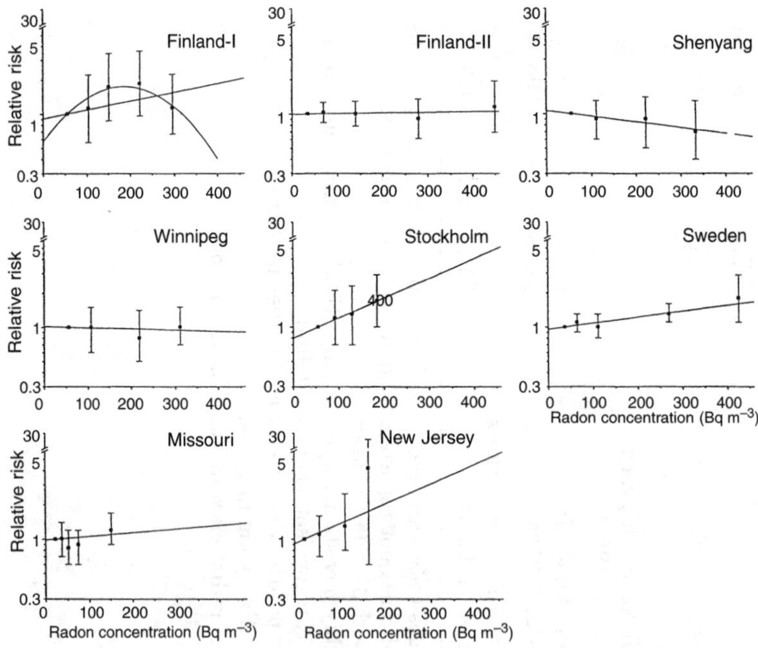

Fig. 9.4. RR by radon concentration and fitted exposure-response trends for eight case-control studies of lung cancer and residential radon levels (Lubin and Boice, 1997).

however, is their limited statistical precision and power to detect small risks. A meta-analysis of these studies (Lubin and Boice, 1997) found a statistically significant positive slope, such that the RR at 150 Bq^{-3} was 1.14 (95 percent CI = 1.01 to 1.30), which is almost identical to the corresponding projection from the miner studies (namely, 1.13). One caveat is that the authors found statistically significant heterogeneity in the risk estimates, which they could not account for by examining various methodological and substantive differences between the studies; this reflects the fact that the slopes differed appreciably in the various studies, with the estimated RRs at 150 Bq^{-3} ranging from 0.84 to 1.83.

In the aggregate, the studies are consistent with risks extrapolated from higher-dose miner studies (Figure 9.5), but they likewise appear to be compatible with the null hypothesis (Lubin, 1994). Lubin *et al.* (1995c), by the use of realistic simulated data, have shown how studies with about 2,000 cases would be needed to have an adequate statistical power to detect effects of the magnitude predicted by

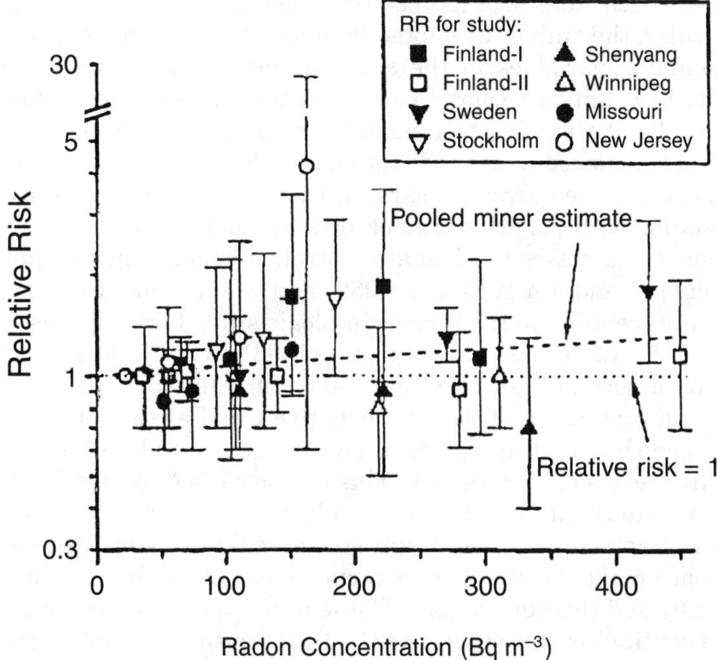

Fig. 9.5. RR from eight lung cancer case-control studies of indoor radon levels. Dashed line is the extrapolation of RR from miners (Lubin *et al.*, 1995a), while the dotted line depicts a RR of one (adapted from Lubin and Boice, 1997).

BEIR IV risk estimates *if* there were no exposure measurement error. But given realistic conditions in which there is exposure measurement error due to both technical measurement factors and things like residential mobility and temporal changes in home insulation or ventilation, they find that the required numbers are more likely to be 5,000 to 13,000 cases with twice that number of controls. Even if one could mount a study of this size, an effect of the magnitude being expected might well be smaller than the confounding and biasing effects in the study, so that a definitive result may not be possible.

It is of interest that none of the case-control studies showed a statistically significant result in the negative direction, in view of the negative slope that Cohen found in an ecologic study (Cohen, 1995) (see Figure 9.6). His analysis was based on lung cancer mortality rates in most United States counties in relation to measurements of samples of residential radon levels from those counties. (The sampling of homes was not strictly random, however). Because this was an ecologic study, it is potentially subject to the largely unresolvable biases of such studies, as has been detailed above (Section 9.1.1.2). In particular, the author attempted to control for a number of potential confounding variables in the analysis but did not have available county-by-county information on the frequency distribution of smoking levels, the single most important variable to control for. He instead was forced to use indirect and crude surrogates for smoking frequencies. The surrogate variables probably have only a modest correlation with the desired smoking frequency variables; hence the authors could not exert adequate control for smoking in the analysis (Greenland and Morgenstern, 1989) and would make the analysis highly susceptible to the types of ecologic-study biases discussed in Section 9.1.1.2. Furthermore, Lubin (1998) recently showed that a positive association between radon and lung cancer among individuals could reverse and became a negative trend at the county level, even if one has accurate information on mean radon levels and mean smoking levels in each county. This can be caused by small correlations within counties between radon levels and smoking or age. The intrinsic epidemiologic and statistical weaknesses of the ecologic approach make the results questionable, even though the study was carefully and thoroughly done. These issues have been examined in more depth elsewhere (NAS/NRC, 1999; Smith *et al.*, 1998; Stidley and Samet, 1994).

In summary, some epidemiological data suggest that for low-LET radiation there is little or no lung cancer risk associated with fractionated or protracted exposures, even though the acute exposure data from the Japanese atomic-bomb study appear linear above

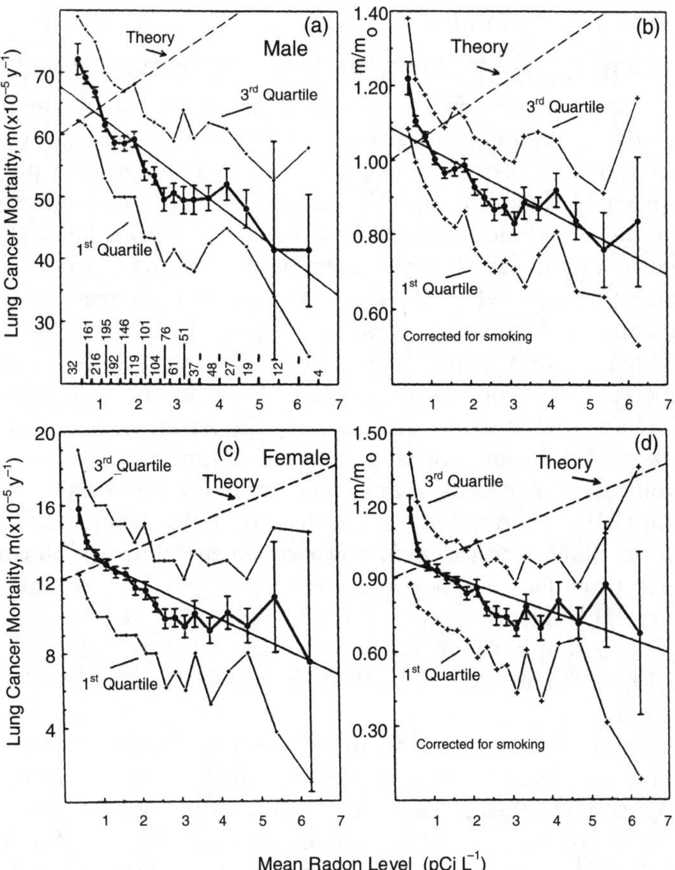

Fig. 9.6. Lung cancer mortality rates versus mean radon level in homes for 1,601 United States counties. Data points shown are average of ordinates for all counties within the range of values shown on the baseline. The number of counties within that range is also shown in Figure 9.6a. Error bars are one standard deviation of the mean, and the first and third quartiles of the distribution are also shown. Figure 9.6c and Figure 9.6d are lung cancer rates corrected for smoking prevalence. Theory lines are arbitrarily normalized lines increasing at a rate $+7.3$ percent (pCi $L^{-1})^{-1}$ (Cohen, 1995).

5 mSv. For high-LET (radon) exposure, the consensus of the best available data, albeit limited as they are, suggests some elevation in risk at fairly low-dose levels. Although the ecologic study by Cohen (1995) was carefully done, intrinsic limitations of the ecologic methodology, particularly in conjunction with the potential confounding by smoking and age, mean the result cannot be relied upon.

9.3.6 *In Utero Irradiation*

Most of the important information concerning the risk of *in utero* ionizing radiation exposure for the development of cancer during childhood has been derived from observations on children who had received prenatal x-ray exposure for medical diagnostic purposes from the early 1940s through the late 1970s. Little new information of this type has developed since then because of the marked reduction in the use of diagnostic x ray during pregnancy due to the extensive medical reporting in the 1960s and 1970s of an increased occurrence of leukemia and other cancers in children during the first 10 to 15 y of life following *in utero* exposure. Most of the studies have been retrospective case-control or cohort in type, in the sense that the events of interest preceded initiation of the study. Several excellent reviews of the consequences of *in utero* diagnostic x-ray exposure have appeared in recent years (Boice and Inskip, 1996; Doll and Wakeford, 1997; NAS/NRC, 1990; NRPB, 1995; UNSCEAR, 1994; Wakeford, 1995). Additional *in utero* exposure follow-up information has been provided in reports of possible effects from atomic-bomb exposure (Delongchamp *et al.*, 1997; Yoshimoto *et al.*, 1988) and various types of radioactive fallout (Darby *et al.*, 1992; Ivanov *et al.*, 1998; Michaelis *et al.*, 1997; Petridou *et al.*, 1996; Stevens *et al.*, 1990).

The Oxford Survey of Childhood Cancer (OSCC), which commenced in the early 1950s, has provided the largest volume of information regarding the consequences of *in utero* exposure to diagnostic x rays for the development of both leukemia and all types of childhood cancer (Bithell, 1989; Bithell and Stewart, 1975; Bithell and Stiller, 1988; Gilman *et al.*, 1988; Knox *et al.*, 1987; Mole, 1974; 1990; Stewart, 1961; Stewart and Kneale, 1970; Stewart *et al.*, 1956; 1958). Cancer and leukemia cases for the OSCC study were selected from the official statistics of mortality in England, Scotland and Wales. All OSCC reports with the exception of a retrospective study of twins (Mole, 1974), have been case-control in type, following which many other smaller case-control studies have evolved throughout the world. The OSCC study and the majority (Ager *et al.*, 1965; Dyer *et al.*, 1969; Graham *et al.*, 1966; Gunz and Atkinson, 1964; Harvey *et al.*, 1985; Hopton *et al.*, 1985; Kaplan, 1958; Polhemus and Koch, 1959; Rodvall *et al.*, 1990; Salonen, 1976) but not all (Kjeldsberg, 1957; Mills *et al.*, 1958; Murray *et al.*, 1959; Rabinowitch, 1956; Wells and Steer, 1961) of the other case-control studies have supported the contention that maternal diagnostic abdominal x-ray exposures during pregnancy carries an increased risk for the subsequent development of leukemia during the first 10 to 15 y of life (Table 9.15). The risk estimates are significant in all reports of the OSCC study, one other case-control study (Kaplan, 1958) and a single large case-cohort study (Monson and MacMahon, 1984). Most of the case-control studies and the case-cohort study of Monson and MacMahon in 1984 have demonstrated an increased risk for the development of childhood cancer, including leukemia, following maternal abdominal diagnostic x-ray exposure during pregnancy, but the results of many of the smaller studies are not significant (Table 9.16). None of a limited number of cohort studies, however, has confirmed a significant increased risk from *in utero* exposure to diagnostic x rays for the development of either childhood leukemia (Table 9.17) or childhood cancer of all types (Table 9.18).

The RR of childhood cancer, all types combined, following prenatal diagnostic pelvimetry or other diagnostic abdominal x-ray studies was estimated to be two or more, based on the early OSCC study reports (Stewart *et al.*, 1956; 1958). The most prevalent current opinion, which is based mostly on OSCC reports, is that the risks for developing either childhood leukemia or other cancers are about the same following diagnostic abdominal x-ray exposures, which generally have been reported to be in the range of 5 to 50 mGy with an average value of about 10 mGy (Doll and Wakeford, 1997; Gilman *et al.*, 1988; Monson and MacMahon, 1984; NAS/NRC, 1990; NRPB, 1995; UNSCEAR, 1994). It should be noted, however, that the 1962 MacMahon study of solid tumors showed a risk ratio of 1.45 in comparison to the risk ratio of only 1.06 for solid tumors in the extended study (Monson and MacMahon, 1984). They attributed this nonsignificant difference to chance, but the low RR for cancer other than leukemia raises some questions about this interpretation. Several analyses of some of the larger studies have suggested that the sensitivity to carcinogenic effects is greater in the first trimester than in the later trimesters of pregnancy (Bithell, 1989; Gilman *et al.*, 1988; Stewart and Kneale, 1970; Stewart *et al.*, 1958). This interpretation, however, has been challenged through consideration of the larger first trimester x-ray doses that probably were given in the early years (Doll and Wakeford, 1997; Mole, 1990; NRPB, 1995). OSCC reports strongly suggest a dose-response relationship as determined by an increase in cancer risk during childhood with increasing number of x-ray exposures during fetal life (Bithell, 1989; Bithell and Stewart, 1975; Bithell and Stiller, 1988; Doll and Wakeford, 1997; Knox *et al.*, 1987; Stewart and Kneale, 1970). Although these dose-response relationships have been interpreted to be linear, they have wide CI and are of marginal statistical significance (Bithell and Stewart, 1975). The observation, however, that cancer risk has

TABLE 9.15—*RR of developing childhood leukemia following in utero x-ray exposure, as determined by case-control studies.*

Reference	Study Years	Ages	Types of Exposure	Leukemia Cases		Controls		RR (95% CI)[a]	P-Value
				Total Number	Number Exposed	Total Number	Number Exposed		
Stewart *et al.* (1956)	1953–55	<10	Abdominal	269	42	269	24	1.89 (1.11, 3.22)	0.02
Kjeldsberg (1957)	1946–56	Not stated	Abdominal	55	5	55	8	0.59 (0.18, 1.92)	0.38
Kaplan (1958)	1955–57	Not stated	Pelvimetry	147	37	147 (sibs)	24	1.72 (0.97, 3.06)	0.06
				125	34	175 (playmates)	27	2.05 (1.16, 3.62)	0.01
Stewart *et al.* (1958)	1953–57	<10	Abdominal	619	79	1,299	93	1.90 (1.38, 2.60)	<0.01
Mills *et al.* (1958)	1947–57	<10	Pelvimetry	140	0	None	—	—	—
Polhemus and Koch (1959)	1950–57	<18	Pelvimetry	251	66	251	58	1.19 (0.79, 1.78)	0.41
Ford *et al.* (1959)	1951–55	<10	Abdominal	78	21	306	56	1.64 (0.92, 2.93)	0.09
Murray *et al.* (1959)	1940–57	<20	Abdominal & pelvic	65	9	240	12	0.92 (0.25, 3.36)	0.89
Stewart (1961)	1953–55	<10	Abdominal	780	96	1,638	117	1.82 (1.37, 2.43)	<0.01
Wells and Steer (1961)	Not stated	"children"	Abdominal	77	4	156	11	0.72 (0.22, 2.35)	0.59
MacMahon (1962)[b]	1947–60	<13	Abdominal & pelvic	304	47	7,242	770	1.54 (1.12, 2.12)[c] 1.36 (0.98, 1.90)	<0.01 0.07
Gunz and Atkinson (1964)	1958–61	<15	Abdominal	102	14	89	11	1.13 (0.48, 2.63)	0.78

Ager *et al.* (1965)	1953–57	<5	Abdominal or pelvic	107	20	210	32	1.28 (0.69, 2.36)	0.43
Graham *et al.* (1966)	1945–62	<15	Abdominal	313	27	854	54	1.40 (0.86, 2.26)	0.17
Mole (1974)[d]	1945–69	<13	Abdominal & pelvic	3,466 (singletons)	511	14,771,901 (singletons)	1,447,190[e]	1.5 (1.36, 1.60)	<0.01
				70 (twins)	51	353,114 (twins)	194,213[e]	1.32 (1.15, 1.53)	<0.01
Bithell and Stewart (1975)	1953–67	<10	Abdominal	4,052	569	4,052	406	1.47 (1.28, 1.68)	<0.01
Salonen (1976)[f]	1959–68	<15	Pelvic	373	19	373	14	1.38 (0.68, 2.79)	0.38
Monson and MacMahon (1984)[g]	1947–67	<10	Abdominal & pelvic	597	86	14,276	1,342[e]	1.62 (1.28, 2.05)	<0.01
Hopton *et al.* (1985)[f]	1980–83	<15	Abdominal & pelvic	245	37	490	57	1.35 (0.87, 2.11)	0.19
Harvey *et al.* (1985)[d]	1930–69	<15	Abdominal & pelvic	13	5	109	28	1.81 (0.55, 5.99)	0.33
Rodvall *et al.* (1990)[d]	1952–83	<16	Abdominal & pelvic	29	10	58	13	1.82 (0.68, 4.87)	0.23

[a]RR calculated as unadjusted odds ratio, unless otherwise designated.
[b]A case-cohort study with a full cohort of 734,243 live births.
[c]Risk adjusted for variables.
[d]Twin study.
[e]Number of exposed controls calculated from percent values in table.
[f]Includes lymphoma.
[g]This study represents an expansion of the 1962 MacMahon case-cohort study to a full cohort of 1,429,400 live births.

TABLE 9.16—*RR of developing all types of childhood cancer (including leukemia) following in utero x-ray exposure, as determined by case-control studies.*

Reference	Study Years	Ages	Types of Exposure	Cancer Cases: Total Number	Cancer Cases: Number Exposed	Controls: Total Number	Controls: Number Exposed	RR (95% CI)[a]	P-Value
Stewart *et al.* (1956)	1953–55	<10	Abdominal	547	85	547	45	2.05 (1.40, 3.01)	<0.01
Stewart *et al.* (1958)	1953–57	<10	Abdominal	1,299	178	1,299	93	2.06 (1.58, 2.68)	<0.01
Ford *et al.* (1959)	1951–55	<10	Abdominal	152	42	306	56	1.70 (1.08, 2.70)	0.02
Stewart (1961)	1953–60	<10	Abdominal	1,638	211	1,638	117	1.92 (1.52, 2.74)	<0.01
MacMahon (1962)[b]	1947–60	<13	Abdominal & pelvic	556	85	7,242	770	1.58 (1.24, 2.02) 1.42 (0.55, 3.66)	0.47 <0.01
Mole (1974)[c]	1945–69	<13	Abdominal & pelvic	7,528 (singletons)	1,091	14,771,901 (singletons)	1,477,190[d]	1.5 (1.37, 1.60)	<0.01
				161 (twins)	111	353,114 (twins)	194,213[d]	1.25 (1.13, 1.39)	<0.01
Bithell and Stewart (1975)	1953–67	<10	Abdominal	8,513	1,181	8,513	840	1.47 (1.34, 1.62)	<0.01
Salonen (1976)	1959–68	<15	Pelvic	972	52	972	45	1.16 (0.77, 1.75)	0.47
Monson and MacMahon (1984)[b]	1947–67	<10	Abdominal & pelvic	1,104	146	14,276	1,344[d]	1.47 (1.22, 1.76)	<0.01
Hopton *et al.* (1985)	1980–83	<15	Abdominal & pelvic	555	72	1,110	120	1.23 (0.9, 1.68)	0.19

Harvey *et al*. (1985)[c]	1930–69	<15	Abdominal & pelvic	31	12	109	28	1.83 (0.79, 4.24)	0.16
								2.4 (1.0, 5.9)[e]	0.04
Rodvall *et al*. (1990)[f]	1952–83	<16	Abdominal & pelvic	95	25	190	39	1.38 (0.78, 2.46)	0.27

[a]RR calculated as unadjusted odds ratio, unless otherwise designated.
[b]A case-cohort study with a full cohort of 734,243 live births.
[c]Twin study.
[d]Number of exposed controls calculated from percent values in the tables.
[e]RR adjusted for variables.
[f]This study represents an expansion of the 1962 MacMahon case-cohort study to a full cohort of 1,429,400 live births.

TABLE 9.17—*RR of developing childhood leukemia following in utero exposure to ionizing radiation, as determined by cohort studies.*

Reference	Study Years	Ages	Types of Exposure	Exposed: Total Number	Exposed: Number of Leukemias	Controls (not exposed): Total Number	Controls (not exposed): Number of Leukemias	RR (95% CI)	P-Value
Court-Brown *et al.* (1960)	1945–58	<15	Abdominal x ray	39,166	9	39,166	10.5[a]	0.86 (0.35, 2.09)	0.73
Lewis (1960)[b]	1943–58	<10	Abdominal x ray	4,291	1	12,657	7	0.42 (0.05, 3.42)	0.42
LeJeune *et al.* (1960)	1947–52	<12	Abdominal x ray	491	2	468	0	—	—
Magnin (1962)	1948–56	<11	Abdominal x ray	5,353	0	5,353	0	—	—
Griem *et al.* (1967)	1948–66	<15	Pelvimetry x ray	998	1	1,776	2	0.89 (0.08, 9.87)	0.93
Hagstrom *et al.* (1969)	1945–67	<15	Radioactive iron	634	1	655	0	—	—
Diamond *et al.* (1973)	1947–59	<10	Abdomen, spine & pelvic x ray	19,889	6	35,753	7	1.54 (0.52, 4.59)	0.44
Oppenheim *et al.* (1974)[c]	1948–67	<19	Pelvimetry x ray	857	1	1,129	2	0.66 (0.06, 7.25)	0.73

Yoshimoto *et al.* (1988)	1950–60	<15	Atomic bomb	920	0	710	0	—	—
Petridou *et al.* (1996)	1986–88	<1	Chernobyl fallout	163,337	12	1,112,566	31[d]	2.6 (1.35, 5.13)	<0.01
	1987–90	>1 & <4		163,337	43	1,112,566	266[d]	1.10 (0.8, 1.5)	0.56
Michaelis *et al.* (1997)	1986–88	<1	Chernobyl fallout	928,649	35	5,630,789	143[e]	1.48 (1.03, 2.15)	0.01
Ivanov *et al.* (1998)	1986–88	<1	Chernobyl	242,065	17	1,600,964	89[g]	1.26 (0.75, 2.12)	0.38
	1987–90	>1 & <4	fallout	729,301[f]	53	5,065,281[f]	346[g]	1.064 (0.80, 1.42)	0.68

[a]Expected number of control cases of leukemias based on national rates.
[b]Years and ages of study are implied but are not stated. Study overlaps with study of Court-Brown *et al.* (1960).
[c]Children studied were slightly different from those in the Griem *et al.* (1967) study.
[d]Expected number of control cases of leukemia derived from population statistics and leukemia records (1980 to 1985 and 1988 to 1990).
[e]Expected number of control cases of leukemia derived from population and tumor registry statistics (1980 to 1985 and 1988 to 1990).
[f]Child years.
[g]Expected number of control cases of leukemia derived from population statistics and leukemia records (1982 to 1985).

TABLE 9.18—*RR of developing all types of childhood cancer (including leukemia) following in utero exposure to ionizing radiation, as determined by cohort studies.*

				Exposed		Controls (not exposed)			
Reference	Study Years	Ages	Types of Exposure	Total Number	Number of Cancers	Total Number	Number of Cancers	RR (95% CI)	P-Value
Magnin (1962)	1948–56	<15	Abdominal & pelvic x ray	5,353	1	5,353	1	1.0 (0.06, 15.98)	0.99
Griem *et al.* (1967)	1948–66	<15	Pelvimetry x ray	998	4	1,776	6	1.19 (0.34, 4.19)	0.79
Hagstrom *et al.* (1969)	1945–67	<15	Radioactive iron	634	3[a]	655	0.65[b]	4.77 (0.33, 69.5)	0.25
Diamond *et al.* (1973)	1947–59	<10	Abdomen, spine & pelvic x ray	19,889	13	35,753	23	1.02 (0.51, 2.01)	0.96
Yoshimoto *et al.* (1988)	1950–60	<15	Atomic bomb	920	2	710	0.73[c]	2.11 (0.15, 30.79)	0.58

[a]A fourth case of liver cancer at age 11 is not included as it also occurred in two of his older unexposed siblings and was believed to be familial in origin.
[b]State of Tennessee mortality estimate for malignancies from population statistics.
[c]Japanese national statistics estimate for malignancies.

declined over the years with a concomitant decrease in fetal dose per x-ray exposure further strengthens a dose-response relationship (Bithell, 1989; Doll and Wakeford, 1997) (Table 9.19). UNSCEAR has summarized the best estimate for the absolute risk of cancer before age 15 following prenatal exposure to x rays at a high-dose rate to be about 5×10^{-2} Gy^{-1}, similar to the lifetime exposure for adults (UNSCEAR, 1994). Mole has estimated that the risk coefficient for irradiation in the third trimester for childhood cancer incidence and mortality during the first 15 y of life is 4 to 5×10^{-2} Gy^{-1} fetal whole-body dose (Mole, 1990). The overall results of the OSCC study suggest that the risk for childhood cancer is increased by a dose of the order of 10 mGy, resulting in a risk coefficient of about 6×10^{-2} Gy^{-1} (Doll and Wakeford, 1997).

There have been a number of criticisms of the results of the *in utero* radiation case-control studies over the years. Reassurances for many of the criticisms have been presented (Doll and Wakeford, 1997), but some are still of concern. One of the major criticisms of the early OSCC studies was that there may have been recall bias by the mothers regarding the number of abdominal diagnostic x-ray exposures that they received during pregnancy. This criticism was largely discounted by the Monson and MacMahon study which was based on examination of medical records in many hospitals in northeastern United States (MacMahon, 1962; Monson and MacMahon, 1984). Cross-check of a significant number of OSCC exposure histories with hospital records also has been effective in disclaiming recall bias (Knox *et al.*, 1987). Another early criticism of the OSCC study was that there may have been selection bias for a possible increased

TABLE 9.19—*Estimated EAR of incidence of all cancers in children 0 to 14 y old associated with prenatal x-ray examinations in the OSCC.*

	EAR (10^{-4} Gy^{-1}) based on	
Year of Birth	Deaths During 1964–1979 with Adjustment for Pregnancy, Illnesses and Drugs	Deaths During 1953–1979 with No Adjustment for Pregnancy, Illnesses and Drugs[a]
1946	203	185 (98–304)
1952	100	96 (50–152)
1957	49	56 (21–98)
1962	27	36 (6–73)

[a]Approximate 95 percent CI given in parentheses (UNSCEAR, 1994).

cancer potential in the cases being studied (Totter and MacPherson, 1981). This argument has been somewhat reduced by virtue of the outcome of abdominal x-ray studies of twins, who were mostly examined for their *in utero* position rather than for some possible medical reason (Harvey *et al.*, 1985; Mole, 1974; Rodvall *et al.*, 1992). The RR estimates for leukemia and cancer in twins and singletons who were x rayed *in utero* were similar. Studies of twin births to resolve this issue, however, have been criticized on several grounds (Boice and Inskip, 1996; Inskip *et al.*, 1991). Since only about one-third to one-half of twin births were x rayed *in utero* (Inskip *et al.*, 1991; Rodvall *et al.*, 1990; 1992) there may have been ample reason for selection bias to have occurred. The exact amounts of prenatal irradiation are not known, and the proportion of twins irradiated was higher than that of singletons. Furthermore, it appears that childhood leukemia and cancer risks are moderately (Harvey *et al.*, 1985; Inskip *et al.*, 1991; Mole, 1974) to slightly (Rodvall *et al.*, 1992) lower among twins than singletons; so that comparisons between exposed twin births may be no better than comparisons between twin and singleton births (Harvey *et al.*, 1985; Inskip *et al.*, 1991).

Another peculiarity of many of the *in utero* x-ray case-control study reports is the observation that the RR for the development of many different types of solid tumors and leukemia are all about the same (Boice and Inskip, 1996; Boice and Miller, 1999; Doll and Wakeford, 1997; UNSCEAR, 1994). This is quite contrary to the recognized dissimilar biological origins of various childhood cancers, their known distinct variations in occurrence following radiation exposure in adulthood and the fact that the exposed Japanese children who developed radiation-induced leukemia during the first 10 y of life did not demonstrate a concomitant excess of cancers of other types (Boice and Inskip, 1996; Boice and Miller, 1999; Delongchamp *et al.*, 1997; Doll and Wakeford, 1997). Also the logic behind any greater radiation sensitivity to the induction of childhood cancer by exposure late in pregnancy, as opposed to exposure right after birth, has been questioned (Boice and Inskip, 1996). The fact that none of the cohort studies of *in utero* diagnostic x-ray exposure have demonstrated significance for either a childhood leukemogenic or carcinogenic effect contrasts with the results of the diagnostic x-ray case-control studies, but it must be emphasized again that most of the cohort studies have been quite small (Tables 9.17 and 9.18).

Critical to the question of either relative or absolute risk to the fetus from radiation exposure is the validity of the dose estimates from diagnostic x-ray studies. The early OSCC values of mean dose per film were a third or less of those estimated by UNSCEAR in 1972 (Mole, 1990). A mean fetal body total dose of about 6 mGy for obstetric radiography performed in 1958 was estimated by Mole, but he also has questioned the reliability of assigning a fetal dose per x-ray film without many other considerations (Mole, 1990). The uncertainty of doses from the use of fluoroscopy and multiple examinations during pregnancy must be added to the doses from pelvimetry or plain abdominal film examinations which were estimated over 20 y ago in an NCRP report to range from as little as 1.6 mGy to as much as 50 mGy (NCRP, 1977). These and other variables raise some doubt concerning the validity of dose-response risk estimates from diagnostic x-ray studies and the shape of the corresponding dose-response curve.

Risk estimates for childhood leukemia and other cancers derived from the cohort studies of the *in utero* exposed atomic-bomb survivors are difficult to interpret in comparisen to estimates derived from the case-control studies of prenatally x-irradiated children. Only two cases of cancer and no cases of leukemia were reported during the first 15 y of life in the children exposed *in utero* to radiation from the atomic bombs at doses greater than 0.01 Gy (Delongchamp *et al.*, 1997; Yoshimoto *et al.*, 1988) (Tables 9.17 and 9.18). A more recent leukemia and cancer mortality study has followed up to age 46 a 3,000-member *in utero* cohort and a comparison cohort of about 10,000 children who were less than 6 y of age at the time of the Hiroshima and Nagasaki bombings (Delongchamp *et al.*, 1997). The members of both cohorts had experienced a wide range of radiation exposures. The mean uterine radiation dose for mothers with known exposures greater than 0.01 Sv was 0.24 Sv, a relatively large exposure in comparison to those of the *in utero* diagnostic x-ray series. The estimated ERR for cancer mortality during the first 15 y of life for survivors exposed *in utero* was 23 Sv^{-1} (90 percent CI = 1.7 to 88), a result which is not significant ($p = 0.07$). (This high ERR came about because the cancer mortality in the unexposed group was so small.) In the 17 to 46 age group, the ERR per sievert estimate for all cancers was 2.1 (90 percent CI = 0.2 to 6) with a significant dose response for the nine cancer deaths, including two from leukemia. The latter estimate was not significantly different, however, from that for the children exposed postnatally under the age of six. Unusual, however, was the fact that all of the solid tumors occurred only in females. The dose response was negative for leukemia in the *in utero* group, probably due to the absence of any cases in survivors with high-dose exposures. The ERR for cancer deaths was highest for first trimester exposures, but the results were not significantly different from those of the other two trimesters ($p = 0.74$). Caution has been advised in the interpretation of cancer risks for the atomic bomb *in utero* exposed group due to their small numbers, the absence of early leukemia, the gender difference in the occurrence of solid tumors, and the absence of a leukemia dose response (Delongchamp *et al.*, 1997; Miller and Boice, 1997). Delongchamp has noted that only one case missed or added to the exposure group would have markedly changed risk estimates (Delongchamp *et al.*, 1997).

A provocative recent cohort study in Greece has suggested that prenatal exposure from Chernobyl fallout resulted in a dose-related increase in the incidence of infant leukemia (Petridou *et al.*, 1996). The overall increased RR of about 2.6 for children with leukemia up to the age of 1 y was based on records of children born of mothers exposed before, immediately following, and during the next several years following the Chernobyl accident (Table 9.17). Smaller increases in post Chernobyl infant leukemia incidence rates also have been reported in both western Germany (Michaelis *et al.*, 1997) and Belarus (Ivanov *et al.*, 1998) (Table 9.17). Neither increase has been attributed to *in utero* exposure from Chernobyl fallout, however, since the increased rates in Germany were inversely related to levels of ground contamination and, in Belarus, the increased trend was weaker despite ground levels of contamination which were 10 times or more greater than either Greece or Germany. No significant increase in leukemia incidence for the children in these cohorts during the next 2 y of life was observed in either Greece or Belarus (Table 9.17). Studies following *in utero* exposure to fallout from atmospheric nuclear weapons tests in Nordic countries and Utah also have failed to show an excess risk for childhood leukemia (Darby *et al.*, 1992; Stevens *et al.*, 1990). In a relatively small 1945 to 1949 study of iron metabolism during pregnancy, utilizing ^{59}Fe, a nonsignificant increase in cancers was observed in exposed children (Hagstrom *et al.*, 1969) (Table 9.18), in whom fetal doses are uncertain but were estimated to have been in the range of 0.3 to 3.54 mGy (Dyer *et al.*, 1969). A recent case-control study of acute lymphoblastic leukemia (ALL) in children and residential radon exposure have failed to demonstrate any association between mortality from ALL and radon levels in the bedrooms of the mothers during the time when the child was *in utero* (Lubin *et al.*, 1998).

There now is increasing evidence that the primary transformation event for ALL, which appears between the ages of two and five, may occur *in utero* (Greaves, 1997; Ross *et al.*, 1994; Smith *et al.*, 1997). Smith has demonstrated, through interpretation of various mathematical models, that an *in utero* event may induce a preleukemic state at birth which would develop into childhood leukemia after the occurrence of an early childhood second event (Smith *et al.*, 1997). Strong evidence of the prenatal initiation of acute leukemia in young patients has been provided in recent studies which have demonstrated the presence of clonotypic MLL-AF4 genomic fusion sequences in the neonatal blood of two children who developed ALL at the ages of two months and 5 y (Gale *et al.*, 1997). This finding greatly strengthens evidence that maternal exposures during pregnancy to such agents as ionizing radiation, various drugs and chemicals, and possibly leukemogenic viruses may be critical events in the etiology of leukemia in infants or young children (Greaves, 1997; Ross *et al.*, 1994; Shu *et al.*, 1996).

In summary, the preponderance of case-control epidemiologic evidence supports the contention that sensitivity to the leukemia effect and overall carcinogenic effect of low-LET ionizing radiation is higher during the fetal period than postnatally. The data also suggest that the risk may be greatest for exposure during the first trimester of pregnancy. Both of these contentions, however, remain subjects of a great deal of uncertainty (Boice and Miller, 1999). Many of the studies conducted throughout the world concerning the medical outcome of children who had received *in utero* diagnostic x rays are weak due to their small size, uncertainties in dosimetry, poor documentation of doses, possible selection factors, the mostly negative cohort studies, inconsistencies in the types of tumors which have developed, and failure of support from animal studies. The shape of the dose-response curve from the OSCC x-ray reports appears linear, but the wide CI associated with these observations limit the usefulness of the information as applied to low levels of exposure. The amount of information from *in utero* exposed atomic-bomb survivors may be too small for reliable risk estimates for cancer and leukemia during childhood. It presents some unusual problems but suggests that cancer mortality risk over a lifetime following *in utero* exposure may be comparable to that following exposure during the first 5 y of life. The most important overall results of the major *in utero* abdominal x-ray studies of statistical significance, however, provide evidence that the radiation-related risk following fetal exposure probably is at lower dose levels than has been observed following adult exposure.

9.3.7 *Impact of Modifying Factors on the Shape of the Dose-Response Curve*

9.3.7.1 *Host Susceptibility and Radiation Sensitivity to Cancer: Theory.* Modern genetics and molecular biology are rapidly identifying germline mutations and somatic mutations that predispose the host to developing cancer. Variations in endogenous hormones and other biologic factors are also known to affect people's risk of "spontaneous" cancer. An issue of high relevance to this Report is whether persons who are highly susceptible to a cancer in the absence of radiation also have heightened sensitivity to induction of that cancer by radiation. The little human information that is available on this issue will be summarized but first it should be noted that Chakraborty *et al.* (1997) have shown that, over a range of plausible assumptions, either the frequency or penetrance of sensitive genetic predispositions would have to be rather high to influence dose-response relationships much.

A major effect of heightened genetic susceptibility appears to be to cause cancer at an early age. One could hypothesize that high radiation sensitivity would also tend to increase cancer risk especially at an early age, assuming that the radiation exposure occurred early enough in life to do so. This scenario could well occur concomitantly with the other effects of radiosensitivity mentioned in the previous paragraphs.

9.3.7.2 *Host Susceptibility and Radiation Sensitivity to Cancer: Current Information.* A general discussion of radiation sensitivity to cancer induction in genetically susceptible subgroups is available elsewhere (Sankaranarayanan and Chakraborty, 1995). There have been few human studies to date. One of the most interesting studies, by Wong *et al.* (1997), examined mortality from second tumors among 1,604 retinoblastoma survivors, of whom 961 (60 percent) had hereditary retinoblastoma (based on a family history of retinoblastoma or bilateral disease). About 88 percent of the hereditary patients and 18 percent of the nonhereditary patients received radiotherapy. After a median 20 y follow-up, 190 second primary cancers were observed in the hereditary group, when only 6.3 were expected. The nonhereditary group had nine cancers, with 5.6 expected. The second primaries were mainly of bone, connective tissue, nasal cavity, brain, and melanoma. About half the excess tumors in the irradiated hereditary patients were sarcomas, and 75 percent of the sarcomas occurred on the head, in or near the radiation fields. There was a strong effect of the hereditary *Rb1* gene mutation; the SIR in the hereditary retinoblastoma group was 7.3 in the unirradiated group versus 1.3 in the unirradiated nonhereditary retinoblastoma group. There was also some indication of a synergism between radiation and hereditary predisposition. The EAR of radiotherapy in the hereditary group was 6.8 per 1,000 PY versus one in the nonhereditary group, but the statistical significance of the difference is unknown since an evaluation was not performed.

Children with the nevoid basal cell cancer syndrome (NBCCS) are at high risk of developing multiple basal cell cancers and other effects, including medulloblastomas. Among NBCCS patients whose medulloblastomas were treated with radiotherapy the development of a large crop of basal cell skin cancers in and near the radiation field has been observed, beginning as early as six months to 3 y after radiotherapy (Strong, 1977a). Others have reported similar findings in NBCCS patients following irradiation for other conditions (Rayner *et al.*, 1976; Schweisguth *et al.*, 1968). Recent work indicates that NBCCS skin cancers are associated with mutations in the *PTCH* gene (Wicking *et al.*, 1997). Rapid tumor induction following irradiation has also been reported for pre-pubertal ovarian fibromas in NBCCS patients (Heimler *et al.*, 1979; Strong, 1977b). The rapid and pronounced effect of radiation suggests a two-hit mechanism in which the first hit was a germline mutation and radiation supplies the second hit.

There is limited evidence that a germline *p53* mutation consistent with the Li-Fraumeni Syndrome is involved in the associated high prevalence of second primary sarcomas or acute leukemia following irradiation for a first primary tumor (Heyn *et al.*, 1993; Strong and Williams, 1987). This has been confirmed by molecular studies (Malkin *et al.*, 1992) and animal models (Kemp *et al.*, 1994) as well.

The studies of second cancers among those with childhood cancer, which are the context for most of the studies of hereditary cancers, have not suggested notable excesses of the most common adult cancers, such as lung, breast or colon. This may indicate little association, or it may be partly because these cohorts have not yet entered the age ranges in which lung, breast or colon cancers become common.

The Japanese atomic-bomb data on breast cancer suggest the possibility of a highly sensitive subgroup who develop radiogenic breast cancer at an early age. Land *et al.* (1993a) found a particularly high RR for breast cancer before age 35 among those who were irradiated at ages 0 to 19. Although this has not yet been studied, the possibility that this risk is associated with germline mutations in the *BRCA1*, *BRCA2*, *RAD5* or *ATM* genes needs to be investigated. One study has been interpreted as showing a heightened risk of radiation-induced breast cancer among *ATM* heterozygotes (Swift *et al.*, 1991), although others have considered the data not to be sufficiently strong to support that interpretation (Boice and Miller, 1992; Hall *et al.*, 1992b; Land, 1992).

Currently little is known about whether the more common hereditary genes that confer cancer susceptibility in the general population, such as *APC*, *HNPCC* genes, *BRCA1* or *BRCA2*, also confer sensitivity to the induction of cancer by radiation. The answer to this question, including the effects of new cancer susceptibility genes that are yet to be discovered, will be a major determinant of the impact of hereditary predisposition upon radiation risk and upon the shape of dose-response curves.

In summary, there are modest amounts of data to indicate that certain subgroups with high genetic susceptibility to cancer also show high sensitivity for radiation induction of those cancers. However, at this point it is not known whether radiation sensitivity is a common characteristic of genetically susceptible subgroups or whether the exemplars mentioned are exceptional in that regard. It is of interest that there are suggestions that genetic susceptibility may lead to radiation induction of cancer at an early age or within a short time after irradiation. However, a question of principal interest that has not yet been investigated is whether such genetically susceptible persons manifest great sensitivity to low doses of radiation and thereby create an overall supralinear dose-response relationship. The degree to which this might occur would depend on both the degree of genetic susceptibility and the prevalence of such susceptible persons in the population. The studies to date of the rare genetic mutations do not suggest they will have a major impact on total irradiated-population risk or on the shape of the dose-response curve (ICRP, 1998).

9.3.7.3 *Interactions between Radiation and Other Cancer Risk Factors or Exposures.* The joint effects of radiation and some other agent could be pertinent to the issue of low-dose linearity if the other agent in question were to "prime" a person for cancer induction or if radiation were to produce latent initiated cells that another agent would then be able to "promote" to a cancer clone. There are experimental data to suggest that combinations of exposures can change the shape of the dose-response curve, *e.g.*, a sublinear curve for radiation cancer induction can be made more linear by the application of a tumor promoter (Fry and Ullrich, 1986). Any consideration of interactions between radiation and other exposures or risk factors for cancer should also examine the possibility that certain protective factors may make persons insensitive to cancer radiation induction at low doses. A good discussion of this topic is given in the 1982 UNSCEAR report (UNSCEAR, 1982).

The principal interaction that has been examined between radiation and exposure to another agent is that with smoking and lung cancer. In the Japanese atomic-bomb study a test for interaction showed that the joint effect was compatible with additivity. The studies of radon-exposed miners have found that the joint effect of radiation and smoking is greater than additive in conferring lung cancer risk (Lubin *et al.*, 1995a; NAS/NRC, 1990), which is to say that on an absolute risk basis the radiation risk per unit dose is greater among smokers than nonsmokers (although the RR is actually higher among nonsmokers). Among miners, a linear fit to the dose-response curves for radiation-induced lung cancer proved adequate for smokers and nonsmokers considered together, and although there clearly was an excess risk even among nonsmokers (Lubin *et al.*, 1995a), the number of lung cancer cases among nonsmokers appears too small to evaluate low-dose linearity in this subpopulation. In contrast, the atomic-bomb study (Kopecky *et al.*, 1986; Prenctice *et al.*, 1983) and the Hanford occupational study (Petersen *et al.*, 1990) of (mostly) low-LET irradiation have found the effects of smoking and radiation to be approximately additive.

Interactions between breast cancer risk factors and radiation also have been studied, but these have been mainly protective factors. Land *et al.* (1994b) examined several reproductive factors and found that an early age at first full-term pregnancy, multiple births, and lengthy lactation all protected against radiation-induced breast cancer as well as spontaneous breast cancer. Another study found that irradiation at the time of the first childbirth conferred more risk than irradiation at a second or later childbirth (Shore *et al.*, 1980), which is intriguing in view of the epidemiologic finding that the age at first birth is a key factor in breast cancer risk. The biological basis for age-at-first-birth differences in risk is unknown, although an hypothesis has been explored that childbirth increases the differentiation of possible progenitor cells (Russo *et al.*, 1982).

None of the epidemiologic studies of interaction between radiation and risk factors for breast cancer have been able to evaluate whether the factors changed the shape of the dose-response curve. Likewise, other studies of the interaction between radiation and other exposures (*e.g.*, irradiation by both ionizing and ultraviolet radiations) have not had adequate data to evaluate the shape of dose-response curves, so very little is known about the issue from epidemiologic data.

9.3.8 *Status of the Dose-Response Relationship in Epidemiologic Data*

9.3.8.1 *Hormesis.* There is a suggestion that low doses of ionizing radiation protect against cancer rather than conferring cancer risk (Muckerheide *et al.*, 1998), based both on experimental results showing adaptive responses (UNSCEAR, 1994) and on interpretations of epidemiologic studies. Two prime examples of studies cited as supporting hormesis are briefly described below.

It has been argued (Muckerheide, 1995) that a paper by Miller *et al.* (1989b) on the Canadian TB cohort with multiple fluoroscopic examinations showed an apparent deficit of breast cancer mortality in the range of 100 to 300 mGy as compared to the 0 to 90 mGy baseline. However, an analysis performed during the preparation of this Report showed that the nominal deficit was not statistically significant; furthermore, the original authors reported that a simple linear dose-response curve provided an adequate fit to the data. An updated report of breast cancer mortality (Howe and McLaughlin, 1996) did not provide a breakdown of dose categories within the 10 to 500 mGy range so as to provide a comparison to the earlier data. However, the most recent report by Howe (2001), using breast cancer incidence, found that a simple linear model fit the data better than a pure quadratic model, and several additional tests strongly confirmed that the dose-response curve was essentially linear. Furthermore, subjects in the lowest dose group (10 to 190 mGy) had an indication of elevated risk compared to the zero dose group (RR = 1.25, CI = 0.94, 1.67) (Howe, 2001). Thus the alleged deficit shown by the Miller study (Miller *et al.*, 1989b) should be regarded as a statistical anomaly that has now disappeared as more and better data have become available.

A large study by Matanoski (1991; 1993) of United States nuclear shipyard workers purportedly showed hormesis (Muckerheide, 1995), in that there was lower total mortality in the exposed groups (both the <5 mSv and ≥ 5 mSv groups) than in the nonradiation workers. This interpretation ignores the likelihood of occupational selection factors that led some to qualify for radiation work while others did not. The fact that there was a difference for total mortality, and not just for radiosensitive cancers, supports the interpretation that selection factors were operative.

The review of the major studies of a variety of cancer endpoints, as given in preceding sections, indicates there is not strong support for a hormesis interpretation of the radiation epidemiology literature, albeit that many low-dose studies are equivocal because of intrinsic limitations in their precision and statistical power. Because of these limitations, there is a danger in over-interpreting either individual negative studies or individual highly positive studies.

9.3.8.2 *Linearity and Dose Thresholds.* The most recent analyses of the Japanese atomic-bomb data indicate that the data for solid tumors are highly compatible with nonthreshold linearity of the dose-response curve in the aggregate and for major subgroups of cancer types (Pierce *et al.*, 1996; Thompson *et al.*, 1994). On the other hand, Hoel and Li (1998) recently showed that the data for several cancer endpoints could be fit about equally well with a dose-response model having a dose threshold at around 50 to 100 mSv. However, they did not evaluate the fit of a linear-quadratic model versus a threshold model. Even if one brings in other radiation data, there is no clear picture of the shape of the dose-response curve at low doses for a number of tumor sites. Breast cancer and thyroid cancer are perhaps two exceptions, where risks have been demonstrated at a low enough dose or dose-per-fraction, that the data support low-dose linearity. On the other hand, the data on radiation and leukemia seem to support mainly a dose threshold or upward-quadratic curvature (Hoel and Li, 1998; Preston *et al.*, 1994). A variant on a dose threshold, termed a "practical threshold" by Raabe *et al.* (1983) appears to apply to bone cancer induced by ^{226}Ra, in which the median time to malignancy is inversely related to dose, until the mean age of diagnosis exceeds the average lifetime at low doses. Radiation-induced sarcomas also appear to exhibit a threshold response in that they occur primarily following high-dose radiotherapy. Nevertheless, it should be stressed that bone cancer, sarcomas and perhaps leukemia tend to be the exceptions in showing apparent thresholds. Total solid tumors and a number of the major solid-tumor subtypes provide evidence that is compatible with, although not definitive for, low-dose linearity.

9.4 Summary

Epidemiologic data have high validity in that they represent the average risk in human populations resulting from an integration of all the biological processes that are part of the radiation carcinogenic pathway, and that no extrapolation from other species or biological systems is required. On the other hand, epidemiologic data have limited precision because of an unfavorable signal to noise ratio at low dose, uncertainties in dose estimates, heterogeneity in susceptibility to cancer induction, potential exposure to other carcinogenic cofactors or protectors, and potential confounding variables, so that most epidemiologic data have rather limited ability to discriminate between different shapes of the dose-response curve, especially at low doses. All types of radiation epidemiologic study designs have potential problems associated with them, but the cohort (*i.e.*, "follow-up") study is generally regarded as providing the most valid evidence, while data from cluster studies and "ecological" (aggregate) studies are unconvincing because of weaknesses intrinsic to these study designs.

Studies that are useful for evaluating low-dose effects have reasonably accurate dosimetry, substantial numbers of persons in the dose range of interest, a long follow-up period, and a high rate of follow-up. A number of studies with average doses less than 1 Sv and a range of doses, plus those with substantial fractionation or protraction of radiation exposures, are relevant.

For total solid malignant tumors, the Japanese atomic-bomb study provides evidence that acute exposures, even at low doses, increase total cancer risk. Several studies also provide substantial evidence that fractionated or protracted radiation exposures increase total-cancer risk (*e.g.*, therapeutic ^{131}I studies, Techa River cohort, early radiation worker studies). However, when the total doses are low, occupational data do not provide clear evidence of risk. This may be because of their limited precision and the potential masking of small radiation effects by the "healthy worker effect" commonly found in occupational studies.

For leukemia, the Japanese atomic-bomb study has found a curvilinear dose-response relationship (upward concave), with no detectable risk below about 0.2 Sv. Several studies of fractionated or protracted irradiation with relatively high cumulative doses have shown leukemia excesses; most of those with low cumulative doses have not, although a notable exception is a large study of combined radiation-worker cohorts (Cardis *et al.*, 1995). The bulk of the data indicate that the risk per millisievert at low doses is probably less than at higher doses. However, one can find both low-dose studies that suggest there is no risk and others that suggest there is risk. It is probably beyond the capability of epidemiologic data to resolve the issue of low-dose leukemia risk.

A number of studies indicate a risk of thyroid cancer among children exposed to relatively low doses of acute radiation. The limited data available indicate that there is little thyroid cancer risk following irradiation in adult life. Several studies of ^{131}I exposure suggest that the risk from ^{131}I may be lower than that from external, high-dose-rate irradiation, but the studies have thus far involved only small numbers of exposed children. The most extensive data on childhood ^{131}I exposure, that from Chernobyl, indicate an appreciable risk in the low-to-moderate dose range and are clear evidence that protracted exposures increase thyroid cancer risk in children.

A number of studies of breast irradiation have involved fractionated exposures, protracted exposures, and/or subgroups that receive relatively low doses. Most of the studies either have shown significant breast cancer effects or have at least been consistent with such effects. The data are supportive of an interpretation that low doses or fractionated doses are largely additive (*i.e.*, linear) in their effects upon breast cancer risk.

For low-LET radiation the data suggest that there is little or no lung cancer risk associated with fractionated or protracted exposures, even though the acute exposure data from the Japanese atomic-bomb study appear linear. For high-LET (radon) exposure, the consensus of the best available data suggests some elevation in lung cancer risk at fairly low-dose levels, based on both the lower-dose end of the spectrum in the pooled miner data and on a meta-analysis of the main case-control studies of residential radon exposure. Although the ecologic study by Cohen (1995) was carefully done, intrinsic limitations of the ecologic methodology, particularly in conjunction with the potential confounding by smoking and age, mean the negative result is not compelling.

The potential effects of genetic heterogeneity (*i.e.*, intrinsic radiation sensitivity factors) on the shape of the dose-response curve are unknown, although one could plausibly hypothesize that having a subpopulation of persons who are highly sensitive to cancer induction by radiation would tend to "linearize" an otherwise sublinear (upward quadratic) curve. Currently, there is a limited amount of evidence that (1) persons with certain rare heritable mutations are at an unusually high risk of radiation-induced cancers, and (2) a subpopulation of women may be at high risk for breast cancer at an early age following radiation exposure.

The joint effects of radiation and exposure to some other agent could be pertinent to the issue of low-dose linearity if the other exposure in question were to "prime" a person for cancer induction or radiation were to produce latent initiated cells that the other exposure could then "promote" to a cancer clone. However, none of the epidemiologic studies of interaction between radiation and other exposures or risk factors for cancer have evaluated whether the factors changed the shape of the dose-response curve, so little is known about this issue from epidemiologic data.

As irradiated populations are studied for longer times and more cancers accrue, it becomes increasingly possible to detect effects at low doses or with highly fractionated/protracted exposures because of increased statistical power. For instance, in the last few years a statistically significant dose-response relationship for total solid malignant tumors was observed over the dose range 5 to 50 mSv in the Japanese atomic-bomb survivors, a clear excess of thyroid cancer has been observed at doses of about 100 mGy, and breast cancer risk has been demonstrated after highly fractionated doses (on the order of 10 mGy per fraction or less). *In utero* exposures on the order of a few tens of milligray appear to induce childhood leukemia. The data are suggestive, but not yet definitive, that relatively low exposures to radon induce lung cancer. On the other hand, whereas acute doses of low-LET radiation do induce lung cancer, highly fractionated doses of low-LET radiation to the lung do not appear to induce lung cancer. Hence, one cannot necessarily impute the same low-dose sensitivity or the same shape of a dose-response curve to all cancers as a group. Nevertheless, there are a number of studies that indicate there is excess risk for at least some cancer types at fairly low or fractionated radiation doses.

9.5 Research Needs

The research needs in radiation epidemiology relevant to better defining risk at low doses and the shape of the dose-response curve are diverse because there are a number of gaps in our knowledge. Several generic needs are listed below.

- Further follow-up of the Japanese atomic-bomb survivors and other major irradiated cohorts is called for. The addition of more incident and fatal cancers, and further refinements in relevant dose estimates will increase the precision in defining the risk in low-dose range, and, in time, provide information on the risk of cancer in those irradiated under age 20 which, at present, is very inadequately known.
- Information on whether and to what extent persons with hereditary mutations or polymorphisms that confer cancer susceptibility also have increased sensitivity to radiation-induced cancer and whether such sensitivity alters the shape of the dose-response curve in irradiated populations.
- Continued follow-up of the Japanese atomic-bomb cohort irradiated *in utero* has the potential to determine whether *in utero* exposure affects cancer risk throughout the lifetime.
- Further information is needed on the magnitude of risk from highly fractionated or protracted irradiation. Studies of highly fractionated or protracted irradiation in which the cumulative doses are fairly high have the potential to be more informative regarding dose linearity than are low-dose studies.
- Since cancer survival is improving, follow-up studies of cancer patients treated with radiation to evaluate the risk in organs that received low-to-moderate dose scatter will be important.

- A number of cohorts in the former Soviet Union exposed to radiation occupationally (*e.g.*, Chelyabinsk) or through releases (*e.g.*, Chernobyl, Techa River, Semipalatinsk, etc.) have the potential to provide interesting and important new information on radiation risks at both low and high doses, providing exposures can be estimated with sufficient certainty and epidemiologic methods of high quality can be implemented.

10. Adaptive Responses

10.1 Types of Adaptive Responses and Their Dose-Response Relationships

As noted at the outset, there is growing evidence that small doses of radiation can sometimes elicit transient homeostatic responses which may enhance the ability of cells and organisms to withstand subsequent irradiation. Such responses include: (1) heightened capacity for repair of damage to DNA, genes and chromosomes, which has been observed to varying degrees in acutely exposed human, animal and plant cells; (2) acceleration of cellular proliferation, which has been observed in a variety of experimental test systems; (3) enhancement of immune reactions (including the possible immunological surveillance of preneoplastic cells), which has been observed under certain conditions in laboratory rodents; (4) increases of growth and longevity, which have been observed experimentally in various species; and (5) reduction in intercurrent mortality from infectious diseases, which have been observed in some instances in chronically irradiated laboratory mice and rats (UNSCEAR, 1994; Wojcik and Shadley, 2000). Although incompletely understood, the mechanisms of the adaptive responses to radiation appear to have some features in common with those responsible for adaptive responses to other physical and chemical agents (*e.g.*, Calabrese *et al.*, 1987; UNSCEAR, 1994).

Among the adaptive responses to radiation most studied thus far is a heightened capacity for repair of chromosome damage, first reported by Olivieri *et al.* (1984), which has since been amply confirmed (UNSCEAR, 1994). The response is characterized by enhancement of the capacity of cells to repair chromosomal damage for a short time following exposure to an appropriate "priming" or "conditioning" dose of radiation. In the typical case, a "priming" dose of about 10 mGy results in an approximately two-fold reduction in the frequency of chromosome aberrations that is induced by a "challenge" dose of 1 to 3 Gy administered shortly thereafter. The ameliorating effect of the "priming" or "conditioning" dose is thought to result from the activation of an induced repair system. By its very nature, however, the response requires a priming dose the net effects of which are often unknown; *i.e.*, a dose of at least 5 mGy is apparently necessary to evoke the response in lymphocytes (Shadley and Wiencke, 1989), and, while a dose of this magnitude may not suffice to cause a statistically significant increase in the frequency of chromosomal aberrations among the 100 to 200 cells that are typically scored in a cytogenetic assay, such a dose can be expected to damage the DNA of every exposed cell and to produce bds, dsbs and mds in many cells. It is conceivable, in fact, that the adaptive response merely reduces the two-track component of the dose-response curve, without affecting the slope of the linear component of the curve, which predominates in the low-dose range (less than 50 mGy). The failure to observe the response at the lowest doses and dose rates tested suggests that cells must first sustain a certain level of damage before the adaptive response is evoked, as has been suggested by certain data on cell killing (Wouters and Skarsgard, 1997). Hence the data imply that multiple radiation tracks are probably required to induce the response with low-LET radiation, a situation unlikely to be encountered under conditions of natural background irradiation. Although the precise shape of the dose-response curve and the extent to which it may be influenced by variations in dose rate, LET, and other variables remain to be determined, it is noteworthy that the adaptive response of human lymphocytes irradiated with x rays *in vitro* has been observed to vary markedly with the dose, dose rate (*e.g.*, Figure 10.1), interval between successive exposures, and other factors, and that it also varies markedly among individuals, some of whom are apparently nonresponders (UNSCEAR, 1994). Thus, the existing data concerning this response, which is perhaps the best studied of adaptive responses to date, imply that: (1) the response is transitory, lasting only a few hours at most; (2) too low a dose of radiation (*e.g.*, less than 5 mGy) or too low a dose rate (less than 50 mGy min^{-1}) may not suffice to activate the response, while too high a dose (greater than 0.2 Gy) may inhibit it; (3) the level of response to a given radiation dose can be increased or decreased by different chemical agents (*e.g.*, cytotoxic drugs, antimetabolites, enzyme inhibitors), depending on the specific agent and exposure conditions in question; and (4) no response can be detected in the cells from some individuals, for reasons yet to be determined (UNSCEAR, 1994; Wojcik and Shadley, 2000). This limitation casts doubt on its relevance as a factor in the shape of low-dose curves.

Along with enhancement of the capacity to repair chromosome damage, the adaptive response has been observed to reduce the sensitivity of cells to the mutagenic effects of radiation; *i.e.*, low doses of tritiated thymidine, x rays, or gamma rays (approximately 10 mGy) administered to mammalian cells *in vitro* have been observed to decrease by about 50 percent the frequency of mutations

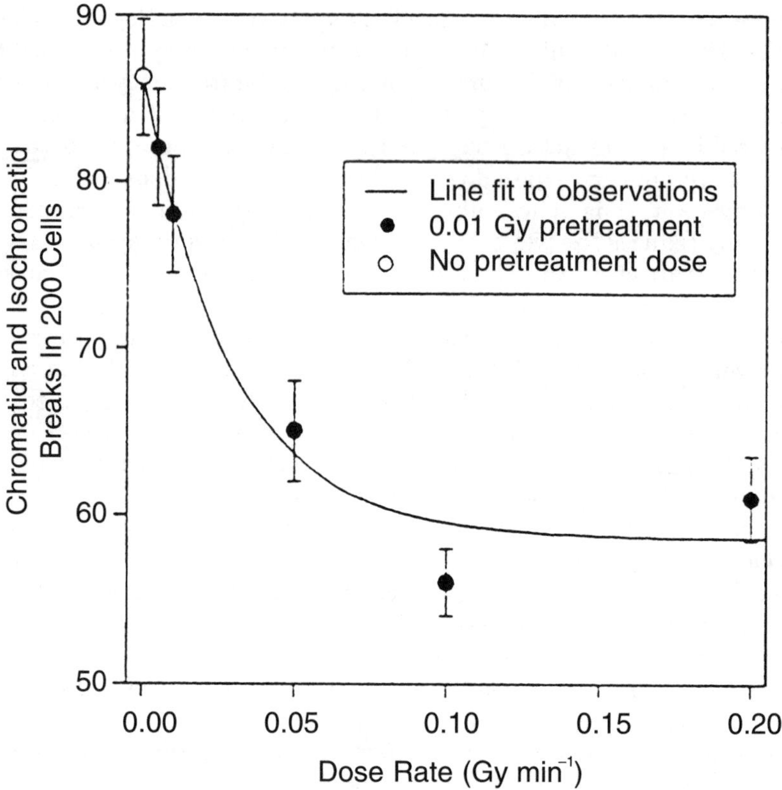

Fig. 10.1. Influence of the dose rate on the effectiveness of a conditioning dose of 10 mGy of x radiation in inducing the adaptive response in human lymphocytes, as measured by the frequency of chromatid and isochromatid breaks produced by a subsequent challenge dose of 1.5 Gy [Puskin (1997) based on the data of Shadley and Weincke (1989)]. (The dose rate plotted on the abscissa is the dose rate during the "conditioning" phase.)

induced in such cells by a subsequent, large, "challenge" dose delivered 5 to 24 h later (Kelsey *et al.*, 1991c; Rigaud *et al.*, 1993; Sanderson and Morley, 1986; Sasaki, 1995; Zhou *et al.*, 1991; 1994). The adaptation is reported to protect selectively against the formation of deletions (Rigaud *et al.*, 1995; Ueno *et al.*, 1996; Zhou *et al.*, 1994). Again, however, there is marked variation among individuals in the expression of the response; *i.e.*, of 18 subjects in one study, only 14 exhibited a detectable response (Bosi and Olivieri, 1989). In the ultimate instance of cancer *in vivo*, the adaptive response may take the form of an immune surveillance of pre-neoplastic cells

(UNSCEAR, 1994). This is certainly a possibility, but it is difficult to design laboratory experiments to test it.

10.2 Implications for the Linear-Nonthreshold Model

Although it is widely acknowledged that adaptive responses may underlie some of the observed dose- and dose-rate-dependent variations in the effectiveness of low-LET radiation, there is no firm evidence thus far that such responses can be expected to operate effectively enough to protect completely against the mutagenic and carcinogenic effects of low-level irradiation. Thus, in spite of some suggestions to the contrary (*e.g.*, Feinendegen *et al.*, 1995; Luckey, 1991; 1994; Muckerheide *et al.*, 1998; Pollycove, 1995), the data are generally interpreted not to exclude the linear-nonthreshold model and thus to provide insufficient grounds for rejecting the linear-nonthreshold dose-response model as a basis for assessing the risks of low-level ionizing radiation in radiation protection (ACRP, 1996; NRPB, 1995; UNSCEAR, 1994).

11. Research Needs

As indicated in foregoing sections of this Report, evaluation of the linear-nonthreshold dose-response model is hampered by limitations in the information that is now available concerning the mutagenic, clastogenic and carcinogenic effects of low-level ionizing radiation and the extent to which such effects may be prevented by adaptive responses to irradiation in the low-dose domain. To provide the data that are needed, various lines of research are called for, which have been discussed in detail elsewhere (NCRP, 1993c). Although no attempt is made to discuss such research needs in depth again at this time, the following deserve to be reemphasized in the context of this Report:

1. Inasmuch as the principal risks to health attributable to low-level irradiation are postulated to result from mutagenic and clastogenic effects on exposed cells, research is urgently needed to define more precisely the shape of the dose-response curve for the mutagenic and clastogenic effects of low doses of low-LET radiation, and the extent to which the production of these effects may be influenced by such variables as the dose rate, genetic background, tissue of origin, stage in the cell cycle, metabolic state, physiological condition, cell communication, damage to neighboring cells, adaptive responses, and other reparative processes.
2. Owing to the uncertainty that is inevitably inherent in cross-species extrapolation, it is essential that further research be conducted to elucidate the dose-response relationships for the carcinogenic effects of low-level ionizing radiation in human populations, including the continued follow-up of Japanese atomic-bomb survivors and the investigation of other populations exposed to low doses at low-dose rates (*e.g.*, radiation workers, populations residing downwind from the Chernobyl accident, selected groups of medically irradiated patients, etc.).
3. To the extent that the dose-response models now being used for estimating the risks of genetic and carcinogenic effects of

radiation are based on incomplete understanding of the cellular and molecular mechanisms of the effects in question, elucidation of the mechanisms of these effects is needed to strengthen the scientific basis for risk assessment.

12. Discussion and Conclusions

The extent to which the existing data on the mutagenic, clastogenic and carcinogenic effects of ionizing radiation are, or are not, compatible with the linear-nonthreshold dose-response hypothesis has been evaluated in the foregoing sections of the Report, taking into account the relevant experimental and epidemiological evidence. The conclusions that may be drawn from the evaluation are necessarily limited by the dearth of quantitative information on dose-response relationships in the low-dose domain, incomplete knowledge of the mechanisms of the effects in question, and uncertainty about the degree to which induction of the effects may be inhibited by adaptive reactions under conditions of low-level irradiation. These limitations notwithstanding, the conclusions that emerge and the rationale underlying them are summarized in the following.

At the outset, it must be noted that radiation imparts its energy to living matter through a stochastic process, such that a single ionizing track has a finite probability of depositing enough energy in traversing a cell to damage a critical molecular target within the cell, such as DNA. Furthermore, the amount of the various types of DNA damage that are known to result from irradiation appears to increase linearly with the dose in the low-to-intermediate dose range. Also, although most such DNA damage is reparable to varying degrees, some types of lesions—namely, dsbs and mds—are often repaired through a process that is error-prone. Because of the vast number of target cells, vanishingly small frequencies of nonlethal, unrepaired or misrepaired lesions may nevertheless result in a finite number of cells undergoing a cancer-initiating event even at low doses, although the possibility of a threshold in the millisievert range cannot be excluded.

Those lesions in DNA that remain unrepaired or are misrepaired may be expressed initially in the form of mutations, the frequency of which increases with the dose of radiation over the dose range in which the effects are amenable to measurement. Although the shape of the dose-response curve varies, depending on the LET of the radiation, the dose rate, the type of mutation, and other variables, it is noteworthy that mutations of types implicated in carcinogenesis—namely, point mutations and partial deletion mutations—have been observed to be inducible at relatively low doses (*e.g.*, <0.01 Gy) with apparently linear-nonthreshold dose-response relationships in various kinds of cells.

Damage to DNA can also give rise to chromosomal alternations, which, in turn, may be linked to the causation of various cancers. Most chromosomal structural changes result from the misrepair of DNA lesions (dsbs, base alterations, cross-links, or more complex lesions) that arise close together in space and time. The frequency of such aberrations therefore typically increases as a linear function of the dose of high-LET radiation. With low-LET radiation, the frequency increases as a linear-quadratic function of the dose in cells exposed acutely, but as the dose rate is reduced, the quadratic component of the response decreases progressively, leaving a response that appears linear in cells exposed chronically. Thus the data imply that traversal of the cell nucleus by a single low-LET radiation track may occasionally suffice to cause a chromosomal aberration. Data from human population monitoring are consistent with this conclusion. At doses in the millisievert range, however, the shape of the dose-response curve is open to question, owing to uncertainty about the fidelity of repair in the low-dose domain and a threshold cannot be excluded.

Cells irradiated in culture have also been observed to undergo dose-dependent neoplastic transformation. The process of transformation appears to involve a succession of steps, during which the affected cells characteristically accumulate a growing number of mutations and/or chromosomal abnormalities, indicating the presence of genomic instability. Although the details of each step remain to be elucidated in full, the activation of oncogenes and/or inactivation or loss of tumor-suppressor genes have been implicated in some instances. Epigenetic changes are also suggested, in view of the fact that the radiation-induced alteration occurs with a frequency that is orders of magnitude above any known mutation rates. Furthermore, susceptibility to transformation varies markedly with the genetic background of the exposed cells and other variables. Not unexpectedly, therefore, the dose-response curve for transformation is complex in shape and subject to variation, depending on the particular experimental conditions investigated. Few data are available as yet on the shape of the curve at low doses, but there is evidence that exposure to a dose involving only one alpha particle traversal per nucleus may suffice to transform a small percentage of exposed cells. The microbeam data, discussed earlier, show that exactly one particle per nucleus is less effective at producing transformation than an average of one with a Poisson distribution. This implies that the cells transformed are those receiving multiple traversals. In the case of low-LET radiations the lowest dose at which a statistically significant increase of transformation over background has been demonstrated is 10 mGy.

In laboratory animals, benign and malignant neoplasms of many types are readily inducible by irradiation. The dose-response curves for such neoplasms vary markedly, depending on the neoplasm in question, the species, strain, sex, and age of the exposed animals, the LET and the dose rate of the radiation, and other variables. In general, the tumorigenic effectiveness of low-LET radiation is appreciably lower than that of high-LET radiation and is reduced at low doses and low-dose rates, whereas the tumorigenic effectiveness of high-LET radiation tends to remain constant or even to increase in some instances with protraction. The available information does not suffice to define the dose-response curve unambiguously for any neoplasm in the dose range below 0.5 Sv, and it indicates the existence of substantial thresholds for the induction of some types of neoplasms. For other types of neoplasms, however, and for the overall life-shortening effects of all radiation-induced neoplasms combined, the data are not inconsistent with a linear-nonthreshold relationship in mice exposed chronically to low-to-intermediate doses of low-LET radiation. The basis for the differences among neoplasms in dose-response relationships remains to be determined. Although the data imply that the initial cellular alteration induced by irradiation *in vivo* typically occurs far more frequently than a mutation at any one genetic locus and that it tends to be followed ultimately by genomic instability in the affected cells, the precise nature and sequence of each of the steps that may be involved in the induction of a particular neoplasm are yet to be fully characterized. Noteworthy, nevertheless, is the fact that various cancer-susceptibility genes, hormones, and other growth-regulating factors have been implicated in a growing number of instances.

Dose-dependent increases in the frequency of many types of benign and malignant neoplasms are also well documented in irradiated human populations. Likewise, it is evident from the available data that the dose-response relationship for such neoplasms may vary, depending on the type of neoplasm, the LET and dose rate of the radiation, the age, sex, and genetic background of the exposed individuals, and other factors. For the most part, moreover, the data come from observations at relatively high doses and dose rates, and they do not suffice to define the shape of the dose-response curve in the low-dose domain. Nevertheless, the following points are noteworthy: (1) in the Japanese atomic-bomb survivors, although the dose-response curve for leukemia appears to be mainly linear-quadratic, the dose-response curve for the overall frequency of solid cancers is not inconsistent with a linear-nonthreshold relationship down to a dose of 50 mSv; (2) prenatal exposure to a dose of only about 10 mGy of x rays appears to increase the risk of cancer in the exposed fetus; (3) analysis of the pooled data from several large cohorts of radiation workers discloses a dose-dependent excess of leukemia (but not solid cancers) in this population that is similar in magnitude to the excess observed in atomic-bomb survivors; (4) a dose of about 100 mSv to the thyroid gland in childhood causes a substantial increase in the risk of thyroid cancer later in life; (5) highly fractionated doses of about 10 mGy per fraction, delivered in multiple fluoroscopic examinations during the treatment of pulmonary TB with artificial pneumothorax, appear to be fully additive in their carcinogenic effects on the female breast, although much less than fully additive in carcinogenic effects on the lung; and (6) certain rare hereditary traits appear to increase sensitivity to radiation-induced cancer, although, there are as yet insufficient data to determine whether the more common hereditary cancer-related gene mutations (*e.g.*, *FAP*, *HNPCC*, *BRCA1*, *BRCA2*, and *ATM* genes) do so. However, some evidence from large low-dose studies has been negative, *e.g.,* there was no dose-response relationship for solid tumors in the large pooled study of workers exposed to radiation (Cardis *et al.*, 1995).

Assessment of the dose-response relationships for low-level irradiation is also complicated by uncertainty about the extent to which adaptive reactions may reduce the effects of radiation in the low-dose domain. Although adaptive reactions may well account in part for dose-dependent and dose-rate-dependent variations in the effectiveness of low-LET radiation at higher doses and higher dose rates, they have yet to be shown to be elicitable in cells or organisms exposed to less than 10 mGy delivered at a dose rate of less than 50 mGy min^{-1}. Furthermore, cells from different individuals vary markedly in their ability to mount such reactions. Given the various lines of evidence that are consistent with the linear-nonthreshold dose-response hypothesis, the existing data on adaptive reactions provide no convincing evidence to the contrary.

In conclusion, although the evidence for linearity is stronger with high-LET radiation than with low-LET radiation, the weight of the evidence, both experimental and theoretical, suggests that the dose-response relationships for many of the biological alterations that are likely precursors to cancer are compatible with linear-nonthreshold functions. The epidemiological evidence, likewise, while necessarily limited to higher doses, suggests that the dose-response relationships for some, but not all, types of cancer may not depart significantly from linear-nonthreshold functions. The existing data do not exclude other dose-response relationships. Further efforts to clarify the relevant dose-response relationships in the low-dose domain are strongly warranted.

References

ABELN, E.C., CORVER, W.E., KUIPERS-DIJKSHOORN, N.J., FLEUREN, G.J. and CORNELISSEM, C.J. (1994). "Molecular genetic analysis of flow-sorted ovarian tumour cells: Improved detection of loss of heterozygosity," Br. J. Cancer **70**, 255–262.

ACRP (1996). Advisory Committee on Radiological Protection. *Biological Effects of Low Doses of Radiation at Low Dose Rate*, ACRP-18, INFO-0654 (Atomic Energy Control Board of Canada, Ottawa).

ADAMS, L.M., ETHIER, S.P. and ULLRICH, R.L. (1984). "Survival of mammary epithelial cells from virgin female BALB/c mice following *in vivo* gamma irradiation," Radiat. Res. **100**, 264–272.

ADAMS, L.M., ETHIER, S.P. and ULLRICH, R.L. (1987). "Enhanced *in vitro* proliferation and *in vivo* tumorigenic potential of mammary epithelium from BALB/c mice exposed *in vivo* to γ-radiation and/or 7,12-dimethylbenz[a]anthracene," Cancer Res. **47**, 4425–4431.

AGER, E.A, SCHUMAN, L.M., WALLACE, H.M., ROSENFIELD, A.B. and GULLEN, W.H. (1965). "An epidemiological study of childhood leukemia," J. Chronic. Dis. **18**, 113–132.

ALAVANJA, M.C., BROWNSON, R.C., LUBIN, J.H., BERGER, E., CHANG, J. and BOICE, J.D., JR. (1994). "Residential radon exposure and lung cancer among nonsmoking women," J. Natl. Cancer Inst. **86**, 1829–1837.

ALBERT, R.E., BURNS, F.J. and BENNETT, P. (1972). "Radiation-induced hair-follicle damage and tumor formation in mouse and rat skin," J. Natl. Cancer Inst. **49**, 1131–1137.

ALEXANDER, P. (1985). "Do cancers arise from a single transformed cell or is monoclonality of tumours a late event in carcinogenesis?" Br. J. Cancer **51**, 453–457.

ALPEN, E.L., POWERS-RISIUS, P., CURTIS, S.B. and DEGUZMAN, R. (1993). "Tumorigenic potential of high-Z, high-LET charged-particle radiations," Radiat. Res. **136**, 382–391.

ALTMAN, P.L. and KATZ, D.D., Eds. (1976). *Cell Biology* (Federation of American Societies for Experimental Biology, Bethesda, Maryland).

AMUNDSON, S.A. and CHEN, D.J. (1996). "Inverse dose-rate effect for mutation induction by gamma-rays in human lymphoblasts," Int. J. Radiat. Biol. **69**, 555–563.

AMUNDSON, S.A., XIA, F., WOLFSON, K. and LIBER, H.L. (1993). "Different cytotoxic and mutagenic responses induced by x-rays in two human lymphoblastoid cell lines derived from a single donor," Mutat. Res. **286**, 233–241.

AMUNDSON, S.A., CHEN, D.J. and OKINAKA, R.T. (1996). "Alpha particle mutagenesis of human lymphoblastoid cell lines," Int. J. Radiat. Biol. **70**, 219–226.

AOYAMA, T. (1989). "Radiation risk of Japanese and Chinese low dose-repeatedly irradiated population," Sangyo Ika Daigaku Zasshi **11** (suppl.), 432–442.

AOYAMA, T., YOSHINAGA, S., YAMAMOTO, Y., KATO, H., SHIMIZU, Y. and SUGAHARA, T. (1998). "Mortality survey of Japanese radiological technologists during the period 1969-1993," Radiat. Prot. Dosim. **77**, 123–128.

ARNOLD, A., COSSMAN, J., BAKSHI, A., JAFFE, E.S., WALDMANN, T.A. and KORSMEYER, S.J. (1983). "Immunoglobulin-gene rearrangements as unique clonal markers in human lymphoid neoplasms," N. Engl. J. Med. **309**, 1593–1599.

ASHBY, J., DOERRER, N.G., FLAMM, F.G., HARRIS, J.E., HUGHES, D.H., JOHANNSEN, F.R., LEWIS, S.C., KRIVANEK, N.D., MCCARTHY, J.F., MOOLENAAR, R.J., RAABE, G., REYNOLDS, R., SMITH, J., STEVENS, J., TETA, M. and WILSON, J. (1990). "A scheme for classifying carcinogens," Regul. Toxicol. Pharmacol. **12**, 270–295.

ASTAKHOVA, L.N., ANSPAUGH, L.R., BEEBE, G.W., BOUVILLE, A., DROZDOVITCH, V.V., GARBER, V., GAVRILIN, Y.I., KHROUCH, V.T., KUVSHINNIKOV, A.V., KUZMENKOV, Y.N., MINENKO, V.P., MOSCHIK, K.V., NALIVKO, A.S., ROBBINS, J., SHEMIAKINA, E.V., SHINKAREV, S., TOCHITSKAYA, S.I. and WACLAWIW, M.A. (1998). "Chernobyl-related thyroid cancer in children of Belarus: A case-control study," Radiat. Res. **150**, 349–356.

AURIAS, A. (1993). "Acquired chromosomal aberrations in normal individuals," in *The Causes and Consequences of Chromosomal Aberrations*, Kirsch, I.R., Ed. (CRC Press, Boca Raton, Florida).

AUVINEN, A., HAKAMA, M., ARVELA, H., HAKULINEN, T., RAHOLA, T., SUOMELA, M., SODERMAN, B. and RYTOMAA, T. (1994). "Fallout from Chernobyl and incidence of childhood leukaemia in Finland, 1976-92," BMJ **309**, 151–154.

AUVINEN, A., MAKELAINEN, I., HAKAMA, M., CASTREN, O., PUKKALA, E., REISBACKA, H. and RYTOMAA, T. (1996). "Indoor radon exposure and risk of lung cancer: A nested case-control study in Finland," J. Natl. Cancer Inst. **88**, 966–972.

AWA, A.A., SOFUNI, T., HONDA, T., ITOH, M., NERIISHI, S. and OTAKE, M. (1978). "Relationship between radiation dose and chromosome aberrations in atomic bomb survivors in Hiroshima and Nagasaki," J. Radiat. Res. (Tokyo) **19**, 126–140.

AZZAM, E.I., RAAPHORST, G.P. and MITCHEL, R.E.J. (1994). "Radiation-induced adaptive response for protection against micronucleus formation and neoplastic transformation in C3H 10T½ mouse embyro cells," Radiat. Res. **138**, S28–S31.

AZZAM, E.I., DE TOLEDO, S.M., RAAPHORST, G.P. and MITCHEL, R.E J. (1996). "Low-dose ionizing radiation decreases the frequency of neoplastic transformation to a level below the spontaneous rate in C3H 10T½ cells," Radiat. Res. **146**, 369–373.

AZZAM, E.I., DE TOLEDO, S.M., GOODING, T. and LITTLE, J.B. (1998). "Intercellular communication is involved in the bystander regulation of gene expression in human cells exposed to very low fluences of alpha particles," Radiat. Res. **150**, 497–504.

BALCER-KUBICZEK, E.K. and HARRISON, G.H. (1988). "Effect of x-ray dose protraction and a tumor promoter on transformation induction *in vitro*," Int. J. Radiat. Biol. **54**, 81–89.

BALCER-KUBICZEK, E.K., HARRISON, G.H. and THOMPSON, B.W. (1987). "Repair time for oncogenic transformation in C3H/10T½ cells subjected to protracted x-irradiation," Int. J. Radiat. Biol. **51**, 219–226.

BALCER-KUBICZEK, E.K., HARRISON, G.H., ZEMAN, G.H., MATTSON, P.J. and KUNSKA, A. (1988). "Lack of inverse dose-rate effect on fission neutron induced transformation of C3H/10T½ cells," Int. J. Radiat. Biol. **54**, 531–536.

BALCER-KUBICZEK, E.K., HARRISON, G.H., HILL, C.K. and BLAKELY, W.F. (1993). "Effects of WR-1065 and WR-151326 on survival and neoplastic transformation in C3H/10T½ cells exposed to TRIGA or JANUS fission neutrons," Int. J. Radiat. Biol. **63**, 37–46.

BALCER-KUBICZEK, E.K., HARRISON, G.H., TORRES, B.A. and MCCREADY, W.A. (1994). "Application of the constant exposure time technique to transformation experiments with fission neutrons: Failure to demonstrate dose-rate dependence," Int. J. Radiat. Biol. **65**, 559–569.

BAO, C.Y., MA, A.H., EVANS, H.H., HORNG, M.F., MENCL, J., HUI, T.E. and SEDWICK, W.D. (1995). "Molecular analysis of hypoxanthine phosphoribosyltransferase gene deletions induced by alpha- and x-radiation in human lymphoblastoid cells," Mutat. Res. **326**, 1–15.

BARLOW, C., HIROTSUNE, S., PAYLOR, R., LIYANAGE, M., ECKHAUS, M., COLLINS, F., SHILOH, Y., CRAWLEY, J.N., RIED, T., TAGLE, D. and WYNSHAW-BORIS, A. (1996). "Atm-deficient mice: A paradigm of ataxia telangiectasia," Cell **86,** 159–171.

BATCHELOR, R.J.S., PHILLIPS, R.J. and SEARLE A.G. (1969). "The ineffectiveness of chronic irradiation with neutrons and gamma rays in inducing mutations in female mice," Br. J. Radiol. **42**, 448–451.

BATES, S. and VOUSDEN, K.H. (1996). "p53 in signaling checkpoint arrest or apoptosis," Curr. Opin. Genet. Dev. **6**, 12–18.

BAUCHINGER, M., SCHMID, E., BRASELMANN, H. and NAHRSTEDT, U. (1989). "Absence of adaptive response to low-level irradiation from tritiated thymidine and x-rays in lymphocytes of two individuals examined in serial experiments," Mutat. Res. **227**, 103–107.

BAVERSTOCK, K.F. and PAPWORTH, D. (1989). "The UK radium luminizer survey," pages 72 to 76 in *Risks from Radium and Thorotrast*, BIR Report 21, Taylor, D.M., Mays, C., Gerber, G. and Thomas, R., Ed. (British Institute of Radiology, London).

BECKMAN, K.B. and AMES, B.N. (1997). "Oxidative decay of DNA," J. Biol. Chem. **272**, 19633–19636.

BENDER, M.A., AWA, A.A., BROOKS, A.L., EVANS, H.J., GROER, P.G., LITTLEFIELD, L.G., PEREIRA, C., PRESTON, R.J. and WACHHOLZ, B.W. (1988). "Current status of cytogenetic procedures to detect and quantify previous exposures to radiation," Mutat. Res. **196**, 103–159.

BENNETT, L.M., HAUGEN-STRANO, A., COCHRAN, C., BROWNLEE, H.A., FIEDOREK, F.T., JR. and WISEMAN, R.W. (1995). "Isolation of the mouse homologue of BRCA1 and genetic mapping to mouse chromosome 11," Genomics **29**, 576–581.

BERENBLUM, I. and TRAININ, N. (1963). "New evidence on the mechanism of radiation leukaemogenesis," pages 41 to 52 in *Cellular Basis and Aetiology of Late Somatic Effects of Ionizing Radiation,* Harris, R.J.C., Ed. (Academic Press, New York).

BETTEGA, D., CALZOLARI, P., POLLARA, P. and LOMBARDI, L.T. (1985). "*In vitro* cell transformations induced by 31 MeV protons," Radiat. Res. **104**, 178–181.

BITHELL, J.F. (1989). "Epidemiological studies of children irradiated *in utero*," pages 77 to 87 in *Low Dose Radiation: Biological Bases of Risk Assessment,* Baverstock, K.F. and Stather, J.R., Eds. (Taylor and Francis, New York).

BITHELL, J.F. and STEWART, A.M. (1975). "Pre-natal irradiation and childhood malignancy: A review of British data from the Oxford survey," Br. J. Cancer **31**, 271–287.

BITHELL, J.F. and STILLER, C.A. (1988). "A new calculation of the carcinogenic risk of obstetric x-raying," Stat. Med. **7**, 857–864.

BLAKELY, W. F., FUCIARELLI, A.F., WEGHER, B.J. and DIZDAROGLU, M. (1990). "Hydrogen peroxide-induced base damage in deoxyribonucleic acid," Radiat. Res. **121**, 338–343.

BLANK, K.R., RUDOLTZ, M.S., KAO, G.D., MUSCHEL, R.J. and MCKENNA, W.G. (1997). "The molecular regulation of apoptosis and implications for radiation oncology," Intl. J. Radiat. Biol. **71**, 455–466.

BLOT, W.J., XU, Z.Y., BOICE, J.D., JR., ZHAO, D.Z., STONE, B.J., SUN, J., JING, L.B. and FRAUMENI, J.F., JR. (1990). "Indoor radon and lung cancer in China," J. Natl. Cancer Inst. **82**, 1025–1030.

BOCHE, R.D. (1967). "Effects of chronic exposure to x-radiation on growth and survival," pages 222 to 252 in *Biological Effects of External Radiation,* Blair, H.A., Ed. (Hafner Publishing Company, New York).

BOICE, J.D., JR. (1991). "Epidemiologic studies of radioactively contaminated environments and cancer clusters," pages 94 to 116 in *Health and Ecological Implications of Radioactively Contaminated Environments,* NCRP Annual Meeting Proceedings No. 12 (National Council on Radiation Protection and Measurements, Bethesda, Maryland).

BOICE, J.D. and INSKIP, P.D. (1996). "Radiation-induced leukemia," pages 195 to 209 in *Leukemia*, sixth ed., Henderson E.S., Lister T.A. and Greaves, M.F., Eds. (W. B. Saunders, Philadelphia).

BOICE, J.D., JR. and MILLER, R.W. (1992). "Risk of breast cancer in ataxia-telangiectasia (letter)," N. Engl. J. Med. **326**, 1357–1358.

BOICE, J.D., JR. and MILLER, R.W. (1999). "Childhood and adult cancer after intrauterine exposure to ionizing radiation," Teratology **59**, 227–233.

BOICE J.D., JR., ROSENSTEIN, M. and TROUT, E.D. (1978). "Estimation of breast doses and breast cancer risk associated with repeated fluoroscopic chest examinations of women with tuberculosis," Radiat. Res. **73**, 373–390.

BOICE, J.D., JR., ENGHOLM, G., KLEINERMAN, R.A., BLETTNER, M., STOVALL, M., LISCO, H., MOLONEY, W.C., AUSTIN, D.F., BOSCH, A., COOKFAIR, D.L., KREMENTZ, E.T., LATOURETTE, H.B., MERRILL, J.A., BJORKHOLM, E., PETERSSON, F., BELL, C.M.J., COLEMAN, M.P., FRASER, P., NEAL, F.E., PRIOR, P., CHOI, N.W., HISLOP, T.G., KOCH, M., KREIGER, N., ROBB, D., ROBSON, D., THOMSON, D.H., LOCHMULLER, H., VON FOURNIER, D., FRISCHKORN, R., KJORSTAD, K.E., RIMPELA, A., PEJOVIC, M.H., KIRN, V.P., STANKUSOVA, H., BERRINO, F., SIGURDSSON, K., HUTCHISON, G.B. and MACMAHON, B. (1988). "Radiation dose and second cancer risk in patients treated for cancer of the cervix," Radiat. Res. **116**, 3–55.

BOICE, J.D., JR., MORIN, M.M., GLASS, A.G., FRIEDMAN, G.D., STOVALL, M., HOOVER, R.N. and FRAUMENI, J.F., JR. (1991a). "Diagnostic x-ray procedures and risk of leukemia, lymphoma, and multiple myeloma," JAMA **265**, 1290–1294.

BOICE, J.D., JR., PRESTON, D., DAVIS, F.G. and MONSON, R.R. (1991b). "Frequent chest x-ray fluoroscopy and breast cancer incidence among tuberculosis patients in Massachusetts," Radiat. Res. **125**, 214–222.

BOICE, J.D., JR., MANDEL, J.S. and DOODY, M.M. (1995). "Breast cancer among radiologic technologists," JAMA **274**, 394–401.

BONASSI, S., ABBONDANDOLO, A., CAMURRI, L., DAL PRA, L., DE FERRARI, M., DEGRASSI, F., FORNI, A., LAMBERTI, L., LANDO, C., PADOVANI, P., SBRANA, I., VECCHIO, D. and PUNTONI, R. (1995). "Are chromosome aberrations in circulating lymphocytes predictive of future cancer onset in humans? Preliminary results of an Italian cohort study," Cancer Genet. Cytogenet. **79**, 133–135.

BOND, V.P., CRONKITE, E.P., LIPPINCOTT, S.W. and SHELLABARGER, C.J. (1960). "Studies on radiation-induced mammary gland neoplasia in the rat. III. Relation of the neoplastic response to dose of total-body radiation," Radiat. Res. **12**, 276–285.

BOND, V.P., VARMA, M.N., SONDHAUS, C.A. and FEINENDEGEN, L.E. (1985). "An alternative to absorbed dose, quality, and RBE at low exposures," Radiat. Res. **8**, S52–S57.

BOOTHMAN, D.A., MEYERS, M., FUKUNAGA, N. and LEE, S.W. (1993). "Isolation of x-ray-inducible transcripts from radioresistant human melanoma cells," Proc. Natl. Acad. Sci. USA **90**, 7200–7204.

BOREK, C. and HALL, E.J. (1973). "Transformation of mammalian cells *in vitro* by low doses of x-rays," Nature **243**, 450–453.

BOSI, A. and OLIVIERI, G. (1989). "Variability of the adaptive response to ionizing radiations in humans," Mutat. Res. **211**, 13–17.

BOUFFLER, S., SILVER, A., PAPWORTH, D., COATES, J. and COX, R. (1993). "Murine radiation myeloid leukaemogenesis: Relationship between interstitial telomere-like sequences and chromosome 2 fragile sites," Genes Chromosomes Cancer **6**, 98–106.

BOUFFLER, S.D., MEIJNE, E.I.M., HUISKAMP, R. and COX, R. (1996). "Chromosomal abnormalities in neutron-induced acute myeloid leukemias in CBA/H mice," Radiat. Res. **146**, 349–352.

BRADLEY, M.O. and ERICKSON, L.C. (1981). "Comparison of the effects of hydrogen peroxide and x-ray irradiation on toxicity, mutation, and DNA damage/repair in mammalian cells (V-79)," Biochim. Biophys. Acta **654**, 135–141.

BRECKON, G., PAPWORTH, D. and COX, R. (1991). "Murine radiation myeloid leukaemogenesis: A possible role for radiation-sensitive sites on chromosome 2," Genes Chromosomes Cancer **3**, 367–375.

BRENNER D.J. (1994). "The significance of dose rate in assessing the hazards of domestic radon exposure," Health Phys. **67**, 76–79.

BRENNER, D.J. and HALL, E.J. (1990). The inverse dose-rate effect for oncogenic transformation by neutrons and charged particles: A plausible interpretation consistent with published data," Int. J. Radiat. Biol. **58**, 745–758.

BRENNER, D.J. and WARD, J.F. (1992). "Constraints on energy deposition and target size of multiply damaged sites associated with DNA double-strand breaks," Int. J. Radiat. Biol. **61**, 737–748.

BRENNER, D.J. and WARD, J.F. (1995). "Restraints on energy deposition / target size of multiply damaged sites associated with DNA DSB," pages 443 to 446 in *Radiation Research 1895-1995, Proceedings of the Tenth International Congress of Radiation Research, Vol. 2* , Hagen, U., Jung, H. and Streffer, C., Eds. (Academic Press, New York).

BRENNER, D.J. and ZAIDER, M. (1984). "The application of track calculations to radiobiology. II. Calculations of microdosimetric quantities," Radiat. Res. **98**, 14–25.

BREWEN, J.G. and BROCK, R.D. (1968). "The exchange hypothesis and chromosome-type aberrations," Mutat. Res. **6**, 245–255.

BREWEN, J.G., PRESTON, R.J., JONES, K.P. and GOSSLEE, D.G. (1973). "Genetic hazards of ionizing radiations: Cytogenetic extrapolations from mouse to man," Mutat. Res. **17**, 245–254.

BROERSE, J.J., HENNEN, L.A. and SOLLEVELD, H.A. (1986). "Actuarial analysis of the hazard for mammary carcinogenesis in different rat strains after x- and neutron irradiation," Leuk. Res. **10**, 749–754.

BROERSE, J.J., HENNEN, L.A., KLAPWIJK, W.M. and SOLLEVELD, H.A. (1987). "Mammary carcinogenesis in different rat strains after irradiation and hormone administration," Int. J. Radiat. Biol. Relat. Stud. Phys. Chem. Med. **51**, 1091–1100.

BUCKTON, K.E. (1983). "Chromosome aberrations in patients treated with irradiation for ankylosing spondylitis," pages 491 to 511 in *Radiation-Induced Chromosome Damage in Man*, Ishihara, T. and Sasaki, M.S., Eds. (Alan R. Liss, New York).

BURNS, F.J. and ALBERT, R.E. (1986a). "Dose-response for radiation-induced cancer in rat skin," pages 51 to 70 in *Radiation Carcinogenesis and DNA Alterations*, Burns, F.J., Upton, A.C. and Silini, G., Eds. (Plenum Press, New York).

BURNS, F.J. and ALBERT, R.E. (1986b). "Radiation carcinogenesis in rat skin," pages 199 to 214 in *Radiation Carcinogenesis*, Upton, A.C., Albert, R.E., Burns, F.J. and Shore, R.E., Eds. (Elsevier, New York).

BURNS, F.J., SARGENT, E.V., MCCORMICK, D. and ALBERT, R.E. (1982). "Retinoid inhibition of radiation-induced rat skin tumors," Am. Assoc. Cancer Res. **23**, 89.

CAGGANA, M., LIBER, H.L., MAUCH, P.M., COLEMAN, C.N. and KELSEY, K.T. (1991). "*In vivo* somatic mutation in the lymphocytes from Hodgkin's disease patients," Environ. Mol. Mutagen. **18**, 6–13.

CAGGANA, M., LIBER, H.L., COLEMAN, C.N., MAUCH, P., CLARK, J.R. and KELSEY, K.T. (1992). "A prospective study of HPRT mutant and mutation frequencies in treated cancer patients," Cancer Epidemiol. Biomarkers Prev. **1**, 573–580.

CALABRESE, E.J., MCCARTHY, M.E. and KENYON, E. (1987). "The occurrence of chemically induced hormesis," Health Phys. **52**, 531–541.

CALDWELL, G.G., KELLEY, D.B. and HEATH, C.W., JR. (1980). "Leukemia among participants in military maneuvers at a nuclear bomb test. A preliminary report," JAMA **244**, 1575–1578.

CALDWELL, G.C., KELLEY, D., ZACK, M., FALK, H. and HEATH, C.W., JR. (1983). "Mortality and cancer frequency among military nuclear test (Smoky) participants, 1957 through 1979," JAMA **250**, 620–624.

CAO, J., WELLS, R.L. and ELKIND, M.M. (1992). "Enhanced sensitivity to neoplastic transformation by ^{137}Cs γ-rays of cells in the G_2-/M-phase age interval," Int. J. Radiat. Biol. **62**, 191–199.

CARATERO, A., COURTADE, M., BONNET, L., PLANEL, H. and CARATERO, C. (1998). "Effect of a continuous gamma irradiation at a very low dose on the life span of mice," Gerontology **44**, 272–276.

CARDIS, E., GILBERT, E.S., CARPENTER, L., HOWE, G., KATO, I., ARMSTRONG, B.K., BERAL, V., COWPER, G., DOUGLAS, A., FIX, J., FRY, S.A., KALDOR, J., LAVE, C., SALMON, L., SMITH, P.G., VOELZ, G.L. and WIGGS, L.D. (1995). "Effects of low doses and low dose rates of external ionizing radiation: Cancer mortality among nuclear industry workers in three countries," Radiat. Res. **142**, 117–132.

CARLSON, L.D., SCHEYER, W.J. and JACKSON, B.H. (1957). "The combined effects of ionizing radiation and low temperature on the metabolism, longevity, and soft tissues of the white rat. I. Metabolism and longevity," Radiat. Res. **7**, 190–197.

CARNEY, J.P., MASER, R.S., OLIVARES, H., DAVIS, E.M., LE BEAU, M., YATES, J.R., III, HAYS, L., MORGAN, W.F. and PETRINI, J.H.J. (1998). "The hMre/hRad50 protein complex and Nijmegen breakage syndrome: Linkage of double-strand break repair to the cellular DNA damage response," Cell **93**, 477–486.

CAVENEE, W.K., DRYJA, T.P., PHILLIPS, R.A., BENEDICT, W.F., GODBOUT, R., GALLIE, B.L., MURPHREE, A.L., STRONG, L.C. and WHITE, R.L. (1983). "Expression of recessive alleles by chromosomal mechanisms in retinoblastoma," Nature **305**, 779–784.

CHAKRABORTY, R., LITTLE, M.P. and SANKARANARAYANAN, K. (1997). "Cancer predisposition, radiosensitivity and the risk of radiation-induced cancers. III. Effects of incomplete penetrance and dose-dependent radiosensitivity on cancer risks in populations," Radiat. Res. **147**, 309–320.

CHANG, W.P. and LITTLE, J.B. (1992). "Persistently elevated frequency of spontaneous mutations in progeny of CHO clones surviving x-irradiation: Association with delayed reproductive death phenotypes," Mutat. Res. **270**, 191–199.

CHARPENTIER, P., OSTFELD, A.M., HADJIMICHAEL, O.C. and HESTER, R. (1993). "The mortality of US nuclear submariners, 1969-1982," J. Occup. Med. **35**, 501–509.

CHAUDHRY, M.A., JIANG, Q., RICANATI, M., HORNG, M.F. and EVANS, H.H. (1996). "Characterization of multilocus lesions in human cells exposed to x radiation and radon," Radiat. Res. **145**, 31–38.

CHECKOWAY, H., PEARCE, N., CRAWFORD-BROWN, D.J. and CRAGLE, D.L. (1988). "Radiation doses and cause-specific mortality among workers at a nuclear materials fabrication plant," Am. J. Epidemiol. **127**, 255–266.

CHEN, K.S., SHEPEL, L.A., HAAG, J.D., HEIL, G.M. and GOULD, M.N. (1996). "Cloning, genetic mapping and expression studies of the rat Brca1 gene," Carcinogenesis **17**, 1561–1566.

CHEN, R.Z., PETTERSSON, U., BEARD, C., JACKSON-GRUSBY, L. and JAENISCH, R. (1998). "DNA hypomethylation leads to elevated mutation rates," Nature **395**, 89–93.

CHERBONNEL-LASSERRE, C., GAUNY, S. and KRONENBERG, A. (1996). "Suppression of apoptosis by Bcl-2 or Bcl-xL promotes susceptibility to mutagenesis," Oncogene **13**, 1489–1497.

CLARK, D.J., MEIJNE, E.I.M., BOUFFLER, S.D., HUISKAMP, R., SKIDMORE, C.J., COX, R. and SILVER, A.R.J. (1996). "Microsatellite analysis of recurrent chromosome 2 deletions in acute myeloid leukaemia induced by radiation in F1 hybrid mice," Genes Chromosomes Cancer **16**, 238–246.

CLEARY, M.L., GALILI, N., TRELA, M., LEVY, R. and SKLAR, J. (1988). "Single cell origin of bigenotypic and biphenotypic B cell proliferations in human follicular lymphomas," J. Exp. Med. **167**, 582–597.

CLIFTON, K.H. (1959). "Problems in experimental tumorigenesis of the pituitary gland, gonads, adrenal cortices and mammary glands: A review," Cancer Res. **19**, 2–22.

CLIFTON, K.H. (1990). "The clonogenic cells of the rat mammary and thyroid glands: Their biology, frequency of initiation, and promotion/progression to neoplasia," pages 1 to 21 in *Scientific Issues in Quantitative Cancer Risk Assessment*, Moolgavkar, S.H., Ed. (Birkhauser, Boston).

CLIFTON, K.H. (1996). "Comments on the evidence in support of the epigenetic nature of radiogenic initiation," Mutat. Res. **350**, 77–80.

CLIFTON, K.H., DEMOTT, R.K., MULCAHY, R.T. and GOULD, M.N. (1978). "Thyroid gland formation from inocula of monodispersed cells: Early results on quantitation, function, neoplasia and radiation effects," Int. J. Radiat. Oncol. Biol. Phys. **4**, 987–990.

CLIFTON, K.H., TANNER, M.A. and GOULD, M.N. (1986). "Assessment of radiogenic cancer initiation frequency per clonogenic rat mammary cell *in vivo*," Cancer Res. **46**, 2390–2395.

COHEN, B.L. (1995). "Test of the linear-no threshold theory of radiation carcinogenesis for inhaled radon decay products," Health Phys. **68**, 157–174.

CONGDON, C.C. (1987). "A review of certain low-level ionizing radiation studies in mice and guinea pigs," Health Phys. **52**, 593–597.

COOKFAIR, D.L., BECK, W., SHY, C., LUSHBAUGH, C. and SOWDER, C. (1983). "Lung cancer among workers at a uranium processing plant," pages 398 to 406 in *Epidemiology Applied to Health Physics* (Lippincott Williams & Wilkins, Baltimore).

COOPER, P.K., NOUSPIKEL, T., CLARKSON, S.G. and LEADON, S.A. (1997). "Defective transcription-coupled repair of oxidative damage in Cockayne patients from XPgroup G," Science **275**, 990–993.

CORNFORTH, M.N. (1990). "Testing the notion of the one-hit exchange," Radiat. Res. **121**, 21–27.

COURT-BROWN, W.M., DOLL, R. and BRADFORD HILL, A.B. (1960). "Incidence of leukemia after exposure to diagnostic radiation *in utero*," BMJ **2**, 1539–1545.

COX, R. and MASSON, W.K. (1979). "Mutation and inactivation of cultured mammalian cells exposed to beams of accelerated heavy ions. III. Human diploid fibroblasts," Int. J. Radiat. Biol. Relat. Stud. Phys. Chem. Med. **36**, 149–160.

CRAGLE, D.L., MCLAIN, R.W., QUALTERS, J.R., HICKEY, J.L., WILKINSON, G.S., TANKERSLEY, W.G. and LUSHBAUGH, C.C. (1988). "Mortality among workers at a nuclear fuels production facility," Am. J. Ind. Med. **14**, 379–401.

CRAGLE, D.L., WATKINS, J.P. and ROBERTSON-DEMERS, K. (1998). *Mortality Among Workers at the Savannah River Nuclear Fuels Production Facility*, Proceedings of the American Statistical Association, Section on Statistics in Epidemiology (Oak Ridge Associated Universities, Oak Ridge, Tennessee).

CRILE, G. and SCHUMACHER, O. (1965). "Radioactive iodine treatment of Graves' disease," Am. J. Dis. Child **110**, 501–504.

CROMPTON, N.E.A., ZOLZER, F., SCHNEIDER, E. and KIEFER, J. (1985). "Increased mutant induction by very low dose-rate gamma-irradiation," Naturwissenschaften **72**, 439–440.

CROMPTON, N.E.A., BARTH, B. and KIEFER, J. (1990). "Inverse dose-rate effect for the induction of 6-thioguanine-resistant mutants in Chinese hamster V79-S cells by ^{60}Co γ rays," Radiat. Res. **124**, 300–308.

CROSS, F.T. (1994). "Residential radon risks from the perspective of experimental animal studies (Commentary)," Am. J. Epidemiol. **140**, 333–339.

DA CRUZ, A.D., VOLPE, J.P., SADDI. V., CURRY, J., CURADOC, M.P. and GLICKMAN, B.W. (1997). "Radiation risk estimation in human populations: Lessons from the radiological accident in Brazil," Mutat. Res. **373**, 207–214.

DAMBER, L., LARSSON, L.G., JOHANSSON, L. and NORIN, T. (1995). "A cohort study with regard to the risk of haematological malignancies in patients treated with x-rays for benign lesions in the locomotor system. I. Epidemiological analyses," Acta Oncol. **34**, 713–719.

DARBY, S.C., OLSEN, J.H., DOLL, R., THAKRAR, B., DENULLY BROWN, P., STORM, H.H., BARLOW, L., LANGMARK, F., TEPPO, L. and TULINIUS, H. (1992). "Trends in childhood leukaemia in the Nordic countries in relation to fallout from atmospheric nuclear weapons testing," BMJ **304**, 1005–1009.

DARBY, S., WHITLEY, E., SILCOCKS, P., THAKRAR, B., GREEN, M., LOMAS, P., MILES, J., REEVES, G., FEARN, T. and DOLL, R. (1998). "Risk of lung cancer associated with residential radon exposure in south-west England: A case-control study," Br. J. Cancer **78**, 394–408.

DAVIS, F.G., BOICE, J.D., JR., KELSEY, J.L. and MONSON, R.R. (1987). "Cancer mortality after multiple fluoroscopic examinations of the chest," J. Natl. Cancer Inst. **78**, 645–652.

DAVIS, F.G., BOICE, J.D., JR., HRUBEC, Z. and MONSON, R.R. (1989). "Cancer mortality in a radiation-exposed cohort of Massachusetts tuberculosis patients," Cancer Res **49**, 6130–6136.

DEININGER, M.W., BOSE, S., GORA-TYBOR, J., YAN, X.H., GOLDMAN, J.M. and MELO, J.V. (1998). "Selective induction of leukemia-associated fusion genes by high-dose ionizing radiation," Cancer Res. **58**, 421–425.

DELONGCHAMP, R.R., MABUCHI, K., YOSHIMOTO, Y. and PRESTON, D.L. (1997). "Cancer mortality among atomic bomb survivors exposed *in utero* or as young children, October 1950-May 1992," Radiat. Res. **147**, 385–395.

DEMIDCHIK, E.P., DROBYSHEVSKAYA, I.M., CHERSTVOY, E.D., ASTAKHOVA, L.N., OKEANOV, A.E., VORONTSOVA, T.V. and GERMENCHUK, M. (1996). "Thyroid cancer in children in Belarus," pages 677 to 682 in *The Radiological Consequences of the Chernobyl Accident*, Karaoglou, A., Desmet, G., Kelly, G.N. and Menzel, H.G., Eds., EUR 16544 EN (European Commission, Brussels).

DEMPLE, B. and HARRISON, L. (1994). "Repair of oxidative damage to DNA: Enzymology and biology," Ann. Rev. Biochem. **63**, 915–948.

DESHPANDE, A., GOODWIN, E.H., BAILEY, S.M., MARRONE, B.L. and LEHNERT, B.E. (1996). "Alpha-particle-induced sister chromatid exchange in normal human lung fibroblasts: Evidence for an extranuclear target," Radiat. Res. **145**, 260–267.

DIAMOND, E.L., SCHMERLER, H. and LILIENFELD, A.M. (1973). "The relationship of intra-uterine radiation to subsequent mortality and development of leukemia in children. A prospective study," Am. J. Epidemiol. **97**, 283–313.

DIMAJO, V., COPPOLA, M., REBESSI, S., SARAN, A., PAZZAGLIA, S., PARISET, L. and COVELLI, V. (1996). "The influence of sex on life shortening and tumor induction in CBA/Cne mice exposed to x rays or fission neutrons," Radiat. Res. **146**, 81–87.

DOBYNS, B.M., SHELINE, G.E., WORKMAN, J.B., TOMPKINS, E.A., MCCONAHEY, W.M. and BECKER, D.Y. (1974). "Malignant and benign neoplasms of the thyroid in patients treated for hyperthyroidism: A report of the cooperative thyrotoxicosis therapy follow-up study," J. Clin. Endocrinol. Metab. **38**, 976–998.

DOLL, R. and WAKEFORD, R. (1997). "Risk of childhood cancer from fetal irradiation," Br. J. Radiol. **70**, 130–139.

DOMANN, F.E., FREITAS, M.A., GOULD, M.N. and CLIFTON, K.H. (1994). "Quantifying the frequency of radiogenic thyroid cancer per clonogenic cell *in vivo*," Radiat. Res. **137**, 330–337.

DONIACH, I. (1963). "Effects including carcinogenesis of ^{131}I and x-rays on the thyroid of experimental animals: A review," Health Phys. **9**, 1357–1362.

DONIACH, I. (1970). "Experimental thyroid tumors," pages 73 to 99 in *Tumours of the Thyroid Gland*, Smithers, D., Ed. (E&S Livingstone, Edinburgh).

DOODY, M.N., MANDEL, J.S., LUBIN, J.H. and BOICE, J.D., JR. (1998). "Mortality among United States radiologic technologists, 1926-90," Cancer Causes Control **9**, 67–75.

DOTTORINI, M.E., LOMUSCIO, G., MAZZUCCHELLI, L., VIGNATI A. and COLOMBO, L. (1995). "Assessment of female fertility and carcinogenesis after iodine-131 therapy for differentiated thyroid carcinoma," J. Nucl. Med. **36**, 21–27.

DULOUT, F.N. and NATARAJAN, A.T. (1987). "A simple and reliable *in vitro* test system for the analysis of induced aneuploidy as well as other cytogenetic end-points using Chinese hamster cells," Mutagenesis **2**, 121–126.

DUMONT, J.E., MALONE, J.F. and VAN HERLE, A.J. (1980). *Irradiation and Thyroid Disease: Dosimetric, Clinical and Carcinogenic Aspects*, EUR 6713 EN (Commission of the European Communities, Luxembourg).

DUNCAN, A.M.V. and EVANS, H.J. (1983). "The exchange hypothesis for the formation of chromatid aberrations: An experimental test using bleomycin," Mutat. Res. **107**, 307–313.

DUPREE, E.A., CRAGLE, D.L., MCLAIN, R.W., CRAWFORD-BROWN, D.J. and TETA, M.J. (1987). "Mortality among workers at a uranium processing facility, the Linde Air Products Company Ceramics Plant, 1943-1949," Scand. J. Work Environ. Health **13**, 100–107.

DYER, N.C., BRILL, A.B., GLASSER, S.R. and GOSS, D.A. (1969). "Maternal-fetal transport and distribution of ^{59}Fe and ^{131}I in humans," Am. J. Obstet. Gyn. **103**, 290–296.

EASTMOND, D.A. and PINKEL, D. (1990). "Detection of aneuploidy and aneuploidy-inducing agents in human lymphocytes using fluorescence *in situ* hybridization with chromosome-specific DNA probes," Mutat. Res. **234**, 303–318.

EDMONDS, C.J. and SMITH, T. (1986). "The long-term hazards of the treatment of thyroid cancer with radioiodine," Br. J. Radiol. **59**, 45–51.

EKER, R. and MOSSIGE, J. (1961). "A dominant gene for renal adenomas in the rat," Nature **189**, 858–859.

ELKIND, M.M. (1991). "Enhanced neoplastic transformation due to protracted exposures of fission-spectrum neutrons: Biophysical model (letter)," Int. J. Radiat. Biol. **59**, 1467–1475.

ENNOW, K., ANDERSSON, M., ENGHOLM, G., JESSEN, K. and STORM, H. (1989). "Epidemiological assessment of the cancer risk among the staff in radiotherapy departments in Denmark," pages 327 to 333 in *Low Dose Radiation. Biological Bases of Risk Assessment*, Baverstock, K.F. and Stather, J.W., Ed. (Taylor and Francis, New York).

ENOCH, T. and NORBURY, C. (1995). "Cellular responses to DNA damage: Cell cycle checkpoints, apoptosis and the roles of *p53* and *ATM*," Trends Biochem. Sci. **20**, 426–430.

ETHIER, S.P. and ULLRICH, R.L. (1982). "Detection of ductal dysplasia in mammary outgrowths derived from carcinogen-treated virgin female BALB/c mice," Cancer Res. **42**, 1753–1760.

EVANS, R.D. (1974). "Radium in man," Health Phys. **27**, 497–510.

EVANS, H.J., BUCKTON, K.E., HAMILTON, G.E. and CAROTHERS, A. (1979). "Radiation-induced chromosome aberrations in nuclear-dockyard workers," Nature **277**, 531–534.

EVANS, H.H., NIELSEN, M., MENCL, J., HORNG, M.F. and RICANTI, M. (1990). "The effect of dose rate on x-radiation-induced mutant frequency and the nature of DNA lesions in mouse lymphoma L5178Y cells," Radiat. Res. **122**, 316–325.

FAS (1995). French Academie des Sciences. *Problemes lies aux Effets des Faibles Doses des Radiatens Ionisantes*, Rappert No. 34 [English translation: Report No. 38] (French Academie des Sciences, Paris).

FEARON, E.R. and VOGELSTEIN, B. (1990). "A genetic model for colorectal tumorigenesis," Cell **61**, 759–767.

FEINBERG, A.P., GEHRKE, C.W., KUO, K.C. and EHRLICH, M. (1988). "Reduced genomic 5-methylcytosine content in human colonic neoplasia," Cancer Res. **48**, 1159–1161.

FEINENDEGEN, L.E., LOKEN, M.K., BOOZ, J., MUHLENSIEPEN, H., SONDHAUS, C.A. and BOND, V.P. (1995). "Cellular mechanisms of protection and repair induced by radiation exposure and their consequences for cell system responses," Stem Cells **13** (suppl.), 7–20.

FEY, M.F., WELLS, R.A., WAINSCOAT, J.S. and THEIN, S.L. (1988). "Assessment of clonality in gastrointestinal cancer by DNA fingerprinting," J. Clin. Invest. **82**, 1532–1537.

FIALKOW, P.J. (1976). "Clonal origin of human tumors," Biochem. Biophys. Acta **458**, 283–321.

FIALKOW, P.J. (1984). "Clonal evolution of human myeloid leukemias," pages 215 to 226 in *Genes and Cancer: UCLA Symposia on Molecular and Cellular Biology*, Bishop, J.M. and Rowley, J.D., Eds. (Alan R. Liss, Inc., New York).

FINKEL, M.P. (1959). "Induction of tumors with internally administered isotopes," pages 322 to 335 in *Radiation Biology and Cancer* (University of Texas Press, Austin).

FINKEL, M.P. and BISKIS, B.O. (1968). "Experimental induction of osteosarcomas," Prog. Exp. Tumor Res. **10**, 72–111.

FINKEL, M.P., REILLY, C.A., JR. and BISKIS, B.O. (1976). "Pathogenesis of radiation and virus-induced bone tumors," Recent Results Cancer Res. **54**, 92–103.

FORD, D.D., PATERSON, J.C. and TREUTING, W.L. (1959). "Fetal exposure to diagnostic x-rays, and leukemia and other malignant diseases in childhood," J. Natl. Cancer Inst. **22**, 1093–1104.

FRENCH, N.R., MAZA, B.G., HILL, H.O., ASCHWADEN, A.P. and KAAZ, H.W. (1974). "A population study of irradiated desert rodents," Ecol. Monogr. **44**, 45–72.

FRIEDBERG, W., HANNEMAN, G.D., FAULKNER, D.N., NEAS, B.R., COSGROVE, G.E., JR. and DARDEN, E.G., JR. (1976). "Fast-neutron irradiation of mouse embryos in the pronuclear zygote stage: Mortality curves and neoplastic diseases in 30-day postnatal survivors," Pro. Soc. Exp. Biol. Med. **151**, 808–810.

FRIGERIO, N.A. and STOWE, R.S. (1983). "Carcinogenic and genetic hazard from background radiation," pages 385 to 391 in *International Symposium on the Biological Effects of Low-Level Radiation with Special Regard to Stochastic and Non-Stochastic Effects*, IAEA-SM-202/805 (International Atomic Energy Agency, Vienna).

FROME, E.L., CRAGLE, D.L. and MCLAIN, R.W. (1990). "Poisson regression analysis of the mortality among a cohort of World War II nuclear industry workers," Radiat. Res. **123**, 138–152.

FROME, E.L., CRAGLE, D.L., WATKINS, J.P., WING, S., SHY, C.M., TANKERSLEY, W.G. and WEST, C.M. (1997). "A mortality study of employees of the nuclear industry in Oak Ridge, Tennessee," Radiat. Res. **148**, 64–80.

FRY, R.J.M. (1994). "Effects of low doses of radiation," Health Phys. **70**, 823–827.

FRY, R.J.M. and ULLRICH, R.L. (1986). "Combined effects of radiation and other agents," pages 437 to 454 in *Radiation Carcinogenesis*, Upton, A.C., Albert, R.E., Burns, F.J. and Shore, R.E., Eds. (Elsevier, New York).

FRY, R.J.M., POWERS-RISIUS, P., ALPEN, E.L. and AINSWORTH, E.J. (1985). "High-LET radiation carcinogenesis," Radiat. Res. **104**, S188–S195.

FRY, S., DUPREE, E., SNIPE, A., SEILER, D. and WALLACE, P. (1996). "A study of mortality and morbidity among persons occupationally exposed to ≥50 mSv in a year: Phase I, mortality through 1984," Appl. Occup. Environ. Hyg. **11**, 1–10.

FRY, R.J.M., GROSOVSKY, A., HANAWALT, P.C., JOSTES, R.F., LITTLE, J.B., MORGAN, W.F., OLEINICK, N.L. and ULLRICH, R.L. (1998). "The impact of biology on risk assessment —Workshop of the National Research Council's Board on Radiation Effects Research, " Radiat. Res. **150**, 695–705.

FULOP, G.M. and PHILLIPS, R.A. (1990). "The scid mutation in mice causes a general defect in DNA repair," Nature **347**, 479–482.

FURST, C.J., LUNDELL, M., HOLM, L.E. and SILFVERSWARD, C. (1988). "Cancer incidence after radiotherapy for skin hemangioma: A retrospective cohort study in Sweden," J. Natl. Cancer Inst. **80**, 1387–1392.

FURST, C.J., LUNDELL, M. and HOLM, L.E. (1990). "Tumors after radiotherapy for skin hemangioma in childhood. A case-control study," Acta Oncol. **29**, 557–562.

FURTH, J. and CLIFTON, K.H. (1958). "Experimental observations on mammotropes and the mammary gland," pages 276 to 282 *Endocrine Aspects of Breast Cancer*, Currie, A.R. and Illingworth, C.F.W., Eds. (E & S Livingstone, Ltd., Edinburgh).

FURUNO-FUKUSHI, I., UENO, A.M. and MATSUDAIRA, H. (1988). "Mutation induction by very low dose rate γ rays in cultured mouse leukemia cells L5178Y," Radiat. Res. **115**, 273–280.

GALE, K.B., FORD, A.M., REPP, R., BORKHARDT, A., KELLER, C., OSBORN, B.E. and GREAVES, M.F. (1997). "Backtracking leukemia to birth: Identification of clonotypic gene fusion sequences in neonatal blood spots," Proc. Natl. Acad. Sci. USA **94**, 13950–13954.

GARTE, S.J., BURNS, F.J., ASHKENAZI-KIMMEL, T., FELBER, M. and SAWEY, M.J. (1990). "Amplification of the c-myc oncogene during progression of radiation-induced rat skin tumors," Cancer Res. **50**, 3073–3077.

GIBSON, R., GRAHAM, S., LILIENFELD, A., SCHUMAN, L., DOWD, J.E. and LEVIN, M.L. (1972). "Irradiation in the epidemiology of leukemia among adults," J. Natl. Cancer Inst. **48**, 301–311.

GILBERT, E.S., OMOHUNDRO, E., BUCHANAN, J.A. and HOLTER, N.A. (1993). "Mortality of workers at the Hanford site: 1945-1986," Health Phys. **64**, 577–590.

GILBERT, E.S., GRIFFITH, W.C., BOECKER, B.B., DAGLE, G.E., GUILMETTE, R.A., HAHN, F.F., MUGGENBURG, B.A., PARK, J.F. and WATSON, C.R. (1998). "Statistical modeling of carcinogenic risks in dogs that inhaled $^{238}PuO_2$," Radiat. Res. **150**, 66–82.

GILLILAND, D.G., BLANCHARD, K.L., LEVY, J., PERRIN, S. and BUNN, H.F. (1991). "Clonality in myeloproliferative disorders: Analysis by means of the polymerase chain reaction," Proc. Natl. Acad. Sci. USA **88**, 6848–6852.

GILMAN, E.A., KNEALE, G.W., KNOX, E.G. and STEWART, A.M. (1988). "Pregnancy x-rays and childhood cancers: Effects of exposure and radiation dose," J. Radiol. Prot. **8**, 3–8.

GOFMAN, J.W. (1996). *Preventing Breast Cancer: The Story of Major, Proven, Preventable Cause of this Disease*, 2nd ed. (Committee for Nuclear Responsibility, San Francisco).

GOLDMAN, M. (1986). "Experimental carcinogenesis in the skeleton," pages 215 to 231 in *Radiation Carcinogenesis,* Upton, A.C., Albert, R.E., Burns, F.J. and Shore, R.E., Eds. (Elsevier, New York).

GOLDMAN, M.B., MALOOF, F., MONSON, R.R., ASCHENGRAU, A., COOPER, D.S. and RIDGWAY, E.C. (1988). "Radioactive iodine therapy and breast cancer: A follow-up study of hyperthyroid women," Am. J. Epidemiol. **127**, 969–980.

GOLDSMITH, R.E. (1972). "Radioisotope therapy for Graves' disease," Mayo Clin. Proc. **47**, 953–961.

GOODHEAD, D.T. (1988). "Spatial and temporal distribution of energy," Health Phys. **55**, 231–240.

GOODHEAD, D.T. (1994). "Initial events in the cellular effects of ionizing radiations: Clustered damage in DNA," Int. J. Radiat. Biol. **65**, 7–17.

GOODHEAD, D.T. (1995). "Molecular and cell models of biological effects of heavy ion radiation," Radiat. Environ. Biophys. **34**, 67–72.

GOODHEAD, D.T. and BRENNER, D.J. (1983). "Estimation of a single property of low LET radiations which correlates with biological effectiveness," Phys. Med. Biol. **28**, 485–492.

GOULD, M.N. and ZHANG, R. (1991). "Genetic regulation of mammary carcinogenesis in the rat by susceptibility and suppressor genes," Environ. Health Perspect. **93**, 161–167.

GOULD, M.N., BIEL, W.F. and CLIFTON, K.H. (1977). "Morphological and quantitative studies of gland formation from inocula of monodispersed rat mammary cells," Exp. Cell. Res. **107**, 405–416.

GOWAN, J.W. and STADLER, J. (1964). "Lifespans of mice as affected by continuing irradiation from cobalt-60 accumulated ancestrally and under direct irradiation," Genetics (suppl.) **50**, 1115–1142.

GRAHAM, S., LEVIN, M.L., LILIENFELD, A., DOWD, J., SCHUMAN, L., GIBSON, R., HEMPELMANN, L. and GERHARDT, P. (1963). "Methodological problems and design of the tristate leukemia survey," Ann. NY Acad. Sci. **107**, 557–569.

GRAHAM, S., LEVIN, M.L., LILIENFELD, A.M., SCHUMAN, L.M., GIBSON, R., DOWD, J.E. and HEMPELMANN, L. (1966). "Preconception, intrauterine and postnatal irradiation as related to leukemia," Natl. Cancer Inst. Monogr. **19**, 347–371.

GRAHN, D. (1970). "Biological effects of protracted low-dose radiation exposure of man and animals," pages 101 to 136 in *Late Effects of Radiation*, Fry, R.J.M., Grahn, D., Griem, M.L., *et al.*, Eds. (Taylor and Francis, New York).

GRAHN, D., FRY, R.J.M. and LEA, R.A. (1972). "Analysis of survival and cause of death statistics for mice under single and duration of life gamma irradiation," pages 175 to 186 in *Life Science and Space Research*, Vol. 10, Vishniac, W., Ed. (Academic Verlag, Berlin).

GREAVES, M.F. (1997). "Aetiology of acute leukemia," Lancet **349**, 344–349.

GREENBERG, E.R., ROSNER, B., HENNEKENS, C., RINSKY, R. and COLTON, T. (1985). "An investigation of bias in a study of nuclear shipyard workers," Am. J. Epidemiol. **121**, 301–308.

GREENLAND, S. and MORGENSTERN, H. (1989). "Ecological bias, confounding, and effect modification," Int. J. Epidemiol. **18**, 269–274.

GREENLAND, S. and ROBINS, J. (1994). "Invited commentary: Ecologic studies—biases, misconceptions, and counterexamples," Am. J. Epidemiol. **139**, 747–760.

GRIEM, M.L., MEIER, P. and DOBBEN, G.D. (1967). "Analysis of the morbidity and mortality of children irradiated in fetal life," Radiology **88**, 347–349.

GRIEM, M.L., KLEINERMAN, R.A., BOICE, J.D., JR., STOVALL, M., SHEFNER, D. and LUBIN, J.H. (1994). "Cancer following radiotherapy for peptic ulcer," J. Natl. Cancer Inst. **86**, 842–849.

GRIFFIN, C.S., MARSDEN, S.J., STEVENS, D.L., SIMPSON, P. and SAVAGE, J.R.K. (1995). "Frequencies of complex chromosome exchange aberrations induced by ^{238}Pu alpha-particles and detected by fluorescence *in situ* hybridization using single chromosome-specific probes," Int. J. Radiat. Biol. **67**, 431–439.

GRIFFIN, C.S., STEVENS, D.L. and SAVAGE, J.R.K. (1996). "Ultrasoft 1.5 keV aluminum K x rays are efficient producers of complex chromosome exchange aberrations as revealed by fluorescence *in situ* hybridization," Radiat. Res. **146**, 144–150.

GROSOVSKY, A.J. (1999). "Radiation-induced mutations in unirradited DNA," Proc. Natl. Acad. Sci. USA **96**, 5346–5347.

GROSOVSKY, A.J. and LITTLE, J.B. (1985). "Evidence for linear response for the induction of mutations in human cells by x-ray exposures below 10 rads," Proc. Natl. Acad. Sci. USA **82**, 2092–2095.

GROSOVSKY, A.J., PARKS, K.K., GIVER, C.R. and NELSON, S.L. (1996). "Clonal analysis of delayed karyotypic abnormalities and gene mutations in radiation-induced genetic instability," Mol. Cell Biol. **16**, 6252–6262.

GUERRERO, I. and PELLICER, A. (1987). "Mutational activation of oncogenes in animal model systems of carcinogenesis," Mutat. Res. **185**, 293–308.

GUNZ, F.W. and ATKINSON, H.R. (1964). "Medical radiations and leukaemia: A retrospective survey," BMJ **1**, 389–393.

HAGMAR, L., BROGGER, A., HANSTEEN, I.L., HEIM, S., HOGSTEDT, B., KNUDSEN, L., LAMBERT, B., LINNAINMAA, K., MITELMAN, F., NORDENSON, I., RENTERWALL, C., SALOMAA, S., SKERVING, S. and SORSA, M. (1994). "Cancer risk in humans predicted by increased levels of chromosomal aberrations in lymphocytes: Nordic study group on the health risk of chromosome damage," Cancer Res. **54**, 2919–2922.

HAGSTROM, R.M., GLASSER, S.R., BRILL, A.B. and HEYSSEL, R.M. (1969). "Long term effects of radioactive iron administered during human pregnancy," Am. J. Epidemiol. **90**, 1–10.

HAHN, P., NEVALDINE, B. and MORGAN, W.F. (1990). "X-ray induction of methotrexate resistence due to *dhfr* gene amplification," Somat. Cell Mol. Genet. **16**, 413–423.

HAKODA, M., AKIYAMA, M., HIRAI, Y., KYOIZUMI, S. and AWA, A.A. (1988). "*In vivo* mutant T cell frequency in atomic bomb survivors carrying outlying values of chromosome aberration frequencies," Mutat. Res. **202**, 203–208.

HALL, E.J. (1994). *Radiobiology for the Radiologist*, 4th ed. (J.B. Lippincott, Philadelphia).

HALL, E.J. and HEI, T.K. (1987). "Oncogenic transformation by radiation and chemicals," pages 507 to 512 in *Radiation Research, Proceedings of the 8th International Congress of Radiation Research*, Fielden, E.M., Fowler, J.F., Hendry, J.H. and Scott, D., Eds. (Taylor and Francis, New York).

HALL, E.J., HEI, T.K. and PIAO, C.Q. (1989). "Transformation by simulated radon daughter alpha particles; interaction with asbestos and modulation by tumor promoters," pages 293 to 299 in *Cell Transformation and Radiation-Induced Cancer*, Chadwick, K.H., Seymour, C. and Barnhart B., Eds. (Adam Hilger, New York).

HALL, P., HOLM, L., LUNDELL, G., BJELKENGREN, G., LARSSON, L., LINDBERG, S., TENNVALL, J., WICKLUND, H. and BOICE, J.D., JR. (1991). "Cancer risks in thyroid cancer patients," Br. J. Cancer **64**, 159–163.

HALL, P., BOICE, J.D., JR., BERG, G., BJELKENGREN, G., ERICSSON, U.B., HALLQUIST, A., LIDBERG, M., LUNDELL, G., MATTSSON, A., TENNVALL, J., WIKLUND, K. and HOLM, L. (1992a). "Leukaemia incidence after iodine-131 exposure," Lancet **340**, 1–4.

HALL, E.J., GEARD, C.R. and BRENNER, D.J. (1992b). "Risk of breast cancer in ataxia-telangiectasia (letter)," N. Engl. J. Med. **326**, 1358–1359.

HALL, P., MATTSSON, A. and BOICE, J.D. JR. (1996). "Thyroid cancer after diagnostic administration of iodine-131," Radiat. Res. **145**, 86–92.

HALLQUIST, A., HARDELL, L., DEGERMAN, A., WINGREN, G. and BOQUIST, L. (1994). "Medical diagnostic and therapeutic ionizing radiation and the risk for thyroid cancer: A case-control study," Eur. J. Cancer Prev. **3**, 259–267.

HAMILTON, P., CHIACCHIERINI, R. and KACZMAREK, R. (1989). *A Follow-up Study of Persons Who Had Iodine-131 and Other Diagnostic Procedures During Childhood and Adolescence* (Center for Devices and Radiological Health, Food and Drug Administration, Rockville, Maryland).

HANCOCK, S.L., TUCKER, M.A. and HOPPE, R.T. (1993). "Breast cancer after treatment of Hodgkin's disease," J. Natl. Cancer Inst. **85**, 25–31.

HARDWICK, J.P., SCHLENKER, R.A. and HUBERMAN, E. (1989). "Alteration of the c-mos locus in "normal" tissues from humans exposed to radium," Cancer Res. **49**, 2668–2673.

HARJULEHTO, T., RAHOLA, T., SUOMELA, M., ARVELA, H. and SAXEN, L. (1991). "Pregnancy outcome in Finland after the Chernobyl accident," Biomed Pharmacother **45**, 263–266.

HARVEY, E.B., BOICE, J.D., JR., HONEYMAN, M. and FLANNERY, J.T. (1985). "Prenatal x-ray exposure and childhood cancer in twins," N. Engl. J. Med. **312**, 541–545.

HAYEK, A., CHAPMAN, E.M. and CRAWFORD, J.D. (1970). "Long-term results of treatment of thyrotoxicosis in children and adolescents with radioactive iodine," N. Engl. J. Med. **283**, 949–953.

HEI, T.K., PIAO, C.Q., WILLEY, J.C., THOMAS, S. and HALL, E.J. (1994). "Malignant transformation of human bronchial epithelial cells by radon-simulated alpha-particles," Carcinogenesis **15**, 431–437.

HEI, T.K., PIAO, C.Q., HAN, E., SUTTER, T. and WILLEY, J.C. (1996). "Radon induced neoplastic transformation of bronchial epithelial cells," Radiat. Oncol. Investment **3**, 398–403.

HEI, TK, WU, L.J., LIU, S.X., VANNAIS, D., WALDREN, C.A. and RANDERS-PHERSON, G. (1997). "Mutagenic effects of a single and an exact number of α particles in mammalian cells," Proc. Natl. Acad. Sci. USA **94**, 3765–3770.

HEIM, S. and MITELMAN, F. (1995). *Cancer Cytogenetics*, 2nd ed. (Wiley-Liss, New York).

HEIMLER, A., FRIEDMAN, E. and ROSENTHAL, A.D. (1979). "Nevoid basal cell carcinoma syndrome and Charcot-Marie-Tooth disease. Two autosomal dominant disorders segregating in a family," J. Med. Genet. **15**, 288–291.

HEISTERKAMP, N., VONCKEN, J.W., VAN SCHAIK, M. and GROFFEN, J. (1993). "PL—positive leukemia," pages 359 to 376 in *The Causes and Consequences of Chromosomal Aberrations*, Kirsch, I.R., Ed. (CRC Press, Boca Raton, Florida).

HELBOCK, H.J., BECKMAN, K.B., SHIGENAGA, M.K., WALTER, P.B., WOODALL, A.A., YEO, H.C. and AMES, B.N. (1998). "DNA oxidation matters: The HPLC-electrochemical detection assay of 8-oxo-deoxyguanosine and 8-oxo-guanine," Proc. Natl. Acad. Sci. USA **95**, 288–293.

HENDERSON, I.C. (1993). "Risk factors for breast cancer development," Cancer **71** (suppl.), 2127–2140.

HEYN, R., HAEBERLEN, V., NEWTON, W.A., RAGAB, A.H., RANEY, R.B., TEFFT, M., WHARAM, M., ENSIGN, L.G. and MAURER, H.M. (1993). "Second malignant neoplasms in children treated for rhabdomyosarcoma," J. Clin. Oncol. **11**, 262–270.

HILDRETH, N.G., SHORE, R.E. and DVORETSKY, P.M. (1989). "Risk of breast cancer after irradiation of the thymus in infancy," N. Engl. J. Med. **321**, 1281–1284.

HILL, C.K., HAN, A. and ELKIND, M.M. (1984). "Fission-spectrum neutrons at a low dose rate enhance neoplastic transformation in the linear, low dose region (0-10 cGy)," Int. J. Radiat. Biol. Relat. Stud. Phys. Chem. Med. **46**, 11–15.

HILL, C.K., CARNES, B.A., HAN, A. and ELKIND, M.M. (1985a). "Neoplastic transformation is enhanced by multiple low doses of fission-spectrum neutrons," Radiat. Res. **102**, 404–410.

HILL, C.K., ELKIND, M.M. and HAN, A. (1985b). "Role of repair processes in neoplastic transformation induced by ionizing radiation in C3H/10T-½ cells," pages 379 to 397 in *Mammalian Cell Transformation: Mechanisms of Carcinogenesis and Assays for Carcinogens, Carcinogenesis, Volume 9*, Barret, J.C. and Tennant, R.W., Eds. (Raven Press, New York).

HINO, O., KLEIN-SZANTO, A.J.P., FREED, J.J., TESTA, J.R., BROWN, D.Q., VILENSKY, M., YEUNG, R.S., TARTOF, K.D. and KNUDSON, A.G. (1993). "Spontaneous and radiation-induced renal tumors in the Eker rat model of dominantly inherited cancer," Proc. Natl. Acad. Sci. USA **90**, 327–331.

HIRAI, Y., KUSUNOKI, Y., KYOIZUMI, S., AWA, A.A., PAWEL, D.J., NAKAMURA, N. and AKIYAMA, M. (1995). "Mutant frequency at the HPRT locus in peripheral blood T-lymphocytes of atomic bomb survivors," Mutat. Res. **329**, 183–196.

HOEL, D.G. and LI, P. (1998). "Threshold models in radiation carcinogenesis," Health Phys. **75**, 241–250.

HOFFMAN, D.A. (1984). "Late effects of I-131 therapy in the United States," pages 273 to 280 in *Radiation Carcinogenesis: Epidemiology and Biological Significance*, Boice, J.D., Jr. and Fraumeni, J.F., Jr., Eds. (Raven Press, New York).

HOFFMAN, D.A. and MCCONAHEY, W.M. (1983). "Breast cancer following iodine-131 therapy for hyperthyroidism," J. Natl. Cancer Inst. **70**, 63–67.

HOFFMAN, D.A., MCCONAHEY, W.M., FRAUMENI, J.F., JR. and KURLAND, L.T. (1982). "Cancer incidence following treatment of hyperthyroidism," Int. J. Epidemiol. **11**, 218–224.

HOFFMAN, D.A., LONSTEIN, J.E., MORIN, M.M., VISSCHER, W., HARRIS, B.S., III and BOICE, J.D., JR. (1989). "Breast cancer in women with scoliosis exposed to multiple diagnostic x rays," J. Natl. Cancer Inst. **81**, 1307–1312.

HOHRYAKOV, V.F. and ROMANOV, S.A. (1994). "Lung cancer in radiochemical industry workers," Sci. Tot. Environ. **142**, 25–28.

HOLLIDAY, R. (1991). "Mutations and epimutations in mammalian cells," Mutat. Res. **250**, 351–363.

HOLM, L.E. (1991). "Cancer risks after diagnostic doses of ^{131}I with special reference to thyroid cancer," Cancer Detect. Prev. **15**, 27–30.

HOLM, L.E., WIKLUND, K.E., LUNDELL, G.E., BERGMAN, N.A., BJELKENGREN, G., ERICSSON, U.B., CEDERQUIST, E.S., LIDBERG, M.E., LINDBERG, R.S., WICKLUND, H.V. and BOICE, J.D., JR. (1989). "Cancer risk in population examined with diagnostic doses of ^{131}I," J. Natl. Cancer Inst. **81**, 302–306.

HOLM, L.E., HALL, P., WIKLUND, K., LUNDELL, G., BERG, G., BJELKENGREN, G., CEDERQUIST, E., ERICSSON, U.B., HALLQUIST, A., LARSSON, L.G., LIDBERG, M.E., LINDBERG, R.S., TENNVALL, J., WICKLUND, H.V. and BOICE, J.D., JR. (1991). "Cancer risk after iodine-131 therapy for hyperthyroidism," J. Natl. Cancer Inst. **83**, 1072–1077.

HOPTON, P.A., MCKINNEY, P.A., CARTWRIGHT, R.A., MANN, J.R., BIRCH, J.M., HARTLEY, A.L., WATERHOUSE, J.A.H., JOHNSTON, H.E., DRAPER, G.J. and STILLER, C.A. (1985). "X-rays in pregnancy and the risk of childhood cancer (letter)," Lancet **2**, 773.

HOSHINO, H. and TANOOKA, H. (1975). "Interval effect of β-irradiation and subsequent 4-nitroquinoline-1-oxide painting on skin tumor induction in mice," Cancer Res. **35**, 3663–3666.

HOWE, G.R. (1995). "Lung cancer mortality between 1950 and 1987 after exposure to fractionated moderate-dose-rate ionizing radiation in the Canadian fluoroscopy cohort study and a comparison with lung cancer mortality in the atomic bomb survivors study," Radiat. Res. **142**, 295–304.

HOWE, G.R. (2001). "Breast cancer incidence (1975-83) following exposure to highly fractionated ionizing radiation in the Candadian fluoroscopy cohort study" (to be published).

HOWE, G.R. and MCLAUGHLIN, J. (1996). "Breast cancer mortality between 1950 and 1987 after exposure to fractionated moderate-dose-rate ionizing radiation in the Canadian fluoroscopy cohort study and a comparison with breast cancer mortality in the atomic bomb survivors study," Radiat. Res. **145**, 694–707.

HUANG, L.C, CLARKIN, K.C. and WAHL, G.M. (1996). "Sensitivity and selectivity of the DNA damage sensor responsible for activating p53-dependent G1 arrest," Proc. Natl. Acad. Sci. USA **93**, 4827–4832.

HUMBLET, C., GREIMERS, R., BONIVER, J. and DEFRESNE, M.P. (1997). "Stages in the development of radiation-induced thymic lymphomas in C57BL/Ka mice: Preleukemic cells become progressively resistant to the tumor preventing effects of a bone marrow graft," Exp. Hematol. **25**, 109–113.

ICRP (1977). International Commission on Radiological Protection. *Recommendations of the International Commission on Radiological Protection*, ICRP Publication 26, Annals of the ICRP **1** (Elsevier Science, New York).

ICRP (1991a). International Commission on Radiological Protection. *1990 Recommendations of the International Commission on Radiological Protection*, ICRP Publication 60, Annals of the ICRP **21** (Elsevier Science, New York).

ICRP (1991b). International Commission on Radiological Protection. *The Biological Basis for Dose Limitation in the Skin*, ICRP Publication 59, Annals of the ICRP **22** (Elsevier Science, New York).

ICRP (1998). International Commission on Radiological Protection. *Genetic Susceptibility to Cancer*, ICRP Publication 79, Annals of the ICRP **28** (Elsevier Science, New York.

ICRU (1983). International Commission on Radiation Units and Measurements. *Microdosimetry*, ICRU Report 36 (International Commission on Radiation Units and Measurements, Bethesda, Maryland).

INSKIP, P.D., MONSON, R.R., WAGONER, J.K., STOVALL, M., DAVIS, F.G., KLEINERMAN, R.A. and BOICE, J.D., JR. (1990). "Leukemia following radiotherapy for uterine bleeding," Radiat. Res. **122**, 107–119.

INSKIP, P.D., HARVEY, E.B., BOICE, J.D., JR., STONE, B.J., MATANOSKI, G., FLANNERY, J.T. and FRAUMENI, J.F., JR. (1991). "Incidence of childhood cancers in twins," Cancer Causes Control **2**, 315–324.

INSKIP, P.D., KLEINERMAN, R.A., STOVALL, M., COOKFAIR, D.L., HADJIMICHAEL, O., MOLONEY, W.C., MONSON, R.R., THOMPSON, W.D., WACTAWSKI-WENDE, J., WAGONER, J. and BOICE, J.D., JR. (1993). "Leukemia, lymphoma, and multiple myeloma after pelvic radiotherapy for benign disease," Radiat. Res. **135**, 108–124.

INSKIP, P.D., STOVALL, M. and FLANNERY, J.T. (1994). "Lung cancer risk and radiation dose among women treated for breast cancer," J. Natl. Cancer Inst. **86**, 983–988.

INSKIP, P.D., EKBOM, A., GALANTI, M.R., GRIMELIUS, L. and BOICE, J.D., JR. (1995). "Medical diagnostic x rays and thyroid cancer," J. Natl. Cancer Inst. **87**, 1613–1621.

ISHIHARA, T. and KUMATORI, T. (1983). "Cytogenetic follow-up studies in Japanese fisherman exposed to fallout radiation," pages 457 to 490 in *Radiation-Induced Chromosome Damage in Man*, Ishihara, T. and Sasaki, M.S., Eds. (Alan R. Liss, New York).

ITO, T., SEYAMA, T., MIZUNO, T., HAYASHI, T., IWAMOTO, K.S., DOHI, K., NAKAMURA, N. and AKIYAMA, M. (1993). "Induction of BCR-ABL fusion genes by *in vitro* x-irradiation," Jpn. J. Cancer Res. **84**, 105–109.

IVANOV, E.P., TOLOCHKO, G.V., SHUVAEVA, L.P., BECKER, S., NEKOLLA, E. and KELLERER, A.M. (1996). "Childhood leukemia in Belarus before and after the Chernobyl accident," Radiat. Environ. Biophys. **35**, 75–80.

IVANOV, E.P., TOLOCHKO, G.V., SHUVAEVA, L.P., IVANOV, V.E. IAROSHEVICH, R.F., BECKER, S., NEKOLLA, E. and KELLERER, A.M. (1998). "Infant leukemia in Belarus after the Chernobyl accident," Radiat. Environ. Biophys. **37**, 53–55.

JABLON, S. and MILLER, R.W. (1978). "Army technologists: 29-year follow up for cause of death," Radiology **126**, 677–679.

JABLON, S., HRUBEC, Z. and BOICE, J.D., JR. (1991). "Cancer in populations living near nuclear facilities. A survey of mortality nationwide and incidence in two states," JAMA **265**, 1403–1408.

JACOB, P., GOULKO, G., HEIDENREICH, W.F., LIKHTAREV, I., KAIRO, I., TRONKO, N.D., BOGDANOVA, T.I., KENIGSBERG, J., BUGLOVA, E., DROZDOVITCH, V., GOLOVNEVA, A., DEMIDCHIK, E.P., BALONOV, M., ZVONOVA, I. and BERAL, V. (1998). "Thyroid cancer risk to children calculated (letter)," Nature **392**, 31–32.

JACOBS, I.J., KOHLER, M.F., WISEMAN, R.W., MARKS, J.R., WHITAKER, R., KERNS, B.A., HUMPHREY, P., BERCHUCK, A., PONDER, B.A. and BAST, R.C., JR. (1992). "Clonal origin of epithelial ovarian carcinoma: Analysis by loss of heterozygosity, p53 mutation, and X-chromosome inactivation," J. Natl. Cancer Inst. **84**, 1793–1798.

JANOWSKI, M., COX, R. and STRAUSS, P.G. (1990). "The molecular biology of radiation-induced carcinogenesis: Thymic lymphoma, myeloid leukaemia and osteosarcoma," Int. J. Radiat. Biol. **57**, 677–691.

JAWOROWSKI, Z. (1995). "Beneficial radiation," Nukleonika **40**, 3–12.

JEGGO, P.A., TACCIOLI, G.E. and JACKSON, S.P. (1995). "Menage a trois: Double strand break repair, V(D)J recombination and DNA-PK," Bioessays **17**, 949–957.

JENSEN, R.H., LANGLOIS, R.G., BIGBEE, W.L., GRANT, S.G., MOORE, D., II, PILINSKAYA, M., VOROBTSOVA, I. and PLESHANOV, P. (1995). "Elevated frequency of glycophorin A mutations in erythrocytes from Chernobyl accident victims," Radiat. Res. **141**, 129–135.

JOHNSON, C.J. (1987). "Cancer incidence patterns in the Denver metropolitan area in relation to the Rocky Flats plant," Am. J. Epidemiol. **126**, 153–155.

KADHIM, M.A., MACDONALD, D.A., GOODHEAD, D.T., LORIMORE, S.A., MARSDEN, S.J. and WRIGHT, E.G. (1992). "Transmission of chromosomal instability after plutonium α-particle irradiation," Nature **355**, 738–740.

KADHIM, M.A., LORIMORE, S.A., HEPBURN, M.D., GOODHEAD, D.T., BUCKLE, V.J. and WRIGHT, E.G. (1994). "α-particle-induced chromosomal instability in human bone marrow cells," Lancet **344**, 987–988.

KADHIM, M.A., MARSDEN, S.J. and WRIGHT, E.G. (1998). "Radiation-induced chromosomal instability in human fibroblasts: Temporal effects and the influence of radiation quality," Int. J. Radiat. Biol. **73**, 143–148.

KAI, M., LUEBECK, E.G. and MOOLGAVKAR, S.H. (1997). "Analysis of the incidence of solid cancer among atomic bomb survivors using a two-stage model of carcinogenesis," Radiat. Res. **148**, 348–358.

KAMIYA, K., YASUKAWA-BARNES, J., MITCHEN, J.M., GOULD, M.N. and CLIFTON, K.H. (1995). "Evidence that carcinogenesis involves an imbalance between epigenetic high-frequency initiation and suppression of promotion," Proc. Natl. Acad. Sci. USA **92**, 1332–1336.

KAMADA, N. and TANAKA, K. (1983). "Cytogenetic studies of hematological disorders in atomic bomb survivors," pages 455 to 474 in *Radiation-Induced Chromosome Damage in Man*, Ishihara, T. and Sasaki, M.S., Eds. (Alan R. Liss, New York).

KAPLAN, H.S. (1958). "An evaluation of the somatic and genetic hazards of the medical uses of radiation," Am. J. Roentg. **80**, 696–706.

KAPLAN, H.S. and BROWN, M.B. (1952). "A quantitative dose-response study of lymphoid tumor development in irradiated C57 black mice," J. Natl. Cancer Inst. **13**, 185–208.

KELLERER, A.M. (1985). "Fundamentals of microdosimetry," pages 77 to 162 in *The Dosimetry of Ionizing Radiation*, Kase, K.R. Bjarngard, B.E. and Attix, F.H., Eds. (Academic Press, New York).

KELLERER, A.M. (1987). "Models of cellular radiation action" pages 305 to 376 in *Kinetics of Nonhomogeneous Processes: A Practical Introduction for Chemists, Biologists, Physicists, and Materials Scientists*, Freeman, G.R., Ed. (John Wiley & Sons, New York).

KELLERER, A.M. and BARCLAY, D. (1992). "Age dependencies in the modeling of radiation carcinogenesis," Radiat. Prot. Dosim. **41**, 273–281.

KELLERER, A.M. and CHMELEVSKY, D. (1975). "Criteria for the applicability of LET," Radiat. Res. **63**, 226–234.

KELLERER, A.M. and NEKOLLA, E. (1997). "Neutron versus gamma-ray risk estimates. Inferences from the cancer incidence and mortality data in Hiroshima," Radiat. Environ. Biophys. **36**, 73–83.

KELLERER, A.M. and ROSSI, H.H. (1972). "The theory of dual radiation action," Curs. Topics Radiat. Res., Q. **8**, 85–158.

KELLERER, A.M. and ROSSI, H.H. (1975). "Biophysical aspects of radiation carcinogenesis," pages 405 to 439 in *Cancer—A Comprehensive Treatise, Vol. I, Etiology-Chemical and Physical Carcinogenesis*, Becker, F.F., Ed. (Plenum Press, New York).

KELSEY, K.T., DONOHOE, K.J., BAXTER, B., MEMISOGLU, A., LITTLE, J.B., CAGGANA, M. and LIBER, H.L. (1991a). "Genotoxic and mutagenic effects of the diagnostic use of thallium-201 in nuclear medicine," Mutat. Res. **260**, 239–246.

KELSEY, K.T., DONOHOE, K.J., MEMISOGLU, A., BAXTER, B., CAGGANA, M. and LIBER, H.L. (1991b). "*In vivo* exposure of human lymphocytes to technetium-99m in nuclear medicine patients does not induce detectable genetic effects," Mutat. Res. **264**, 213–218.

KELSEY, K.T., MEMISOGLU, A., FRENKEL, D. and LIBER, H.L. (1991c). "Human lymphocytes exposed to low doses of x-rays are less susceptible to radiation-induced mutagenesis," Mutat. Res. **263**, 197–201.

KEMP C.J., WHELDON, T. and BALMAIN, A. (1994). "p53-deficient mice are extremely susceptible to radiation-induced tumorigenesis," Natl. Genet. **8**, 66–69.

KENDALL, G.M., MUIRHEAD, C.R., MACGIBBON, B.H., O'HAGAN, J.A., CONQUEST, A.J., GOODILL, A.A., BUTLAND, B.K., FELL, T.P., JACKSON, D.A., WEBB, M.A., HAYLOCK, R.G.E., THOMAS, J.M. and SILK, T.J. (1992). "Mortality and occupational exposure to radiation: First analysis of the National Registry for Radiation Workers," BMJ **304**, 220–225.

KENNEDY, A.R. (1985). "Evidence that the first step leading to carcinogen-induced malignant transformation is a high frequency, common event," Carcinog. Compr. Surv. **9**, 355–364.

KENNEDY, A.R. (1991). "Is there a critical target gene for the first step in carcinogenesis?" Environ. Health Perspect. **93**, 199–203.

KENNEDY, A.R., MURPHY, G. and LITTLE, J.B. (1980a). "Effect of time and duration of exposure to 12-0-tetradecanoylphorbol-13-acetate on x-ray transformation of C3H 10T½ cells," Cancer Res. **40**, 1915–1920.

KENNEDY, A.R., FOX., M., MURPHY, G. and LITTLE, J.B. (1980b). "Relationship between x-ray exposure and malignant transformation in C3H 10T½ cells," Proc. Natl. Acad. Sci. USA **77**, 7262–7266.

KERBER, R.A., TILL, J.E., SIMON, S.L., LYON, J.L., THOMAS, D.C., PRESTON-MARTIN, S., RALLISON, M.L., LLOYD, R.D. and STEVENS, W. (1993). "A cohort study of thyroid disease in relation to fallout from nuclear weapons testing," JAMA **270**, 2076–2082.

KHAIDAKOV, M., YOUNG, D., ERFLE, H., MORTIMER, A., VORONKOV, Y. and GLICKMAN, B.W. (1997). "Molecular analysis of mutations in T-lymphocytes from experienced Soviet cosmonauts," Environ. Mol. Mutagen. **30**, 21–30.

KIM, U. and FURTH, J. (1960). "Relation of mammary tumors to mammotropes. I. Induction of mammary tumors in rats," Proc. Soc. Exp. Biol. Med. **103**, 640–642.

KIM, N.D., OBERLEY, T.D. and CLIFTON, K.H. (1993). "Primary culture of flow cytometry-sorted rat mammary epithelial cell (RMEC) subpopulations in a reconstituted basement membrane, Matrigel," Exp. Cell Res. **209**, 6–20.

KJELDSBERG, H. (1957). "Radioaktiv bestraling og leukemifrekvens hos barn," Tidsskr Norske Laegeforen **77**, 1052–1053.

KLIAUGA, P. and DVORAK, R. (1978). "Microdosimetric measurements of ionization by monoenergetic photons," Radiat. Res. **73**, 1–20,

KNEALE, G.W. and STEWART, A.M. (1993). "Reanalysis of Hanford data: 1944-1986 deaths," Am. J. Ind. Med. **23**, 371–389.

KNOX, E.G., STEWART, A.M., KNEALE, G.W. and GILMAN, E.A. (1987). "Prenatal irradiation and childhood cancer," J. Radiol. Prot., **7**, 177–189.

KNUDSON, A.G., JR., HETHCOTE, H.W. and BROWN, B.W. (1975). "Mutation and childhood cancer: A probabilistic model for the incidence of retinoblastoma," Proc. Natl. Acad. Sci. USA **72**, 5116–5120.

KOBAYASHI, T., URAKAMI, S., HIRAYAMA, Y., YAMAMOTO, T., NISHIZAWA, M., TAKAHARA, T., KUBO, Y. and HINO, O. (1997). "Intragenic *Tsc2* somatic mutations as Knudson's second hit in spontaneous and chemically induced renal carcinomas in the Eker rat model," Jpn. J. Cancer Res. **88**, 254–261.

KODURU, P. and CHAGANTI, R. (1988). "Congenital chromosome breakage clusters within Giemsa-light bands and identifies sites of chromatin instability," Cytogenet. Cell Genet. **49**, 269–274.

KOGUT, M.D., KAPLAN, S.A., COLLIPP, P.J., TIAMSIC, T. and BOYLE, D. (1965). "Treatment of hyperthyroidism in children. Analysis of forty-five patients," N. Engl. J. Med. **272**, 217–221.

KOLODNER, R.D. (1995). "Mismatch repair: Mechanisms and relationship to cancer susceptibility," Trends Biochem. Sci. **20**, 397–401.

KONDO, S. (1993). *Health Effects of Low-Level Radiation* (Medical Physics Publishing, Madison, Wisconsin).

KONIG, F. and KIEFER, J. (1988). "Lack of a dose-rate effect for mutation induction by gamma-rays in human TK6 cells," Int. J. Radiat. Biol. **54**, 891–897.

KOPECKY, K.J., NAKASHIMA, E., YAMAMOTO, T. and KATO, H. (1986). *Lung Cancer, Radiation, and Smoking Among A-Bomb Survivors, Hiroshima and Nagasaki* (Radiation Effects Research Foundation, Hiroshima).

KOSHURNIKOVA, N.A., BULDAKOV, L.A., BYSOGOLOV, G.D., BOLOTNIKOVA, M.G., KOMLEVA, N.S. and PESTERNIKOVA, V.S. (1994). "Mortality from malignancies of the hematopoietic and lymphatic tissues among personnel of the first nuclear plant in the USSR," Sci. Total Environ. **142**, 19–23.

KOSHURNIKOVA, N.A., BYSOGOLOV, G.D., BOLOTNIKOVA, M.G., KHOKHRYAKOV, V.F., KRESLOV, V.V., OKATENKO, P.V., ROMANOV, S.A. and SHILNIKOVA, N.S. (1996). "Mortality among personnel who worked at the Mayak complex in the first years of its operation," Health Phys. **71**, 90–93.

KOSHURNIKOVA, N.A., SHILNIKOVA, N.S., OKATENKO, P.V., KRESLOV, V.V., BOLOTNIKOVA, M.G., ROMANOV, S.A. and SOKOLIKOV, M.E. (1997). "The risk of cancer among nuclear workers at the "Mayak" Production Association: Preliminary results of an epidemiological study," pages 113 to 122 in *Implications of New Data on Radiation Cancer Risk*, NCRP Annual Meeting Proceedings No. 18 (National Council on Radiation Protection and Measurements, Bethesda, Maryland).

KOSSENKO, M.M. (1996). "Cancer mortality among Techa River residents and their offspring," Health Phys. **71**, 77–82.

KOSSENKO, M.M. and DEGTEVA, M.O. (1994). "Cancer mortality and radiation risk evaluation for the Techa River population," Sci. Total Environ. **142**, 73–89.

KRAEMER, S.M. and WALDREN, C.A. (1997). "Chromosomal mutations and chromosome loss measured in a new human-hamster hybrid cell line, ALC: Studies with colcemid, ultraviolet irradiation, and ^{137}Cs gamma-rays," Mutat. Res. **379**, 151–166.

KRANERT, T., SCHNEIDER, E. and KIEFER, J. (1990). "Mutation induction in V79 Chinese hamster cells by very heavy ions," Int. J. Radiat. Biol. **58**, 975–987.

KRONENBERG, A. (1991). "Perspectives on fast-neutron mutagenesis of human lymphoblastoid cells," Radiat. Res. **128**, S87–S93.

KRONENBERG, A. (1994). "Radiation-induced genomic instability," Int. J. Radiat. Biol. **66**, 603–609.

KRONENBERG, A. and LITTLE, J.B. (1989a). "Locus specificity for mutation induction in human cells exposed to accelerated heavy ions," Int. J. Radiat. Biol. **55**, 913–924.

KRONENBERG, A. and LITTLE, J.B. (1989b). "Molecular characterization of thymidine kinase mutants of human cells induced by densely ionizing radiation," Mutat. Res. **211**, 215–224.

KRONENBERG, A., GAUNY, S., CRIDDLE, K., VANNAIS, D., UENO, A., KRAEMER, S. and WALDREN, C.A. (1995). "Heavy ion mutagenesis: Linear energy transfer effects and genetic linkage," Radiat. Environ. Biophy. **34**, 73–78.

KUPRYJANCZYK, J., THOR, A.D., BEAUCHAMP, R., POREMBA, C., SCULLY, R.E. and YANDELL, D.W. (1996). "Ovarian, peritoneal, and endometrial serous carcinoma: Clonal origin of multifocal disease," Mod. Pathol. **9**, 166–173.

LAMBERT, B., HOLMBERG, K., HACKMAN, P. and WENNBORG, A. (1998). "Radiation-induced chromosomal instability in human T-lymphocytes," Mutat. Res. **405**, 161–170.

LAND, C.E. (1980). "Estimating cancer risks from low doses of ionizing radiation," Science **209**, 1197–1203.

LAND C. (1992). "Risk of breast cancer in ataxia-telangiectasia (letter)," N. Engl. J. Med. **326**, 1359–1361.

LAND, C.E., TOKUNAGA, M., TOKUOKA, S. and NAKAMURA, N. (1993a). "Early-onset breast cancer in A-bomb survivors (letter)," Lancet **342**, 237.

LAND, C.E., SHIMOSATO, Y., SACCOMANNO, G., TOKUOKA, S., AUERBACH, O., TATEISHI, R., GREENBERG, S.D., NAMBU, S., CARTER, D., AKIBA, S., KEEHN, R., MADIGAN, P., MASON, T. and TOKUNAGA, M. (1993b). "Radiation-associated lung cancer: A comparison of the histology of lung cancers in uranium miners and survivors of the atomic bombings of Hiroshima and Nagasaki," Radiat. Res. **134**, 234–243.

LAND, C.E. HAYAKAWA, N., MACHADO, S.G., YAMADA, Y., PIKE, M.C., AKIBA, S. and TOKUNAGA, M. (1994a). "A case-control interview study of breast cancer among Japanese A-bomb survivors. I. Main effects," Cancer Causes Control **5**, 157–165.

LAND, C.E., HAYAKAWA, N., MACHADO, S.G., YAMADA, Y., PIKE, M.C., AKIBA, S. and TOKUNAGA, M. (1994b). "A case-control interview study of breast cancer among Japanese A-bomb survivors. II. Interactions with radiation dose," Cancer Causes Control **5**, 167–176.

LANGENDORFF, H. (1963). "Effects of small irradiation doses and low dose outputs," Strahlentherapie **52**, 188–196.

LANGLOIS, R.G., AKIYAMA, M., KUSUNOKI, Y., DUPONT, B.R., MOORE, D.H., II, BIGBEE, W.L., GRANT, S.G. and JENSEN, R.H. (1993). "Analysis of somatic cell mutations at the glycophorin A locus in atomic bomb survivors: A comparative study of assay methods," Radiat. Res. **136**, 111–117.

LE, X.C., XING, J.Z., LEE, J., LEADON, S.A. and WEINFELD, M. (1998). "Inducible repair of thymine glycol detected by an ultrasensitive assay for DNA damage," Science **280**, 1066–1069.

LEADON, S.A. and COOPER, P.K. (1993). "Preferential repair of ionizing radiation-induced damage in the transcribed strand of an active human gene is defective in Cockayne Syndrome," Proc. Natl. Acad. Sci. USA **90**, 10499–10503.

LEADON, S.A., DUNN, A.B. and ROSS, C.E. (1996). "A novel DNA repair response is induced in human cells exposed to ionizing radiation at the G_1/S border," Radiat. Res. **146**, 123–130.

LEE, W. and YOUMANS, H. (1970). *Doses to the Central Nervous System of Children Resulting from X-ray Therapy for Tinea Capitis*, FDA Publication No. BRH/DBE 70-4 (Center for Devices and Radiological Health, Food and Drug Administration, Rockville, Maryland).

LEE, W., SHLEIEN, B., TELLES, N.C. and CHIACCHIERINI, R.P. (1979). "An accurate method of ^{131}I dosimetry in the rat thyroid," Radiat. Res. **79**, 55–62.

LEE, W., CHIACCHIERINI, R.P., SHLEIEN, B. and TELLES, N.C. (1982). "Thyroid tumors following ^{131}I or localized x irradiation to the thyroid and pituitary glands in rats," Radiat. Res. **92**, 307–319.

LEHNERT, B.E. and GOODWIN, E.H. (1997). "Extracellular factor(s) following exposure to alpha particles can cause sister chromatid exchanges in normal human cells," Cancer Res. **57**, 2164–2171.

LEJEUNE, J., TURPIN, R., RETHORE, M.O. and MAYER, M. (1960). "Resultats d'une premiere enquete sur les effets somatiques de l'irradiation foeto-embryonnaire *in utero* (cas particulier des heterochromies iriennes)," Rev Franc Etudes Clin et Biol, **V**, 982–989.

LETOURNEAU, E.G., KREWSKI, D., CHOI, N.W., GODDARD, M.J., MCGREGOR, R.G., ZIELINSKI, J.M. and DU, J. (1994). "Case-control study of residential radon and lung cancer in Winnipeg, Manitoba, Canada," Am. J. Epidemiol. **140**, 310–322.

LEVY, R., WARNKE, R A., DORFMAN, R.F. and HAIMOVICH, J. (1977). "The monoclonality of human B-cell lymphomas," J. Exp. Med. **145**, 1014–1028.

LEWIS, T.L.T. (1960). "Leukaemia in childhood after antenatal exposure to x rays. A survey at Queen Charlotte's hospital," BMJ **2**, 1551–1552.

LIBER, H.L., LEMOTTE, P.K. and LITTLE, J.B. (1983). "Toxicity and mutagenicity of x-rays and 125-IdUrd or 3-H-TdR incorporated in the DNA of human lymphoblast cells," Mutat. Res. **111**, 387–404.

LIBER, H.L., OZAKI, V.H. and LITTLE, J.B. (1985). "Toxicity and mutagenicity of low dose rates of ionizing radiation from tritiated water in human lymphoblasts," Mutat. Res. **157**, 77–85.

LIE, R.T., IRGENS, L.M., SKJAERVEN, R., REITAN, J.B., STRAND, P. and STRAND, T. (1992). "Birth defects in Norway by levels of external and food-based exposure to radiation from Chernobyl," Am. J. Epidemiol. **136**, 377–388.

LIEBERMAN, M., HANSTEEN, G.A., WALLER, E.K., WEISSMAN, I.L. and SEN-MAJUMDAR, A. (1992). "Unexpected effects of the severe combined immunodeficiency mutation on murine lymphomagenesis," J. Exp. Med. **176**, 399–405.

LIMOLI, C.L., CORCORAN, J.J., MILLIGAN, J.R., WARD, J.F. and MORGAN, W.F. (1999). "Critical target and dose and dose-rate responses for the induction of chromosomal instability by ionizing radition," Radiat. Res. **151**, 677–685.

LINDBERG, S., KARLSSON, P., ARVIDSSON, B., HOLMBERG, E., LUNBERG, L.M. and WALLGREN, A. (1995). "Cancer incidence after radiotherapy for skin haemangioma during infancy," Acta Oncol. **34**, 735–740.

LINOS, A., GRAY, J.E., ORVIS, A.L., KYLE, R.A., O'FALLON, M. and KURLAND, L.T. (1980). "Low-dose radiation and leukemia," N. Engl. J. Med. **302**, 1101–1105.

LISCO, H. and CONARD, R.A. (1967). "Chromosome studies on Marshall Islanders exposed to fallout radiation," Science **157**, 445–447.

LITTLE, J.B. (1994). "Failla Memorial Lecture: Changing views of cellular radiosensitivity," Radiat. Res. **140**, 299–311.

LITTLE, J.B. (1998). "Radiation-induced genomic instability," Int. J. Radiat. Biol. **74**, 663–671.

LITTLE, M.P. and MUIRHEAD, C.R. (1996). "Evidence for curvilinearity in the cancer incidence dose-response in the Japanese atomic bomb survivors," Int. J. Radiat. Biol. **70**, 83–94.

LITTLE, M.P. and MUIRHEAD, C.R. (1998). "Curvature in the cancer mortality dose response of Japanese atomic bomb survivors: Absence of evidence of threshold," Int. J. Radiat. Biol. **74**, 471–480.

LITTLE, J.B., NAGASAWA, H., PFENNING, T. and VETROVS, H. (1997). "Radiation-induced genomic instability: Delayed mutagenic and cytogenetic effects of x rays and alpha particles," Radiat. Res. **148**, 299–307.

LLOYD, D.C., PURROTT, R.J. and REEDER, E.J. (1980). "The incidence of unstable chromosome aberrations in peripheral blood lymphocytes from unirradiated and occupationally exposed people," Mutat. Res. **72**, 523–532.

LLOYD, D.C., EDWARDS, A.A., LEONARD, A., DEKNUDT, G.L., VERSCHAEVE, L., NATARAJAN, A.T., DARROUDI, F., OBE, G., PALITTI, F., TANZARELLA, C. and TAWN, E.J. (1992). "Chromosomal aberrations in human lymphocytes induced *in vitro* by very low doses of x-rays," Int. J. Radiat. Biol. **61**, 335–343.

LLOYD, R.D., TAYLOR, G.N., ANGUS, W., BRUENGER, F.W. and MILLER, S.C. (1993). "Bone cancer occurrence among beagles given ^{239}Pu as young adults," Health Phys. **64**, 45–51.

LLOYD, R.D., TAYLOR, G.N., ANGUS, W., MILLER, S.C. and BOECKER, B.B. (1994). "Skeletal malignancies among beagles injected with ^{241}Am," Health Phys. **66**, 172–177.

LOEB, L.A. (1994). "Microsatellite instability: Marker of a mutator phenotype in cancer," Cancer Res. **54**, 5059–5063.

LORENZ, E., JACOBSON, L.O., HESTON, W.E., SHIMKIN, M., ESCHENBRENNER, A.B., DERINGER, M.K,, DONIGER, J. and SCHWEISTAL, R. (1954). "Effects of long-continued total body gamma irradiation on mice, guinea pigs, and rabbits. III, Effect on life span, weight, blood picture, and carcinogenesis, and the role of the intensity of of radiation," pages 24 to 48 in *Biological Effects of External X and Gamma Radiation*, Vol. 1, Zirkle, R.E., Ed. (McGraw-Hill, New York).

LORENZ, R., DEUBEL, W., LEUNER, K., GOLLNER, T., HOCHHAUSER, E. and HEMPEL, K. (1994). "Dose and dose-rate dependence of the frequency of HPRT deficient T lymphocytes in the spleen of the ^{137}Cs gamma-irradiated mouse," Int. J. Radiat. Biol. **66**, 319–326.

LORIMORE, S.A., KADHIM, M.A., POCOCK, D.A., PAPWORTH, D., STEVENS, D.L., GOODHEAD, D.T. and WRIGHT, E.G. (1998).

"Chromosomal instability in the descendants of unirradiated surviving cells after alpha-particle irradiation," Proc. Natl. Acad. Sci. USA **95**, 5730–5733.

LUBIN, J.H. (1994). "Invited commentary: Lung cancer and exposure to residential radon," Am. J. Epidemiol. **140**, 323–332.

LUBIN, J. (1998). "On the discrepancy between epidemiologic studies in individuals of lung cancer and residential radon and Cohen's ecologic regression," Health Phys. **75**, 4–10.

LUBIN, J.H. and BOICE, J.D., JR. (1997). "Lung cancer risk from residential radon: Meta-analysis of eight epidemiologic studies," J. Natl. Cancer Inst. **89**, 49–57.

LUBIN, J.H., BOICE, J.D., JR., EDLING, C., HORNUNG, R.W., HOWE, G.R., KUNZ, E., KUSIAK, R.A., MORRISON, H.I., RADFORD, E.P., SAMET, J.M., TIRMARCHE, M., WOODWARD, A., YAO, S.X. and PIERCE, D.A. (1994). *Lung Cancer and Radon: A Joint Analysis of 11 Studies*, National Institutes of Health Publication No. 94-3644 (Bethesda, Maryland).

LUBIN, J.H., BOICE, J.D., JR., EDLING, C., HORNUNG, R.W., HOWE, G.R., KUNZ, E., KUSIAK, R.A., MORRISON, H.I., RADFORD, E.P., SAMET, J.M., TIRMARCHE, M., WOODWARD, A., YAO, S.X. and PIERCE, D.A. (1995a). "Lung cancer in radon-exposed miners and estimation of risk from indoor exposure," J. Natl. Cancer Inst. **87**, 817–827.

LUBIN, J.H., BOICE, J.D., JR., EDLING, C., HORNUNG, R.W., HOWE, G., KUNZ, E., KUSIAK, R.A., MORRISON, H.I., RADFORD, E.P., SAMET, J.M., TIRMARCHE, M., WOODWARD, A. and YAO, S.X. (1995b). "Radon-exposed underground miners and inverse dose-rate (protraction enhancement) effects," Health Phys. **69**, 494–500.

LUBIN, J.H., BOICE, J.D., JR. and SAMET, J.M. (1995c). "Errors in exposure assessment, statistical power and the interpretation of residential radon studies," Radiat. Res. **144**, 329–341.

LUBIN, J.H., TOMASEK, L., EDLING, C., HORNUNG, R.W., HOWE, G., KUNZ, E., KUSIAK, R.A., MORRISON, H.I., RADFORD, E.P., SAMET, J.M., TIRMARCHE, M., WOODWARD, A. and YAO, S.X. (1997). "Estimating lung cancer mortality from residential radon using data for low exposures of miners," Radiat. Res. **147**, 126–134.

LUBIN, J.H., LINET, M.S., BOICE, J.D., JR., BUCKLEY, J., CONRATH, S.M., HATCH, E.E., KLEINERMAN, R.A., TARONE, R.E., WACHOLDER, S. and ROBISON, L.L. (1998). "Case-control study of childhood acute lymphoblastic leukemia and residential radon exposure," J. Natl. Cancer Inst. **90**, 294–300.

LUCAS, J.N., AWA, A., STRAUME, T., POGGENSEE, M., KODAMA, Y., NAKANO, M., OHTAKI, K., WEIER, H.U., PINKEL, D., GRAY, J. and LITTLEFIELD, L.G. (1992). "Rapid translocation frequency analysis in humans decades after exposure to ionizing radiation," Int. J. Radiat. Biol. **62**, 53–63.

LUCAS, J.N., CHEN, A.M. and SACHS, R.K. (1996). "Theoretical predictions on the equality of radiation-produced dicentrics and translocations detected by chromosome painting," Int. J. Radiat. Biol. **69**, 145–153.

LUCKEY, T.D. (1991). *Radiation Hormesis* (CRC Press, Boca Raton, Florida).

LUCKEY, T.D. (1994). "Radiation hormesis in cancer mortality," Int. J. Occup. Med. Toxicol. **3**, 175–191.

LUEBECK, E.G., CURTIS, S.B., CROSS, F.T. and MOOLGAVKAR, S.H. (1996). "Two-stage model of radon-induced malignant lung tumors in rats: Effects of cell killing," Radiat. Res. **145**, 163–173.

LUNDELL, M. and HOLM, L.E. (1996). "Mortality from leukemia after irradiation in infancy for skin hemangioma," Radiat. Res. **145**, 595–601.

LUNDELL, M., HAKULINEN, T., LINDELL, B. and HOLM, L.E. (1994). "Thyroid cancer after radiotherapy for skin hemangioma in infancy," Radiat. Res. **140**, 334–339.

LUNDELL, M., MATTSSON, A., HAKULINEN, T. and HOLM, L.E. (1996). "Breast cancer after radiotherapy for skin hemangioma in infancy," Radiat. Res. **145**, 225–230.

LUNING, K.G. (1960). "Studies of irradiated mouse populations. I. Plans and report of first generation," Hereditas **46**, 668–674.

LUNING, G., SCHEER, J., SCHMIDT, M. and ZIGGEL, H. (1989). "Early infant mortality in West Germany before and after Chernobyl," Lancet **2**, 1081–1083.

LUXIN, W. (1980). "Health survey in high background radiation areas in China," Science **209**, 877–880.

MACMAHON, B. (1962). "Prenatal x-ray exposure and childhood cancer," J. Natl. Cancer Inst. **28**, 1173–1191.

MAGNIN, P. (1962). "L'Avenir des enfants irradies *in utero*. Etude d'une enquete portant sur 5353 observations," Presse Med, **70**, 1199–1202.

MAISIN, J.R., WAMBERSIE, A., GERBER, G.B., MATTELIN, G., LAMBIET-COLLIER, M., DE COSTER, B. and GUEULETTE, J. (1988). "Life-shortening and disease incidence in C57B1 mice after single and fractionated gamma and high-energy neutron exposure," Radiat. Res. **113**, 300–317.

MAISIN, J.R., GERBER, G.B., VANKERKOM, J. and WAMBERSIE, A. (1996). "Survival and diseases in C57BL mice exposed to x rays or 3.1 Mev neutrons at an age of 7 or 21 days," Radiat. Res. **146**, 453–460.

MALKIN, D., JOLLY, K.W., BARBIER, N., LOOK, A.T., FRIEND, S.H., GEBHARDT, M.C., ANDERSEN, T.I., BORRESEN, A.L., LI, F.P., GARBER, J. and STRONG, L.C. (1992). "Germline mutations of the p53 tumor-suppressor gene in children and young adults with second malignant neoplasms," N. Engl. J. Med. **326**, 1309–1315.

MALUMBRES, M., DE CASTRO, I.P., SANTOS, J., MELENDEZ, B., MANGUES, R., SERRANO, M., PELLICER, A. and FERNANDEZ-PIQUERAS, J. (1997). "Inactivation of the cyclin-dependent Chinese inhibitor p15 by deletion and de novo methylation with independence of p16 alterations in murine primary T-cell lymphomas," Oncogene **14**, 1361–1370.

MALYAPA, R.S., BI, C., AHERN, E.W. and ROTI ROTI, J.L. (1998). "Detection of DNA damage by the alkaline comet assay after exposure to low-dose gamma radiation," Radiat. Res. **149**, 396–400.

MARSHALL, J.H. and GROER, P.G. (1977). "A theory of the induction of bone cancer by alpha radiation," Radiat. Res. **71**, 149–192.

MARTINS, M.B., SABATIER, L., RICOUL, M., PINTON, A. and DUTRILLAUX, B. (1993). "Specific chromosome instability induced by heavy ions: A step towards transformation of human fibroblasts?" Mutat. Res. **285**, 229–237.

MATANOSKI, G.M. (1991). *Health Effects of Low-Level Radiation in Shipyard Workers, Final Report*, DOE/EV/10095-T2 (National Technical Information Service, Springfield, Virginia).

MATANOSKI, G. (1993). "Nuclear shipyard workers study," Radiat. Res. **133**, 126–127.

MATANOSKI, G.M., SARTWELL, P., ELLIOTT, E., TONASCIA, J. and STERNBERG, A. (1984). "Cancer risks in radiologists and radiation workers," pages 83 to 96 in *Radiation Carcinogenesis: Epidemiology and Biological Significance*, Boice, J.D., Jr. and Fraumeni, J.F., Jr., Eds. (Raven Press, New York).

MATANOSKI, G.M., STERNBERG, A. and ELLIOTT, E.A. (1987). "Does radiation exposure produce a protective effect among radiologists?" Health Phys. **52**, 637–643.

MATTSSON, A., RUDEN, B.I., HALL, P., WILKING, N. and RUTQVIST, L.E. (1993). "Radiation-induced breast cancer: Long-term follow-up of radiation therapy for benign breast disease," J. Natl. Cancer Inst. **85**, 1679–1685.

MCMORROW, L.E., NEWCOMB, E.W. and PELLICER, A. (1988). "Identification of a specific marker chromosome early in tumor development in γ-irradiated C57BL/6J mice," Leukemia **2**, 115–119.

MCTIERNAN, A.M., WEISS, N.S. and DALING, J.R. (1984). "Incidence of thyroid cancer in women in relation to previous exposure to radiation therapy and history of thyroid disease," J. Natl. Cancer Inst. **73**, 575–581.

MEIJNE, E.I.M., SILVER, A.R.J., BOUFFLER, S.D., MORRIS, D.J., VAN KAMPEN, E.W., SPANJER, S., HUISKAMP, R. and COX, R. (1996). "Role of telomeric sequences in murine radiation-induced myeloid leukaemia," Genes Chromosomes Cancer **16**, 230–237.

MENDELSOHN, M.L. (1995). "A simple reductionist model for cancer risk in atom bomb survivors," pages 185 to 192 in *Modeling of Biological Effects and Risks of Radiation Exposure*, Inaba, J. and Kobayshi, S., Eds. (Japan National Institute of Radiological Sciences, Chiba, Japan).

MENDONCA, M.S., HOWARD, K., FASCHING, C.L., FARRINGTON, D.L., DESMOND, L.A., STANBRIDGE, E.J. and REDPATH, J.L. (1998). "Loss of suppressor loci on chromosomes 11 and 14 may be required for radiation-induced neoplastic transformation of HeLa x skin fibroblast human cell hybrids," Radiat. Res. **149**, 246–255.

MESSING, K. and BRADLEY, W.E.C. (1985). "*In vivo* mutant frequency rises among breast cancer patients after exposure to high doses of gamma-irradiation," Mutat. Res. **152**, 107–112.

MESSING, K., FERRARIS, J., BRADLEY, W.E.C., SWARTZ, J. and SEIFERT, A.M. (1989). "Mutant frequency of radiotherapy technicians appears to be associated with recent dose of ionizing radiation," Health Phys. **57**, 537–544.

METTING, N.F., PALAYOOR, S.T., MACKLIS, R.M., ATCHER, R.W., LIBER, H.L. and LITTLE, J.B. (1992). "Induction of mutations by bismuth-212 alpha particles at two genetic loci in human B-lymphoblasts," Radiat. Res. **132**, 339–345.

MICHAELIS, J., KALETSCH, U., BURKART, W. and GROSCHE, B. (1997). "Infant leukaemia after the Chernobyl accident (letter)," Nature **387**, 246.

MILL, A.J., FRANKENBERG, D., BETTEGA, D., HIEBER, L., SARAN, A., ALLEN, L.A., CALZOLARI, P., FRANKENBERG-SCHWAGER, M., LEHANE, M.M., MORGAN, G.R., PARISET, L., PAZZAGLIA, S., ROBERTS, C.J. and TALLONE, L. (1998). "Transformation of C3H 10T½ cells by low doses of ionising radiation: A collaborative study by six European laboratories strongly supporting a linear dose-response relationship," J. Radiol. Prot. **18**, 79–100.

MILLER, R.W. and BOICE, J.D., JR. (1997). "Cancer after intrauterine exposure to the atomic bomb," Radiat. Res. **147**, 396–397.

MILLER, R.C. and HALL, E.J. (1978). X-ray dose fractionation and oncogenic transformation in cultured mouse embryo cells," Nature **272**, 58–60.

MILLER, R.C., HALL, E.J. and ROSSI, H.H. (1979). "Oncogenic transformation of mammalian cells *in vitro* with split doses of x-rays," Proc. Natl. Acad. Sci. USA **76**, 5755–5758.

MILLER, R.C., OSMAK, R.O., ZIMMERMAN, M. and HALL, E.J. (1982). Sensitizers, protectors and oncogenic transformation *in vitro*," Int. J. Radiat. Oncol. Biol. Phys. **8**, 771–775.

MILLER, R.C., GEARD, C.R., BRENNER, D.J., KOMATSU, K., MARINO, S.A. and HALL, E.J. (1989a). "Neutron-energy-dependent oncogenic transformation of C3H 10T½ mouse cells," Radiat. Res. **117**, 114–127.

MILLER, A.B., HOWE, G.R., SHERMAN, G.J., LINDSAY, J.P., YAFFE, M.J., DINNER, P.J., RISCH, H.A. and PRESTON, D.L. (1989b). "Mortality from breast cancer after irradiation during fluoroscopic examinations in patients being treated for tuberculosis," N. Engl. J. Med. **321**, 1285–1289.

MILLER, R.C., GEARD, C.R., GEARD, M.J. and HALL, E.J. (1992). "Cell-cycle-dependent radiation-induced oncogenic transformation of C3H 10T½ cells," Radiat. Res. **130**, 129–133.

MILLER, R.C., RANDERS-PEHRSON, G., HIEBER, L., MARINO, S.A., RICHARDS, M. and HALL, E.J. (1993). "The inverse dose-rate effect for oncogenic transformation by charged particles is dependent on linear energy transfer," Radiat. Res. **133**, 360–364.

MILLER, R.C., MARINO, S.A., BRENNER, D.J., MARTIN, S.G., RICHARDS, M., RANDERS-PEHRSON, G. and HALL, E.J. (1995). "The biological effectiveness of radon-progeny alpha particles. II. Oncogenic transformation as a function of linear energy transfer," Radiat. Res. **142**, 54–60.

MILLER, R.C., RANDERS-PEHRSON, G., GEARD, C.R., HALL, E.J. and BRENNER, D.J. (1999). "The oncogenic transforming potential of the passage of single alpha particles through mammalian cell nuclei," Proc. Natl. Acad. Sci. USA **96**, 19–22.

MILLS, S.D., BRUWER, A.J., BANNER, E.A, BANNER, E.A., DAVIS, G.D. and GAGE, R.P. (1958). "Effects of irradiation of the fetus: A ten year follow-up of pelvimetry during pregnancy," Minnesota Med. **41**, 339–341.

MINDEN, M.D., TOYONAGA, B., HA, K., YANAGI, Y., CHIN, B., GELFANT, E. and MAK, T. (1985). "Somatic rearrangement of T-cell antigen receptor gene in human T-cell malignancies," Proc. Natl. Acad. Sci. USA **82**, 1224–1227.

MITELMAN, F. (1994). *Catalog of Chromosome Aberrations in Cancer*, 5th ed. (John Wiley & Sons, New York).

MIYAO, N., TSAI, Y.C., LERNER, S.P., OLUMI, A.F., SPRUCK, C.H., III, GONZALEZ-ZULUETA, M., NICHOLS, P.W., SKINNER, D.G. and JONES, P.A. (1993). "Role of chromosome 9 in human bladder cancer," Cancer Res. **53**, 4066–4070.

MOLE, R.H. (1974). "Antenatal irradiation and childhood cancer: Causation or coincidence?" Br. J. Cancer **30**, 199–208.

MOLE, R.H. (1990). "Childhood cancer after prenatal exposure to diagnostic x-ray examinations in Britain," Br. J. Cancer **62**, 152–168.

MOLE, R.H. and DAVIDS, J.A.G. (1982). "Induction of myeloid leukemia and other tumors in mice by irradiation with fission neutrons," pages 31 to 43 in *Neutron Carcinogenesis*, Broerse, J.J. and Gerber, G.H., Eds. (Commission of the European Communities, Luxembourg).

MOLE, R.H. and MAJOR, I.R. (1983). "Myeloid leukaemia frequency after protracted exposure to ionizing radiation: Experimental confirmation of the flat dose-response found in ankylosing spondylylitis after a single treatment course with x-rays," Leuk. Res. **7**, 295–300.

MOLE, R.H., PAPWORTH, D.G. and CORP, M.J. (1983). "The dose response for x-ray induction of myeloid leukaemia in male CBA/H mice," Br. J. Cancer **47**, 285–291.

MONSON, R.R. and MACMAHON, B. (1984). "Prenatal x-ray exposure and cancer in children," page 97 to 105 in *Radiation Carcinogenesis: Epidemiology and Biological Significance*, Boice, J.D., Jr. and Fraumeni, J.F., Jr., Eds. (Raven Press, New York).

MOORE, M.M., AMTOWER, A., STRAUSS, G.H.S. and DOERR, C. (1986). "Genotoxicity of γ-irradiation in L5178Y mouse lymphoma cells," Mutat. Res. **174**, 149–154.

MORGAN, W.F., DAY, J.P., KAPLAN, M.I., MCGHEE, E.M. and LIMOLI, C.L. (1996). "Genomic instability induced by ionizing radiation," Radiat. Res. **146**, 247–258.

MOTHERSILL, C. and SEYMOUR, C.B. (1997). "Medium from irradiated human epithelial cells but not human fibroblasts reduces the clonogenic survival of unirradiated cells," Int. J. Radiat. Biol. **71**, 421–427.

MOTHERSILL, C. and SEYMOUR, C.B. (1998). "Cell-cell contact during gamma irradiation is not required to induce a bystander effect in normal human keratinocytes: Evidence for release during irradiation of a signal controlling survival into the medium," Radiat. Res. **149**, 256–262.

MUCKERHEIDE, J. (1995). "The health effects of low-level radiation: Science, data, and corrective action," Nuclear News **38**, 26–34.

MUCKERHEIDE, J., BECKER, K., CIHAK, R., COHEN, B., JAWOROWSKI, Z., JOVANOVICH, J., KONDO, S., LIU, Z.S., LUCKEY, T.D., MUCKERHEIDE, W.A., PATTERSON, H.W., POLLYCOVE, M., ROCKWELL, T., SEILER, F., TSCHAECHE, A., WALINDER, G. and WEI, L. (1998). *Low Level Radiation Health Effects: Compiling the Data* (Radiation, Science, and Health, Inc., Needham, Massachusetts).

MUGGENBURG, B.A., HAHN, F.F., GRIFFITH, W.C., JR., LLOYD, R.D. and BOECKER, B.B. (1996). "The biological effects of radium-224 injected into dogs," Radiat. Res. **146**, 171–186.

MUIRHEAD, C.R., BUTLAND, B.K., GREEN, B.M. and DRAPER, G.J. (1991). "Childhood leukaemia and natural radiation (letter)," Lancet **335**, 1008–1012.

MULCAHY, R.T., GOULD, M.N. and CLIFTON, K.H. (1984). "Radiogenic initiation of thyroid cancer: A common cellular event," Int. J. Radiat. Biol. Relat. Stud. Phys. Chem. Med. **45**, 419–426.

MURRAY, R., HECKEL, P. and HEMPELMANN, L.H. (1959). "Leukemia in children exposed to ionizing radiation," N. Engl. J. Med. **261**, 585–589.

NAGASAWA, H. and LITTLE, J.B. (1992). "Induction of sister chromatid exchanges by extremely low doses of alpha-particles," Cancer Res. **52**, 6394–6396.

NAGASAWA, H. and LITTLE, J.B. (1999). "Unexpected sensitivity to the induction of mutations by very low doses of alpha-particle radiation: Evidence for a bystander effect," Radiat. Res. **152**, 552–557.

NAGASAWA, H., LI, C. and LITTLE, J.B. (1993). "Analysis of alpha-particle induced HPRT mutation by PCR amplification," Proceedings of the 1993 Annual Meeting of the Radiation Research Society.

NAJARIAN, T. and COLTON, T. (1978). "Mortality from leukaemia and cancer in shipyard nuclear workers," Lancet **1**, 1018–1020.

NAKAMURA, N. and OKADA, S. (1981). "Dose-rate effects of gamma-ray induced mutations in cultured mammalian cells," Mutat. Res. **83**, 127–135.

NAKAMURA, N., SUZUKI, S., ITO, A. and OKADA, S. (1982). "Mutations induced by gamma-rays and fast neutrons in cultured mammalian cells. Differences in dose response and RBE with methotrexate-and 6-thioguanine-resistant systems," Mutat. Res. **104**, 383–387.

NAMBA, M., NISHITANI, K., FUKUSHIMA, F., KIMOTO, T. and NOSE, K. (1986). "Multistep process of neoplastic transformation of normal human fibroblasts by ^{60}Co gamma rays and Harvey sarcoma viruses," Int. J. Cancer **37**, 419–423.

NAMBI, K.S. and SOMAN, S. (1987). "Environmental radiation and cancer in India," Health Phys. **52**, 653–657.

NAS/NRC (1988). National Academy of Sciences/National Research Council. *Health Risks of Radon and Other Internally Deposited Alpha-Emitters*, Committee on the Biological Effects of Ionizing Radiations, BEIR IV (National Academy Press, Washington).

NAS/NRC (1990). National Academy of Sciences/National Research Council. *Health Effects of Exposure to Low Levels of Ionizing Radiation*, Committee on the Biological Effects of Ionizing Radiations, BEIR V (National Academy Press, Washington).

NAS/NRC (1999). National Academy of Sciences/National Research Council. *Health Effects of Exposure to Radon*, Committee on the Biological Effects of Ionizing Radiations, BEIR VI (National Academy Press, Washington).

NCRP (1977). National Council on Radiation Protection and Measurements. *Medical Radiation Exposure of Pregnant and Potentially Pregnant Women,* NCRP Report No. 54 (National Council on Radiation Protection and Measurements, Bethesda, Maryland).

NCRP (1980). National Council on Radiation Protection and Measurements. *Influence of Dose and Its Distribution in Time on Dose-Response Relationships for Low-LET Radiations*, NCRP Report No. 64 (National Council on Radiation Protection and Measurements, Bethesda, Maryland).

NCRP (1985). National Council on Radiation Protection and Measurements. *Induction of Thyroid Cancer by Ionizing Radiation*, NCRP Report No. 80 (National Council on Radiation Protection and Measurements, Bethesda, Maryland).

NCRP (1987). National Council on Radiation Protection and Measurements. *Exposure of the Population in the United States and Canada from Natural Background Radiation*, NCRP Report No. 94 (National Council on Radiation Protection and Measurements, Bethesda, Maryland).

NCRP (1993a). National Council on Radiation Protection and Measurements. *Limitation of Exposure to Ionizing Radiation*, NCRP Report No. 116 (National Council on Radiation Protection and Measurements, Bethesda, Maryland).

NCRP (1993b). National Council on Radiation Protection and Measurements. *Risk Estimates for Radiation Protection*, NCRP Report No. 115 (National Council on Radiation Protection and Measurements, Bethesda, Maryland).

NCRP (1993c). National Council on Radiation Protection and Measurements. *Research Needs for Radiation Protection,* NCRP Report No. 117 (National Council on Radiation Protection and Measurements, Bethesda, Maryland).

NCRP (1995). National Council on Radiation Protection and Measurements. *Principles and Application of Collective Dose in Radiation Protection,* NCRP Report No. 121 (National Council on Radiation Protection and Measurements, Bethesda, Maryland).

NCRP (1997). National Council on Radiation Protection and Measurements. *Uncertainties in Fatal Cancer Risk Estimates Used in Radiation Protection*, NCRP Report 126 (National Council on Radiation Protection and Measurements, Bethesda, Maryland).

NCRP (1998). National Council on Radiation Protection and Measurements. *Operational Radiation Safety Program*, NCRP Report No. 127 (National Council on Radiation Protection and Measurements, Bethesda, Maryland).

NEEL, J.V., SATOH, C., GORIKI, K., ASAKAWA, J., FUJITA, M., TAKAHASHI, N., KAGEOKA, T. and HAZAMA, R. (1988). "Search for mutations altering protein charge and/or function in children of atomic bomb survivors: Final report," Am. J. Hum. Genet. **42**, 663–676.

NICHOLSON, D.W. and THORNBERRY, N.A. (1997). "Caspases: Killer proteases," Trends Biochem. Sci. **22,** 299–306.

NICKLAS, J.A., O'NEILL, J.P., HUNTER, T.C., FALTA, M.T., LIPPERT, M.J., JACOBSON-KRAM, D., WILLIAMS, J.R. and ALBERTINI, R.J. (1991). "*In vivo* ionizing irradiations produce deletions in the HPRT gene of human T-lymphocytes," Mutat. Res. **250**, 383–396.

NIKIFOROV, Y.E. and FAGIN, I.A. (1998). "Radiation-induced thyroid cancer in children after the Chernobyl accident," Thyroid Today **21**, 1–11.

NILSSON, A. and BROOME-KARLSSON, A. (1976). "Influence of steroid hormones on the carcinogenicity of ^{90}Sr," Acta Radiol. Ther. Phys. Biol. **15**, 417–426.

NISHIO, K. (1969). "Effects of ^{137}Cs and ^{90}Sr administered continuously upon mice. VII. Lifespan of mice from the 11th to 20th generation," Ann. Rept. Radiat. Cent. Osaka Prefectr. **10**, 90–91.

NOGUCHI, S., MOTOMURA, K., INAJI, H., IMAOKA, S. and KOYAMA, H. (1992). "Clonal analysis of human breast cancer by means of the polymerase chain reaction," Cancer Res. **52**, 6594–6597.

NOWELL, P.C. (1976). "The clonal evolution of tumor cell populations," Science **194**, 23–28.

NOWELL, P.C. and HUNGERFORD, D.A. (1960). "A minute chromosome in human chronic granuloeylic leukemia," Science **132**, 1497.

NRPB (1995). National Radiological Protection Board. *Risk of Radiation-Induced Cancer at Low Doses and Low Dose Rates for Radiation Protection Purposes*. Documents of the NRPB **6**, No 1 (National Radiological Protection Board, Chilton, United Kingdom).

OECD/NEA (1998). Organisation for Economic Co-operation and Development/Nuclear Energy Agency. *Developments in Radiation Health Science and Their Impact on Radiation Protection* (Organisation for Economic Co-operation and Development, Nuclear Energy Agency, Paris).

OHNISHI, T., WANG, X., TAKAHASHI, A., OHNISHI, K. and EJIMA, Y. (1999). "Low-dose-rate radiation attenuates the response of the tumor suppressor TP53," Radiat. Res. **151**, 368–372.

OHTAKI, K. (1992). "G-banding analysis of radiation-induced chromosome change in lymphocytes of Hiroshima A-bomb survivors," Jpn. J. Hum. Genet. **37**, 245–262.

OKUMOTO, M., NISHIKAWA, R., IWAI, M., IWAI, Y., TAKAMORI, Y., NIWA, O. and YOKORO, K. (1990). "Lack of evidence for the involvement of type-C and type-B retroviruses in radiation leukaemogenesis of NFS mice," Radiat. Res. **121**, 267–273.

OLEINICK, N.L., CHIU, S.M., FRIEDMAN, L.R., XUE, L.Y. and RAMAKRISHNAN, N. (1990). "DNA-protein crosslinks: Windows into chromatin damage and repair," pages 59 to 67 in *Ionizing Radiation Damage to DNA: Molecular Aspects*, Wallace, S.S. and Painter, R.B., Eds. (Wiley-Liss, New York).

OLIVIERI, G., BODYCOTE, J. and WOLFF, S. (1984). "Adaptive response of human lymphocytes to low concentrations of radioactive thymidine," Science **233**, 594–597.

OLIVIERI, G., BOSI, A., GRILLO, R. and SALONE, B. (1994). "Synergism and adaptive response in the interaction of low dose irradiation with subsequent mutagenic treatment in G2 phase human lymphocytes," pages 150 to 159 in *Chromosomal Aberrations: Origin and Significance,* Obe, G. and Natarajan, A.T., Eds. (Springer-Verlag, Berlin).

OOTSUYAMA, A. and TANOOKA, H. (1988). "One hundred percent tumor induction in mouse skin after repeated β irradiation in a limited dose range," Radiat. Res. **115**, 488–494.

OOTSUYAMA, A. and TANOOKA, H. (1993). "Zero tumor incidence in mice after repeated lifetime exposures to 0.5 Gy of beta radiation," Radiat. Res. **134**, 244–246.

OPPENHEIM, B.E., GRIEM, M.L. and MEIER, P. (1974). "Effects of low-dose prenatal irradiation in humans: Analysis of Chicago lying-in data and comparison with other studies," Radiat. Res. **57**, 508–544.

PAINTER, R.B. (1981). "Use of mutant yields to compare the carcinogenic risk of DNA-damaging agents," pages 345 to 355 in *Health Risk Analysis*, Richmond, C.R., Walsh, P.J. and Copenhaver, E.D., Eds. (Franklin Institute Press, Philadelphia).

PARETZKE, H.G. (1987). "Radiation track structure theory," pages 89 to 170 in *Kinetics of Nonhomogeneous Processes*, Freeman, G.R., Ed. (John Wiley & Sons, New York).

PARKER, S.L., TONG, T., BOLDEN, S. and WINGO, P.A. (1997). "Cancer statistics, 1997," CA Cancer J. Clin. **47**, 5–27.

PARKIN, D.M., CLAYTON, D., BLACK, R.J., MASUYER, E., FRIEDL, H.P., IVANOV, E., SINNAEVE, J., TZVETANSKY, C.G., GERYK, E., STORM, H.H., RAHU, M., PUKKALA, E., BERNARD, J.L., CARLI, P.M., L'HUILLIER, M.C., MENEGOZ, F., SCHAFFER, P., SCHRAUB, S., KAATSCH, P., MICHAELIS, J., APJOK, E., SCHULER, D., CROSIGNANI, P., MAGNANI, C., BENNETT, B.G., TERRACINI, B., STENGREVICS, A., KRIAUCIUNAS, R., COEBERGH, J., LANGMARK, F., ZATONSKI, W., TULBURE, R., BOUKHNY, A., MERABISHVILI, V., PLESKO, I., KRAMAROVA, E., POMPE-KIRN, V., BARLOW, L., ENDERLIN, F., LEVI, F., SCHULER, L.R.G., TORHORST, J., STILLER, C., SHARP, L. and BENNETT, B. (1996). "Childhood leukaemia in Europe after Chernobyl: 5 year follow-up," Br. J. Cancer **73**, 1006–1012.

PERSHAGEN, G., LIANG, Z.H., HRUBEC, Z., SVENSSON, C. and BOICE, J.D., JR. (1992). "Residential radon exposure and lung cancer in Swedish women," Health Phys. **63**, 179–186.

PERSHAGEN, G., AKERBLOM, G., AXELSON, O., CLAVENSJO, B., DAMBER, L., DESAI, G., ENFLO, A., LAGARDE, F., MELLANDER, H., SVARTENGREN, M. and SWEDJEMARK, G. (1994). "Residential radon exposure and lung cancer in Sweden," N. Engl. J. Med. **330**, 159–164.

PETERSEN, G.R., GILBERT, E.S., BUCHANAN, J.A. and STEVENS, R.G. (1990). "A case-cohort study of lung cancer, ionizing radiation, and tobacco smoking among males at the Hanford site," Health Phys. **58**, 3–11.

PETRIDOU, E., TRICHOPOULAS, D., DESSYPRIS, N., FLYTZANI, V., HAIDAS, S., KALMANTI, M., KOLIOUSKAS, D., KOSMIDIS, H., PIPEROPOULOU, F. and TZORTZATOU, F. (1996). "Infant leukaemia after *in utero* exposure to radiation from Chernobyl," Nature **382**, 352–353.

PETRINI, J.H.J., BRESSAN, D.A. and YAO, M.S. (1997). "The *RAD52* epistasis group in mammalian double strand break repair," Semin. Immunol. **9**, 181–188.

PIANTADOSI, S. (1994). "Invited commentary: Ecologic biases," Am. J. Epidemiol. **139**, 761–764.

PIERCE, D.A., PRESTON, D.L., STRAM, D.O. and VAETH, M. (1991). "Allowing for dose-estimation errors for the A-bomb survivor data," J. Radiat. Res. (Tokyo) **32** (suppl.), 108–121.

PIERCE, D.A., SHIMIZU, Y., PRESTON, D.L., VAETH, M. and MABUCHI, K. (1996). "Studies of the mortality of atomic bomb survivors. Report 12, Part 1. Cancer: 1950-1990," Radiat. Res. **146**, 1–27.

POHL-RULING, J. and FISCHER, P. (1983). "Chromosome aberrations in inhabitants of areas with elevated natural radioactivity," pages 527 to 560 in *Radiation-Induced Chromosome Damage in Man*, Ishihara, T. and Sasaki, M.S., Eds. (Alan R. Liss, New York).

POHL-RULING, J., FISCHER, P., HAAS, O., OBE, G., NATARAJAN, A.T., VAN BUUL, P.P.W., BUCKTON, K.E., BIANCHI, N.O., LARRAMENDY, M., KUCEROVA, M., POLIKOVA, Z., LEONARD, A., FABRY, L., PALITTI, F., SHARMA, T., BINDER, W., MUKHERJEE, R.N. and MUKHERJEE, U. (1983). "Effect of low-dose acute x-irradiation on the frequencies of chromosomal aberrations in human peripheral lymphocytes *in vitro*," Mutat. Res. **110**, 71–82.

POHL-RULING, J., FISCHER, P., LLOYD, D.C., EDWARDS, A.A., NATARAJAN, A.T., OBE, G., BUCKTON, K.E., BIANCHI, N.O., VAN BUUL, P.P.W., DAS, B.C., DASCHIL, F., FABRY, L., KUCEROVA, M., LEONARD, A., MUKHERHEE, R.N., MUKHERJEE, U., NOWOTNY, R., PALITTI, P., POLIVKOVA, Z., SHARMA, T. and SCHMIDT, W. (1986). "Chromosomal damage induced in human lymphocytes by low doses of D-T neutrons," Mutat. Res. **173**, 267–272.

POLHEMUS, D.W. and KOCH, R. (1959). "Leukemia and medical radiation," Pediatrics **23**, 453–461.

POLLYCOVE, M. (1995). "The issue of the decade: Hormesis," Eur. J. Nucl. Med. **22**, 399–401.

PONNAIYA, B., CORNFORTH, M.N. and ULLRICH, R.L. (1997a). "Radiation-induced chromosomal instability in BALB/c and C57BL/6 mice: The difference is as clear as black and white," Radiat. Res. **147**, 121–125.

PONNAIYA, B., CORNFORTH, M.N. and ULLRICH, R.L. (1997b). "Induction of chromosomal instability in human mammary cells by neutrons and gamma rays," Radiat. Res. **147**, 288–294.

POTTERN, L.M., KAPLAN, M.M., LARSEN, P.R., SILVA, J.E., KOENIG, R.J., LUBIN, J.H., STOVALL, M. and BOICE, J.D., JR. (1990). "Thyroid nodularity after childhood irradiation for lymphoid hyperplasia: A comparison of questionnaire and clinical findings," J. Clin. Epidemiol. **43**, 449–460.

PRENTICE, R.L., YOSHIMOTO, Y. and MASON, M.W. (1983). "Relationship of cigarette smoking and radiation exposure to cancer mortality in Hiroshima and Nagasaki," J. Natl. Cancer Inst. **70**, 611–622.

PRESTON, R.J. (1992). "A consideration of the mechanisms of induction of mutations in mammalian cells by low doses and dose rates of ionizing radiation," Adv. Radiat. Biol. **16**, 125–135.

PRESTON, R.J. (1997). "Telomeres, telomerase and chromosome stability," Radiat. Res. **147**, 529–534.

PRESTON, D.L., KUSUMI, S., TOMONAGA, M., IZUMI, S., RON, E., KURAMOTO, A., KAMADA, N., DOHY, H., MATSUO, T., NONAKA, H., THOMPSON, D., SODA, M. and MABUCHI, K. (1994). "Cancer incidence in atomic bomb survivors. Part III: Leukemia, lymphoma and multiple myeloma, 1950-1987," Radiat. Res. **137**, S68–S97.

PRESTON-MARTIN, S., BERNSTEIN, L., MALDONADO, A.A., HENDERSON, B.E. and WHITE, S.C. (1985a). "A dental x-ray validation study. Comparison of information from patient interviews and dental charts," Am. J. Epidemiol. **121**, 430–439.

PRESTON-MARTIN, S., HENDERSON, B.E. and BERNSTEIN, L. (1985b). "Medical and dental x rays as risk factors for recently diagnosed tumors of the head," Natl. Cancer Inst. Monogr. **69**, 175–179.

PRESTON-MARTIN, S., THOMAS, D.C., YU, M.C. and HENDERSON, B.E. (1989). "Diagnostic radiography as a risk factor for chronic myeloid and monocytic leukaemia (CML)," Br. J. Cancer **59**, 639–644.

PUSKIN, J.S. (1997). "Are low doses of radiation protective?" pages 211 to 213 in *Health Effects of Low-Dose Radiation: Challenges of the 21st Century* (British Nuclear Energy Society, London).

RAABE, O.G., PARKS, N.J. and BOOK, S.A. (1981). "Dose-response relationships for bone tumors in beagles exposed to ^{226}Ra and ^{90}Sr," Health Phys. **40**, 863–880.

RAABE, O.G., BOOK, S.A. and PARKS, N.J. (1983). "Lifetime bone cancer dose-response relationships in beagles and people from skeletal burdens of ^{226}Ra and ^{90}Sr," Health Phys. **44** (suppl.), 33–48.

RAABE, O.G., ROSENBLATT, L.S. and SCHLENKER, R.A. (1990). "Interspecies scaling of risk for radiation-induced bone cancer," Int. J. Radiat. Biol. **57**, 1047–1061.

RABBITTS, T.H. (1994). "Chromosomal translocations in human cancer," Nature **372**, 143–149.

RABINOWITCH, J. (1956). "X rays and leukemia," Lancet **2**, 1261–1262.

RANDERS-PHERSON, G., MARINO, S.A., JOHNSON, G. and GEARD, C.R. (1995). "Single particle irradiation facility," pages 35 to 39 in *Annuenter for Radiological Research*, Kliauga, P., Ed.

RAYNER, C.R., TOWERS, J.F. and WILSON, J.S. (1977). "What is Gorlin's syndrome? The diagnosis and management of the basal cell naevus syndrome, based on a study of thirty-seven patients," Br. J. Plastic Surg. **30**, 62–67.

REDPATH, J.L., SUN, C., COLMAN, M. and STANBRIDGE, E.J. (1987). "Neoplastic transformation of human hybrid cells by gamma radiation: A quantitative assay," Radiat. Res. **110**, 468–472.

REITMAIR, A.H., SCHMITS, R., EWEL, A., BAPAT, B., REDSTON, M., MITRI, A., WATERHOUSE, P., MITTRUCKER, H.W., WAKEHAM, A., LIU, B. *et al.* (1995). "MSH2 deficient mice are viable and susceptible to lymphoid tumours," Natl. Genet. **11**, 64–70.

REVELL, S.H. (1974). "The breakage-and-reunion theory and the exchange theory for chromosomal aberrations induced by ionizing radiations: A short history," Adv. Radiat. Biol. **4**, 367–416.

RHIM, J.S. and DRITSCHILO, A. (1991). "Neoplastic transformation in human cell system—an overview," pages xi to xxxi in *Neoplastic Transformation in Human Cell Culture: Mechanisms of Carcinogenesis*, Rhim, J.S. and Dritschilo, A. Eds. (Humana Press, Totowa, New Jersey).

RIGAUD, O., PAPADOPOULO, D. and MOUSTACCHI, E. (1993). "Decreased deletion mutation in radioadapted human lymphoblasts," Radiat. Res. **133**, 94–101.

RIGAUD, O., LAQUERBE, A. and MOUSTACCHI, E. (1995). "DNA sequence analysis of HPRT-mutants induced in human lymphoblastoid cells adapted to ionizing radiation," Radiat. Res. **144**, 181–189.

RINSKY, R.A., ZUMWALDE, R.D., WAXWEILER, R.J., MURRAY, W.E., JR., BIERBAUM, P.J., LANDRIGAN, P.J., TERPILAK, M. and COX, C. (1981). "Cancer mortality at a Naval nuclear shipyard," Lancet **1**, 231–235.

RINSKY, R.A., MELIUS, J.M., HORNUNG, R.W., ZUMWALDE, R.D., WAXWEILER, R.J., LANDRIGAN, P.J., BIERBAUM, P.J. and MURRAY, W.E., JR. (1988). "Case-control study of lung cancer in civilian employees at the Portsmouth Naval Shipyard, Kittery, Maine," Am. J. Epidemiol. **127**, 55–64.

ROBBINS, J. and ADAMS, W. (1989). "Radiation effects in the Marshall Islands," pages 11 to 24 in *Radiation and the Thyroid*, Nagataki, S., Ed. (Excerpta Medica, Amsterdam).

RODVALL, Y., PERSHAGEN, G., HRUBEC, Z., AHLBOM, A., PEDERSEN, N.L. and BOICE, J.D., JR. (1990). "Prenatal x-ray exposure and childhood cancer in Swedish twins," Int. J. Cancer **46**, 362–365.

RODVALL, Y., HRUBEC, Z., PERSHAGEN, G., AHLBOM, A., BJURMAN, A. and BOICE, J.D., JR. (1992). "Childhood cancer among Swedish twins," Cancer Causes Control **3**, 527–532.

RON, E., KLEINERMAN, R.R.., BOICE, J.D., JR., LIVOLSI, V.A., FLANNERY, J.T. and FRAUMENI, J.F., JR. (1987). "A population-based case-control study of thyroid cancer," J. Natl. Cancer Inst. **79**, 1–12.

RON, E., MODAN, B., PRESTON, D., ALFANDARY, E., STOVALL, M. and BOICE, J.D., JR. (1989). "Thyroid neoplasia following low-dose radiation in childhood," Radiat. Res. **120**, 516–531.

RON, E., LUBIN, J.H., SHORE, R.E., MABUCHI, K., MODAN, B., POTTERN, L.M., SCHNEIDER, A.B., TUCKER, M.A. and BOICE, J.D., JR. (1995). "Thyroid cancer after exposure to external radiation: A pooled analysis of seven studies," Radiat. Res. **141**, 259–277.

RON, E., DOODY, M.M., BECKER, D.V., BRILL, A.B., CURTIS, R.E., GOLDMAN, M.B., HARRIS, B.S.H., III, HOFFMAN, D.A., MCCONAHEY, W.M., MAXON, H.R., PRESTON-MARTIN, S., WARSHAUER,

M.E., WONG, F.L. and BOICE, J.D., JR. (1998). "Cancer mortality following treatment for hyperthyroidism," JAMA **280**, 347–355.

ROOTS, R. and OKADA, S. (1975). "Estimation of life times and diffusion distances of radicals involved in x-ray-induced DNA strand breaks or killing of mammalian cells," Radiat. Res. **64**, 306–320.

ROSS, J.A., DAVIS, S.M., POTTER, J.D. and ROBISON, L.L. (1994). "Epidemiology of childhood leukemia, with a focus on infants," Epidemiol. Rev. **16**, 243–272.

ROSSI, H.H. (1967). "Energy distribution in the absorption of radiation," Adv. Biol. Med. Phys. **11**, 27–85.

ROSSI, H.H. (1980). "Microdosimetry and its application to biology," pages 75 to 82 in *Advances in Radiation Protection and Dosimetry in Medicine*, Thomas, R.H. and Perez-Mendez, V., Eds. (Plenum Press, New York).

ROSSI, H.H. and KELLERER, A.M. (1986). "The dose rate dependence of oncogenic transformation by neutrons may be due to variation of response during the cell cycle," Int. J. Radiat. Biol. Relat. Stud. Phys. Chem. Med. **50**, 353–361.

ROTHMAN, K.J. (1990). "A sobering start for the cluster busters' conference," Am. J. Epidemiol. **132**, S6–S13.

ROWLAND, R.E. (1997). "Bone sarcoma in humans induced by radium: A threshold response?" Radioprotection **32**, C1-331–C1-338.

RUOSTEENOJA, E. (1991). *Indoor Radon and Risk of Lung Cancer: An Epidemiological Study in Finland*, STUK-A99 (Finnish Centre for Radiation and Nuclear Safety, Helsinki).

RUSSELL, W.L. and KELLY, E.M. (1982a). "Specific-locus mutation frequencies in mouse stem-cell spermatogonia at very low radiation dose rates," Proc. Natl. Acad. Sci. USA **79**, 539–541.

RUSSELL, W.L. and KELLY, E.M. (1982b). "Mutation frequencies in male mice and the estimation of genetic hazards of radiation in men," Proc. Natl. Acad. Sci. USA **79**, 542–544.

RUSSO, J. and RUSSO, I.H. (1987). "Biology of disease: Biological and molecular bases of mammary carcinogenesis," Lab. Invest. **57**, 112–137.

RUSSO, J., TAY, L.K. and RUSSO, I.H. (1982). "Differentiation of the mammary gland and susceptibility to carcinogenesis," Br. Cancer Res. Treat. **2**, 5–73.

SACHER, G.A. (1966). "The Gompertz transformation in the study of the injury-mortality relationship: Application to late radiation effects and aging," pages 411 to 441 in *Radiation and Aging*, Lindop, P.J. and Sacher, G.A., Eds. (Taylor and Francis, New York).

SACHER, G.A. (1973). "Dose dependence for life-shortening by x-rays, gamma-rays, and fast neutrons," pages 1425 to 1432 in *Advances in Radiation Research, Vol. 3*, Duplan, J.F. and Chapiro, A., Eds. (Gordon and Breach, New York).

SADO, T. (1992). "Experimental radiation carcinogenesis studies at NIRS," pages 36 to 42 in *Proceedings of the International Conference on Radiation Effects and Protection* (Mito, Ibaraki, Japan).

SADO, T., KAMISAKU, H. and KUBO, E. (1991). "Bone marrow-thymus interactions during thymic lymphomagenesis induced by fractionated radiation exposure in B10 mice: Analysis using bone marrow transplantation between Thy 1 congenic mice," J. Radiat. Res. (Tokyo) **32** (suppl.), 168–180.

SAENGER, E.L., THOMAS, G.E. and TOMPKINS, E.A. (1968). "Incidence of leukemia following treatment of hyperthyroidism," JAMA **205**, 855–862.

SAFA, A.M., SCHUMACHER, O.P. and RODRIGUEZ-ANTUNEZ, A. (1975). "Long-term follow-up results in children and adolescents treated with radioactive iodine (I-131) for hyperthyroidism," N. Engl. J. Med. **292**, 167–171.

SALA-TREPAT, M., COLE, J., GREEN, M.H.L., RIGAUD, O., VILCOQ, J.R. and MOUSTACCHI, E. (1990). "Genotoxic effects of radiotherapy and chemotherapy on the circulating lymphocytes of breast cancer patients. III. Measurement of mutant frequency to 6-thioguanine resistance," Mutagenesis **5**, 593–598.

SALONEN, T. (1976). "Prenatal and perinatal factors in childhood cancer," Ann. Clin. Res. **8**, 27–42.

SANDBERG, A.A. (1993). "Chromosome changes in leukemia and cancer and their molecular lining," pages 141 to 163 in *The Causes and Consequences of Chromosomal Aberrations*, Kirsch, I.R., Ed. (CRC Press, Boca Raton, Florida).

SANDERS, C.L. and LUNDGREN, D.L. (1995). "Pulmonary carcinogenesis in the F344 and Wister rat after inhalation of plutonium dioxide," Radiat. Res. **144**, 206–214.

SANDERSON, B.J.S. and MORLEY, A.A. (1986). "Exposure of human lymphocytes to ionizing radiation reduces mutagenesis induced by subsequent ionizing radiation," Mutat. Res. **164**, 347–351.

SANDERSON, B.J.S, DEMPSEY, J.L. and MORLEY, A.A. (1984). "Mutations in human lymphocytes: Effect of x- and UV-irradiation," Mutat. Res. **140**, 223–227.

SANKARANARAYANAN, K. and CHAKRABORTY, R. (1995). "Cancer predisposition, radiosensitivity and the risk of radiation-induced cancers. I. Background," Radiat. Res. **143**, 121–143.

SANKARANARAYANAN, K., VON DUYN, A., LOOS, M.J. and NATARAJAN, A.T. (1989). "Adaptive response of human lymphocytes to low-level radiation from radiosotopes or x-rays," Mutat. Res. **211**, 7–12.

SASAKI, M.S. (1995). "On the reaction kinetics of the radioadaptive response in cultured mouse cells," Int. J. Radiat. Biol. **68**, 281–291.

SASAKI, S. and KASUGA, T. (1981). "Life-shortening and carcinogenesis in mice irradiated neonatally with x rays," Radiat. Res. **88**, 313–325.

SAVAGE, J.R.K. (1977). "Assignment of aberration breakpoints in banded chromosomes," Nature **270**, 513–514.

SAVAGE, J.R.K. (1979). "Annotation: Classification and relationship of induced chromosomal structural changes," J. Med. Genet. **12**, 103–122.

SAVAGE, J.R.K. (1996). "Insight into sites," Mutat. Res. **366**, 81–95.

SAWEY, M.J., HOOD, A.T., BURNS, F.J. and GARTE, S.J. (1987). "Activation of c-myc and c-K-ras oncogenes in primary rat tumors induced by ionizing radiation," Mol. Cell Biol. **7**, 932–935.

SAX, K. (1938). "Induction by x rays of chromosome aberrations in Tradescantia microspores," Genetics **23**, 494–516.

SCHIESTL, R.H., KHOGALI, F. and CARLS, N. (1994). "Reversion of the mouse pink-eyed unstable mutation induced by low doses of x-rays," Science **266**, 1573–1576.

SCHMID, E. and BAUCHINGER, M. (1975). "Chromosome aberrations in human lymphocytes after irradiation with 15.0-MeV neutrons *in vitro*. II. Analysis of the number of absorption events and the interaction distance in the formation of dicentric chromosomes," Mutat. Res. **27**, 111–117.

SCHNEIDER, A.B., RON, E., LUBIN, J., STOVALL, M. and GIERLOWSKI, T.C. (1993). "Dose-response relationships for radiation-induced thyroid cancer and thyroid nodules: Evidence for the prolonged effects of radiation on the thyroid," J. Clin. Endocrinol. Metab. **77**, 362–369.

SCHOENBERG, J.B., KLOTZ, J.B., WILCOX, H.B., NICHOLLS, G.P., GIL-DEL-REAL, M.T., STEMHAGEN, A. and MASON, T.J. (1990). "Case-control study of residential radon and lung cancer among New Jersey women," Cancer Res. **50**, 6520–6524.

SCHULTE-FROHLINDE, D. and VON SONNTAG, C. (1990). "Sugar lesions in radiobiology," pages 31 to 42 in *Ionizing Radiation Damage to DNA*, Wallace, S.S. and Painter, R.B., Eds. (Wiley-Liss, New York).

SCHULZ, R.J. and ALBERT, R.E. (1968). "Follow-up study of patients treated by x-ray epilation for tinea capitis. 3. Dose to organs of the head from the x-ray treatment of tinea capitis," Arch. Environ. Health **17**, 935–950.

SCHWARTZ, J.L., JORDAN, R., SEDITA, B.A., SWENNINGSON, M.J., BANATH, J.P. and OLIVE, P.L. (1995). "Different sensitivity to cell killing and chromosome mutation induction by gamma rays in two human lymphoblastoid cell lines derived from a single donor: Possible role of apoptosis," Mutagenesis **10**, 227–233.

SCHWEISGUTH, O., GERARD-MARCHANT, R. and LEMERLE, J. (1968). "Basal cell envus syndrome. Association with congenital rhabdomyosarcoma," Arch. Fr. Pediatr. **25**, 1083–1093.

SEARLE, A.G. and BEECHEY, C.V. (1974). "Sperm-count, egg-fertilization and dominant lethality after x-irradiation of mice," Mutat. Res. **22**, 63–72.

SEIFERT, A.M., BRADLEY, W.E.C. and MESSING, K. (1987). "Exposure of nuclear medicine patients to ionizing radiation is associated with rises in HPRT-mutant frequency in peripheral T-lymphocytes," Mutat. Res. **191**, 57–63.

SEIFERT, A.M., DEMERS, C., DUBEAU, H. and MESSING, K. (1993). "HPRT-mutant frequency and lymphocyte characteristics of workers exposed to ionizing radiation on a sporadic basis: A comparison of two exposure indicators, job title and dose," Mutat. Res. **319**, 61–70.

SELVANAYAGAM, C.S., DAVIS, C.M., CORNFORTH, M.N. and ULLRICH, R.L. (1995). "Latent expression of *p53* mutations and radiation-induced mammary cancer," Cancer Res. **55**, 3310–3317.

SEMOV, A.B., IOFA, E.L., AKAEVA, E.A. and SHEVCHENKO, V.A. (1994). "The dose dependence of the induction of chromosome aberrations in those who worked in the cleanup of the Chernobyl accident," Radiat. Biol. Radioecol. **34**, 865–871.

SHADLEY, J.D. and WIENCKE, J.K. (1989). "Induction of the adaptive response by x-rays is dependent on radiation intensity," Int. J. Radiat. Biol. **56**, 107–118.

SHELINE, G.E., LINDSAY, S., MCCORMACK, K.R. and GALANTE, M. (1962). "Thyroid nodules occurring late after treatment of thyrotoxicosis with radioiodine," J. Clin. Endocr. **22**, 8–18.

SHELLABARGER, C.J. and BROWN, R.D. (1972). "Rat mammary neoplasia following ^{60}Co irradiation at 0.03 R or 10 R per minute," Radiat. Res. **51**, 493–494.

SHELLABARGER, C.J., BOND, V.P., APONTE, G.E. and CRONKITE, E.P. (1966). "Results of fractionation and protraction of total-body radiation on rat mammary neoplasia," Cancer Res. **26**, 509–513.

SHELLABARGER, C.J., STONE, J.P. and HOLTZMAN, S. (1978). "Rat differences in mammary tumor induction with estrogen and neutron radiation," J. Natl. Cancer Inst. **61**, 1505–1508.

SHELLABARGER, C.J., CHMELEVSKY, D. and KELLERER, A.M. (1980). "Induction of mammary neoplasms in the Sprague-Dawley rat by 430 keV neutrons and x-rays," J. Natl. Cancer Inst. **64**, 821–833.

SHELLABARGER, C.J., CHMELEVSKY, D., KELLERER, A.M., STONE, J.P. and HOLTZMAN, S. (1982). "Induction of mammary neoplasms in the ACI rat by 430-keV neutrons, x-rays and diethylstilbestrol," J. Natl. Cancer Inst. **69**, 1135–1146.

SHIMIZU, Y., KATO, H. and SCHULL, W.J. (1990). "Studies of the mortality of A-bomb survivors. 9. Mortality, 1950-1985: Part 2. Cancer mortality based on the recently revised doses (DS86)," Radiat. Res. **121**, 120–141.

SHORE, R.E. (1992). "Issues and epidemiological evidence regarding radiation-induced thyroid cancer," Radiat. Res. **131**, 98–111.

SHORE, R.E. (1996). "Epidemiological issues related to dose reconstruction," pages 245 to 260 in *Environmental Dose Reconstruction and Risk Implications*, NCRP Annual Meeting Proceedings No. 17 (National Council on Radiation Protection and Measurements, Bethesda, Maryland).

SHORE, R.E., WOODARD, E.D., HEMPELMANN, L.H. and PASTERNACK, B.S. (1980). "Synergism between radiation and other risk factors for breast cancer," Prev. Med. **9**, 815–822.

SHORE, R.E., HILDRETH, N., DVORETSKY, P., ANDRESEN, E., MOSESON, M. and PASTERNACK, B. (1993). "Thyroid cancer among persons given x-ray treatment in infancy for an enlarged thymus gland," Am. J. Epidemiol. **137**, 1068–1080.

SHOWE, L.C. and CROCE, C.M. (1987). "The role of chromosomal translocations in B- and T-cell neoplasia," Annu. Rev. Immunol. **5**, 253–277.

SHU, X.O., ROSS, J.A., PENDERGRASS, T.W., REAMAN, G.H., LAMPKIN, B. and ROBISON, L.L. (1996). "Parental alcohol consumption, cigarette smoking, and risk of infant leukemia: A childrens cancer group study," J. Natl. Cancer Inst. **88**, 24–31.

SIGG, M., CROMPTON, N.E.A. and BURKART, W. (1997). "Enhanced neoplastic transformation in an inhomogeneous radiation field: An effect of the presence of heavily damaged cells," Radiat. Res. **148**, 543–547.

SIMPSON, P.J. and SAVAGE, J.R.K. (1996). "Dose-response curves for simple and complex chromosome aberrations induced by x-rays and detected using fluorescence *in situ* hybridization," Int. J. Radiat. Biol. **69**, 429–436.

SINCLAIR, W.K. (1998). "The linear no-threshold response: Why not linearity," Med. Phys. **25**, 285–290.

SINGH, N.P., GRAHAM, M.M., SINGH, V. and KHAN, A. (1995). "Induction of DNA single-strand breaks in human lymphocytes after low doses of gamma-rays," Int. J. Radiat. Biol. **68**, 563–569.

SMITH, P.G. and DOLL, R. (1981). "Mortality from cancer and all causes among British radiologists," Br. J. Radiol. **54**, 187–194.

SMITH, M.A., CHEN, T. and SIMON, R. (1997). "Age-specific incidence of acute lymphoblastic leukemia in U.S. children: *In utero* initiation model," J. Natl. Cancer Inst. **89**, 1542–1544.

SMITH, B.J., FIELD, R.W. and LYNCH, C. (1998). "Residential ^{222}Rn exposure and lung cancer: Testing the linear no-threshold theory with ecologic data," Health Phys. **75**, 11–17.

SPALDING, J.R., THOMAS, R.G. and TIETSEN, G.L. (1982). *Life Span of C57 Mice as Influenced by Radition Dose, Dose Rate and Age at Exposure*, Los Alamos National Laboratory (National Technical Information Service, Springfield, Virginia).

SPARROW, A.H., UNDERBRINK, A.G. and ROSSI, H.H. (1972). "Mutations induced in trandescantia by small doses of x-rays and neutrons: Analysis of dose-response curves," Science **176**, 916–918.

SRDOC, D. and MARINO, S.A. (1996). "Microdosimetry of monoenergetic neutrons," Radiat. Res. **146**, 466–474.

STANNARD, J.N. (1988). *Radioactivity and Health: A History*, U.S. Department of Energy RL/01830-T59 (National Technical Information Service, Springfield, Virginia).

STARR, P., JAFFE, H.L. and OETTINGER, L., JR. (1969). "Later results of ^{131}I treatment of hyperthyroidism in 73 children and adolescents: 1967 followup," J. Nuc. Med. **10**, 586–590.

STEBBINGS, J.H., LUCAS, H.F. and STEHNEY, A.F. (1983). "Multiple myeloma, leukemia, and breast cancer among the U.S. radium dial workers," pages 298 to 307 in *Epidemiology Applied to Health Pysics* (Health Physics Society, Albuquerque, New Mexico).

STEBBINGS, J.H., LUCAS, H.F. and STEHNEY, A.F. (1984). "Mortality from cancers of major sites in female radium dial workers," Am. J. Ind. Med. **5**, 435–459.

STERN, F.B., WAXWEILER, R.A., BEAUMONT, J.J., LEE, S.T., RINSKY, R.A., ZUMWALDE, R.D., HALPERIN, W.E., BIERBAUM, P.J., LANDRIGAN, P.J. and MURRAY, W.E., JR. (1986). "A case-control study of leukemia at a Naval nuclear shipyard," Am. J. Epidemiol. **123**, 980–992.

STEVENS, W., THOMAS, D.C., LYON, J.L., TILL, J.E., KERBER, R.A., SIMON, S.L., LLOYD, R.D., ELGHANY, N.A. and PRESTON-MARTIN, S. (1990). "Leukemia in Utah and radioactive fallout from the Nevada test site. A case-control study," JAMA **264**, 585–591.

STEWART, A. (1961). "Aetiology of childhood malignancies," BMJ **1**, 452–460.

STEWART, A. and KNEALE, G.W. (1970). "Radiation dose effects in relation to obstetric x-rays and childhood cancer," Lancet **1,** 1185–1188.

STEWART, A., WEBB, J., GILES, D. and HEWITT, D. (1956). "Malignant disease in childhood and diagnostic irradiation *in utero*," Lancet **2**, 447–448.

STEWART, A., WEBB, J. and HEWITT, D. (1958). "A survey of childhood malignancies," BMJ, 1495–1508.

STEWART, A., PENNYBACKER, W. and BARBER, R. (1962). "Adult leukaemias and diagnostic x rays," BMJ, 882–890.

STIDLEY, C.A. and SAMET, J.M. (1994). "Assessment of ecologic regression in the study of lung cancer and indoor radon," Am. J. Epidemiol. **139**, 312–322.

STOLL, U., SCHMIDT, A., SCHNEIDER, E. and KIEFER, J. (1995). "Killing and mutation of Chinese hamster V79 cells exposed to accelerated oxygen and neon ions," Radiat. Res. **142**, 288–294.

STONE, J.P., HOLTZMAN, S. and SHELLABARGER, C.J. (1979). Neoplastic responses and correlated plasma prolactin levels in diethylstilbestrol-treated ACI and Sprague-Dawley rats," Cancer Res. **39**, 773–778.

STORER, J.B., SERRANO, L.J., DARDEN, E.B., JR., JERNIGAN, M.C. and ULLRICH, R.L. (1979). "Life shortening in RFM and BALB/c mice as a function of radiation quality, dose, and dose rate," Radiat. Res. **78**, 122–161.

STORER, R.D., KRAYNAK, A.R., MCKELVEY, T.W., ELIA, M.C., GOODROW, T.L. and DELUCA, J.G. (1997). "The mouse lymphoma L5178Y Tk+/- cell line is heterozygous for a codon 170 mutation in the p53 tumor suppressor gene," Mutat. Res. **373**, 157–165.

STRAM, D.O., SPOSTO, R., PRESTON, D., ABRAHAMSON, S., HONDA, T. and AWA, A.A. (1993). "Stable chromosome aberrations among A-bomb survivors: An update," Radiat. Res. **136**, 29–36.

STRAUSS, P.G., MITREITER, K., ZITZELSBERGER, H., LUZ, A., SCHMIDT, J., ERFLE, V. and HOFLER, H. (1992). "Elevated p53 RNA expression correlates with incomplete osteogenic differentiation of radiation-induced murine osteosarcomas," Int. J. Cancer **50**, 252–258.

STRONG, L.C. (1977a). "Theories of pathogenesis: Mutation and cancer," pages 401 to 415 in *Genetics of Human Cancer,* Mulvihill, J.J., Miller, R.W. and Fraumeni, J.F., Jr., Eds. (Raven Press, New York).

STRONG, L.C. (1977b). "Genetic and environmental interactions," Cancer **40** (suppl.), 1861–1866.

STRONG, L.C. and WILLIAMS, W.R. (1987). "The genetic implications of long-term survival of childhood cancer. A conceptual framework," Am. J. Pediatr. Hematol. Oncol. **9**, 99–103.

SUGAHARA, T., SAGAN, L.A. and AOYAMA, T., Eds. (1992). *Low Dose Irradiation and Biological Defense Mechanisms* (Excerta Medica, New York).

SWIFT, M., MORRELL, D., MASSEY, R.B. and CHASE, C.L. (1991). "Incidence of cancer in 161 families affected by ataxia-telangiectasia," N. Engl. J. Med. **325**, 1831–1836.

TABOCCHINI, M.A., LITTLE, J.B. and LIBER, H.L. (1989). "Mutation induction in human lymphoblasts after protracted exposure to low doses of tritiated water," pages 439 to 445 in *Low Dose Radiation: Biological Bases of Risk Assessment*. Baverstock, K.F. and Stather, J.W., Eds. (Taylor and Francis, New York).

TERZAGHI, M. and LITTLE, J.B. (1976). "X-radiation-induced transformation in a C3H mouse embryo-derived cell line," Cancer Res. **36**, 1367–1374.

TERZAGHI, M., NETTESHEIM, P. and WILLIAMS, M.L. (1978). "Repopulation of denuded tracheal grafts with normal, preneoplastic and neoplastic epithelial cell populations," Cancer Res. **38**, 4546–4553.

TERZAGHI-HOWE, M. and FORD, J. (1994). "Effects of radiation on rat respiratory epithelial cells: Critical target cell populations and the importance of cell-cell interactions," Adv. Space Res. **14**, 565–572.

THACKER, J., STRETCH, A. and STEPHENS, M.A. (1979). "Mutation and inactivation of cultured mammalian cells exposed to beams of accelerated heavy ions. II. Chinese hamster V79 cells," Int. J. Radiat. Biol. Relat. Stud. Phys. Chem. Med. **36**, 137–148.

THOMPSON, D.E., MABUCHI, K., RON, E., SODA, M., TOKUNAGA, M., OCHIKUBO, S., SUGIMOTO, S., IKEDA, T., TERASAKI, M., IZUMI, S. and PRESTON, D.L. (1994). "Cancer incidence in atomic bomb survivors. Part II: Solid tumors, 1958-1987," Radiat. Res. **137**, S17–S67.

THOMSON, J.F., WILLIAMSON, F.S., GRAHN, D. and AINSWORTH, E.J. (1981a). "Life shortening in mice exposed to fission neutrons and γ rays. I. Single and short-term fractionated exposures," Radiat. Res. **86**, 559–572.

THOMSON, J.F., WILLIAMSON, F.S., GRAHN, D. and AINSWORTH, E.J. (1981b). "Life shortening in mice exposed to fission neutrons and γ rays. II. Duration-of-life and long-term fractionated exposures," Radiat. Res. **86**, 573–579.

THOMSON, J.F., LOMBARD, L.S., GRAHN, D., WILLIAMSON, F.S. and FRITZ, T.F. (1982). "RBE of fission neutrons for life shortening and tumorigenesis," pages 75 to 94 in *Neutron Carcinogenesis*, Broerse, J.J. and Gerber, G.B., Eds. (Commission of European Communities, Luxembourg).

THRAVES, P., SALEHI, Z., DRITSCHILO, A. and RHIM, J.S. (1990). "Neoplastic transformation of immortalized human epidermal keratinocytes by ionizing radiation," Proc. Natl. Acad. Sci. USA **87**, 1174–1177.

TOKARSKAYA, Z.B., OKLADNIKOVA, N.D., BELYAEVA, Z.D. and DROZHKO, E.G. (1997). "Multifactorial analysis of lung cancer dose-response relationships for workers at the Mayak nuclear enterprise," Health Phys. **73**, 899–905.

TOKUNAGA, M., LAND, C.E., TOKUOKA, S., NISHIMORI, I., SODA, M. and AKIBA, S. (1994). "Incidence of female breast cancer among atomic bomb survivors, 1950-1985," Radiat. Res. **138**, 209–223.

TOTTER, J.R. and MACPHERSON, H.G. (1981). "Do childhood cancers result from prenatal x-rays?" Health Phys. **40**, 511–524.

TROTT, K.R., JAMALI, M., MANTI, L. and TEIBE, A. (1998). "Manifestations and mechanisms of radiation-induced genomic instability in V-79 Chinese hamster cells," Int. J. Radiat. Biol. **74**, 787–791.

TSUBOI, K., YANG T.C. and CHEN, D.J. (1992). "Charged-particle mutagenesis. 1. Cytotoxic and mutagenic effects of high-LET charged ion particles on human skin fibroblasts," Radiat. Res. **129**, 171–176.

TUCKER, J.D., SORENSEN, K.J., CHU, C.S., NELSON, D.O., RAMSEY, M.J., URLANDO, C. and HEDDLE, J.A. (1998). "The accumulation of chromosome aberrations and *Dlb-1* mutations in mice with highly fractionated exposure to gamma radiation," Mutat. Res. **400**, 321–335.

UENO, A.M., VANNAIS, D.B., GUSTAFSON, D.L., WONG, J.C. and WALDREN, C.A. (1996). "A low, adaptive dose of gamma-rays reduced the number and altered the spectrum of S1-mutants in human-hamster hybrid A_L cells," Mutat. Res. **358**, 161–169.

ULLRICH, R.L. (1983). "Tumor induction in BALB/c female mice after fission neutron or gamma irradiation," Radiat. Res. **93**, 506–515.

ULLRICH, R.L. (1984). "Tumor induction in BALB/c mice after fractionated or protracted exposures to fission-spectrum neutrons," Radiat. Res. **97**, 587–597.

ULLRICH, R.L. and DAVIS, C.M. (1999). "Radiation-induced cytogenetic instability *in vivo*," Radiat. Res. **152**, 170–173.

ULLRICH, R.L. and PONNAIYA, B. (1998). "Radiation-induced instability and its relation to radiation carcinogenesis," Int. J. Radiat. Biol. **74**, 747–754.

ULLRICH, R.L. and PRESTON, R.J. (1987). "Myeloid leukemia incidence in male RFM mice following irradiation with fission spectrum neutrons or gamma rays," Radiat. Res. **109**, 165–170.

ULLRICH, R.L. and STORER, J.B. (1979). "Influence of γ irradiation on the development of neoplastic disease in mice. III. Dose-rate effects," Radiat. Res. **80**, 325–342.

ULLRICH, R.L., JERNIGAN, M.C., SATTERFIELD, L.C. and BOWLES, N.D. (1987). "Radiation carcinogenesis: Time-dose relationships," Radiat. Res. **111**, 179–184.

ULLRICH, R.L., BOWLES, N.D., SATTERFIELD, L.C. and DAVIS, C.M. (1996). "Strain-dependent susceptibility to radiation-induced mammary cancer is a result of differences in epithelial cell sensitivity to transformation," Radiat. Res. **146**, 353–355.

UNSCEAR (1977). United Nations Scientific Committee on the Effects of Atomic Radiation. *Sources and Effects of Ionizing Radiation, UNSCEAR 1977 Report to the General Assembly, with annexes*, No. E.77.IX.1 (United Nations, New York).

UNSCEAR (1982). United Nations Scientific Committee on the Effects of Atomic Radiation. *Ionizing Radiation: Sources and Biological Effects, UNSCEAR 1982 Report to the General Assembly, with annexes*, No. E.82.IX.8 (United Nations, New York).

UNSCEAR (1986). United Nations Scientific Committee on the Effects of Atomic Radiation. *Genetic and Somatic Effects of Ionizing Radiation, UNSCEAR 1986 Report to the General Assembly, with annexes*, No. E.96.IX.9 (United Nations, New York).

UNSCEAR (1988). United Nations Scientific Committee on the Effects of Atomic Radiation. *Sources, Effects and Risks of Ionizing Radiation,*

UNSCEAR 1988 Report to the General Assembly with annexes, No. E.88.IX.7 (United Nations, New York).

UNSCEAR (1993). United Nations Scientific Committee on the Effects of Atomic Radiation. *Sources and Effects of Ionizing Radiation, UNSCEAR 1993 Report to the General Assembly, with scientific annexes*, No. E.94.IX.2 (United Nations, New York).

UNSCEAR (1994). United Nations Scientific Committee on the Effects of Atomic Radiation. *Sources and Effects of Ionizing Radiation, UNSCEAR 1994 Report to the General Assembly, with scientific annexes*, No. E.94.IX.11 (United Nations, New York).

UPTON, A.C. (1959). "Studies on the mechanism of leukaemogenesis by ionizing radiation," pages 249 to 268 in *Ciba Foundation Symposium on Carcinogenesis: Mechanisms of Action* (Little, Brown Company, Boston).

UPTON, A.C., KIMBALL, A.W., FURTH, J., CHRISTENBERRY, K.W. and BENEDICT, W.H. (1960). "Some delayed effects of atom-bomb radiations in mice," Cancer Res. **20**, 1–60.

UPTON, A.C., RANDOLPH, M.L. and CONKLIN, J.W. (1967). "Late effects of fast neutrons and gamma rays in mice as influenced by the dose rate of irradiation: Life shortening," Radiat. Res. **32**, 493–509.

UPTON, A.C., RANDOLPH, M.L., CONKLIN, J.W., KASTENBAUM, M.A., SLATER, M., MELVILLE, G.S., JR., CONTE, F.P. and SPROUL, J.A., JR. (1970). "Late effects of fast neutrons and gamma-rays in mice as influenced by the dose rate of irradiation: Induction of neoplasia," Radiat. Res. **41**, 467–491.

UPTON, A.C., ALBERT, R.E., BURNS, F.J. and SHORE, R.E., Eds. (1986). *Radiation Carcinogenesis* (Elsevier, New York).

VAETH, M., PRESTON, D. and MABUCHI, K. (1992). "The shape of the cancer incidence dose-response curve for the A-bomb survivors," pages 75 to 78 in *Low Dose Irradiation and Biological Defense Mechanisms*, Sugahara, T., Sagan, L.A. and Aoyama, T., Eds. (Excerta Medica, New York).

VARON, R., VISSINGA, C., PLATZER, M., CEROSALETTI, K.M., CHRZANOWSKA, K.H., SAAR, K., BECKMANN, G., SEEMANOVA, E., COOPER, P.R., NOWAK, N.J., STUMM, M., WEEMAES, C.M.R., GATTI, R.A., WILSON, R.K., DIGWEED, M., ROSENTHAL, A., SPERLING, K., CONCANNON, P. and REIS, A. (1998). "Nibrin, a novel DNA double-strand break repair protein, is mutated in Nijmegen breakage syndrome," Cell **93**, 467–476.

VIJAYALAXMI and EVANS, H.J. (1984). "Measurement of spontaneous and x-irradiation-induced 6-thioguanine-resistant human blood lymphocytes using a T-cell cloning technique," Mutat. Res. **125**, 87–94.

VIJAYALAXMI, LEAL, B.Z., DEAHL, T.S. and MELTZ, M.L. (1995). "Variability in adaptive response to low dose radiation in human blood lymphocytes: Consistent results from chromosome aberrations and micronuclei," Mutat. Res. **348**, 45–50.

VIRSIK, R.P., SCHAFER, C., HARDER, D., GOODHEAD, D.T., COX, R. and THACKER, J. (1980). "Chromosome aberrations induced in human lymphocytes by ultrasoft Al (K) and C (K) x-rays," Int. J. Radiat. Biol. Relat. Stud. Phys. Chem. Med. **38**, 545–557.

VOGELSTEIN, B., FEARON, E.R., HAMILTON, S.R. and FEINBERG, A.P. (1984). "Use of restriction fragment length polymorphisms to determine the clonal origin of human tumors," Science **227**, 642–645.

VOGELSTEIN, B., FEARON, E.R., HAMILTON, S.R., PREISINGER, A.C., WILLARD, H.F., MICHELSON, A.M., RIGGS, A.D. and ORKIN, S.H. (1987). "Clonal analysis using recombinant DNA probes from the X-chromosome," Cancer Res. **47**, 4806–4813.

WAINSCOAT, J.S. and FEY, M.F. (1990). "Assessment of clonality in human tumors: A review," Cancer Res. **50**, 1355–1360.

WAKEFORD, R. (1995). "The risk of childhood cancer from intrauterine and preconceptional exposure to ionizing radiation," Environ. Health Perspect. **103**, 1018–1025.

WALBURG, H.E., JR. (1975). "Radiation-induced life-shortening and premature aging," Adv. Radiat. Biol. **7**, 145–179.

WALBURG, H.E., JR., COSGROVE, G.E. and UPTON, A.C. (1968). "Influence of microbial environment on development of myeloid leukemia in x-irradiated RFM mice," Int. J. Cancer **3**, 150–154.

WALDREN, C., CORRELL, L., SOGNIER, M.A. and PUCK, T.T. (1986). "Measurement of low levels of x-ray mutagenesis in relation to human disease," Proc. Natl. Acad. Sci USA **83**, 4839–4843.

WALDREN, C, UENO, A., VANNAIS, D., BEDFORD, J. and HEI, T. (1992). "Molecular analysis of chromosomal mutations induced in mammalian cells by high and low dose rate ^{137}Cs γ rays," pages 339 to 342 in *Low Dose Irradiation and Biological Defense Mechanisms*, Sugahara, T., Sagan, L.A. and Aoyama, T. Eds. (Excerta Medica, New York).

WALLACE, S.S. (1998). "Enzymatic processing of radiation-induced free radical damage in DNA," Radiat. Res. **150**, S60–S79.

WALLACE, C., BERNSTEIN, R. and PINTO, M.R. (1984). "Non-random *in vitro* 7:14 translocations detected in routine cytogenetic series. 12 examples and their possible significance," Hum. Genet. **66**, 157–161.

WANG, Y., PARKS, W.C., WIGLE, J.C., MAHER, V.M. and MCCORMICK, J.J. (1986). "Fibroblasts from patients with inherited predisposition to retinoblastoma exhibit normal sensitivity to the mutagenic effects of ionizing radiation," Mutat. Res. **175**, 107–114.

WANG, Z.Y., BOICE, J.D., JR., WEI, L.X., BEEBE, G.W., ZHA, Y.R., KAPLAN, M.M., TAO, Z.F., MAXON, H.R., III, ZHANG, S., SCHNEIDER, A.B., TAN, B., WESSELER, T.A., CHEN, D., ERSHOW, A.G., KLEINERMAN, R.A., LITTLEFIELD, L.G. and PRESTON, D. (1990a). "Thyroid nodularity and chromosome aberrations among women in areas of high background radiation in China," J. Natl. Cancer Inst. **82**, 478–485.

WANG, J.X., INSKIP, P.D., BOICE, J.D., JR., LI, B.X., ZHANG, J.Y. and FRAUMENI, J.F., JR. (1990b). "Cancer incidence among medical diagnostic x-ray workers in China, 1950 to 1985," Int. J. Cancer **45**, 889–895.

WARD, J.F. (1988). "DNA damage produced by ionizing radiation in mammalian cells: Identities, mechanisms of formation and repairability," Prog. Nucleic Acid Res. Mol. Biol. **35**, 95–125.

WARD, J.F. (1990). "The yield of double-strand breaks produced intracellularly by ionizing radiation: A review," Int. J. Radiat. Biol. **57**, 1141–1150.

WARD, J.F. (1991). "Letter to the editor: Response to commentary by D. Billen," Radiat. Res. **126**, 385–387

WARD, J.F. (1995). "Radiation mutagenesis: The initial DNA lesions responsible," Radiat. Res. **142**, 362–368.

WARD, J.F., BLAKELY, W.F. and JONER, E.I. (1985). " Mammalian cells are not killed by DNA single-strand breaks caused by hydroxyl radicals from hydrogen peroxide," Radiat. Res. **103**, 383–392.

WATANABE, H., TANNER, M.A., DOMANN, F.E., GOULD, M.N. and CLIFTON, K.H. (1988). "Inhibition of carcinoma formation and of vascular invasion in grafts of radiation-initiated thyroid clonogens by unirradiated thyroid cells," Carcinogenesis **9**, 1329–1335.

WATSON, G.E., LORIMORE, S.A., CLUTTON, S.M., KADHIM, M.A. and WRIGHT, E.G. (1997). "Genetic factors influencing alpha-particle-induced chromosomal instability," Int. J. Radiat. Biol. **71**, 497–503.

WEI, L.X., ZHA, Y.R., TAO, Z.F., HE, W.H., CHEN, D.Q. and YUAN, Y.L. (1990). "Epidemiological investigation of radiological effects in high background areas of Yanjiang, China," J. Radiat. Res. (Tokyo) **31**, 119–136.

WEINBERG, C.R., BROWN, K.G. and HOEL, D.G. (1987). "Altitude, radiation, and mortality from cancer and heart disease," Radiat. Res. **112**, 381–390.

WEISS, H.A., DARBY, S.C. and DOLL, R. (1994). "Cancer mortality following x-ray treatment for ankylosing spondylitis," Int. J. Cancer **59**, 327–338.

WELCH, J.P., LEE, C.L.Y., BEATTY-DESANA, J.W., HOGGARD, M.J., COOLEDGE, J.W., HECHT, F., MCCAW, B.K., PEAKMAN, D. and ROBINSON, A. (1975). "Non-random occurrence of 7-14 translocations in human lymphocyte cultures," Nature **255**, 241–245.

WELLS, J. and STEER, C.M. (1961). "Relationship of leukemia in children to abdominal irradiation of mothers during pregnancy," Am. J. Obstet. Gynec. **81**, 1059–1063.

WERNER, A., MODAN, B. and DAVIDOFF, D. (1968). "Doses to brain, skull and thyroid, following x-ray therapy for tinea capitis," Phys. Med. Biol. **13**, 247–258.

WHITTEMORE, A.S. and MCMILLAN, A. (1982). "Osteosarcomas in beagles exposed to ^{232}Pu," Radiat. Res. **90**, 41–56.

WICKING, C., SHANLEY, S., SMYTH, I., GILLIES, S., NEGUS, K., GRAHAM, S., SUTHERS, G., HAITES, N., EDWARDS, M., WAINWRIGHT, B. and CHENEVIX-TRENCH, G. (1997). "Most germ-line mutations in the nevoid basal cell carcinoma syndrome lead to a premature termination of the PATCHED protein, and no genotype-phenotype correlations are evident," Am. J. Hum. Genet. **60**, 21–26.

WIGGS, L.D., COX-DEVORE, C.A., WILKINSON, G.S. and REYES, M. (1991). "Mortality among workers exposed to external ionizing radiation at a nuclear facility in Ohio," J. Occup. Med. **33**, 632–637.

WIGGS, L.D., JOHNSON, E.R., COX-DEVORE, C.A. and VOELZ, G.L. (1994). "Mortality through 1990 among white male workers at the Los

Alamos National Laboratory: Considering exposures to plutonium and external ionizing radiation," Health Phys. **67**, 577–588.

WILLEY, J.C., BROUSSOUND, A., SLEEMI, A., BENNETT, W.P., CERUTTI, P. and CURTIS, C.C. (1991). "Immortalization of human bronchial epithelial cells by human papillomaviruses 16 or 18," Cancer Res. **51**, 5370–5377.

WING, S., SHY, C.M., WOOD, J.L., WOLF, S., CRAGLE, D.L. and FROME, E.L. (1991). "Mortality among workers at Oak Ridge National Laboratory. Evidence of radiation effects in follow-up through 1984," JAMA **265**, 1397–1402.

WOJCIK, A. and SHADLEY, J.D. (2000). "The current status of the adaptive response to ionizing radiation in mammalian cells," Hum. Ecol. Risk Assessm. **6**, 281–300.

WOLFF, S. (1996). "Aspects of the adaptive response to very low doses of radiation and other agents," Mutat. Res. **358**, 135–142.

WONG, F.L., BOICE, J.D., JR., ABRAMSON, D.H., TARONE, R.E., KLEINERMAN, R.A., STOVALL, M., GOLDMAN, M.B., SEDDON, J.M., TARBELL, N., FRAUMENI, J.F., JR. and LI, J.F. (1997). "Cancer incidence after retinoblastoma. Radiation dose and sarcoma risk," JAMA **278**, 1262–1267.

WORSHAM, M.J., WOLMAN, S.R. and ZARBO, R.J. (1996). "Molecular genetic fingerprints: Clues to monoclonal origin of multifocal disease," Mod. Pathol. **9**, 163–165.

WOUTERS, B.G. and SKARSGARD, L.D. (1997). "Low-dose radiation sensitivity and induced radioresistance to cell killing in HT-29 cells is distinct from the "adaptive response" and cannot be explained by a subpopulation of sensitive cells," Radiat. Res. **148**, 435–442.

WRIGHT, E.G. (1998). "Radiation-induced genomic instability in haemopoietic cells," Int. J. Radiat. Biol. **74**, 681–687.

XIA, F., WANG, X., WANG, Y.H., TSANG, N.M., YANDELL, D.W., KELSEY, K.T. and LIBER, H.L. (1995). "Altered p53 status correlates with differences in sensitivity to radiation-induced mutation and apoptosis in two closely related human lymphoblast lines," Cancer Res. **55**, 12–15.

XU, D.B., FENG, J., HEI, T.K., PIAO, C.Q. and KAHN, S.M. (1997). "Absence of the microsatellite mutator phenotype in human bronchial epithelial cells transformed by alpha particles," Int. J. of Oncology **10**, 921–925.

YANG, T.C.H., CRAISE, L.M., MEI, M.T. and TOBIAS, C.A. (1985). "Neoplastic cell transformation by heavy charged particles," Radiat. Res. **104**, S177–S187.

YEUNG, R.S., BUETOW, K.H., TESTA, J.R. and KNUDSON, A.G., JR. (1993). "Susceptibility to renal carcinoma in the Eker rat involves a tumor suppressor gene on chromosome 10," Proc. Natl. Acad. Sci. USA **90**, 8038–8042.

YOKORO, K., NAKANO, M., ITO, A., NAGAO, K. and KODAMA, Y. (1977). "Role of prolaction in rat mammary carcinogenesis: Detection of carcinogenicity of low-dose carcinogens and of persisting dormant cancer cells," J. Natl. Cancer Inst. **58**, 1777–1783.

YOSHIMOTO, Y., KATO, H. and SCHULL, W.J. (1988). "Risk of cancer among children exposed *in utero* to A-bomb radiations, 1950-84," Lancet **2**, 665–669.

ZAICHKINA, S.I., APTIKAEVA, G.F., ROZANOVA, O.M., AKHMADIEVA, A.K., SMIRNOVA, E.N. and GANASSI, E.E. (1997). "Action of chronic irradiation on the cytogenetic damage of human lymphocyte culture," Environ. Health Perspect. **105** (suppl.), 1441–1443.

ZHOU, P.K., LIU, X.Y., SUN, W.Z., ZHANG, Y.P. and WEI, K. (1991). "Cultured mouse SR-1 cells exposed to low doses of γ-rays become less susceptible to the induction of mutagenesis by radiation as well as bleomycin," Mutagenesis **8**, 109–111.

ZHOU, P.K., ZHIANG, X.Q., SUN, W.Z., LIU, X.Y., ZHANG, Y.P. and WEI, K. (1994). "Adaptive response to mutagenesis and its molecular basis in a human T-cell leukemia line primed with a low dose of gamma rays," Radiat. Env. Biophys. **33**, 211–217.

The NCRP

The National Council on Radiation Protection and Measurements is a nonprofit corporation chartered by Congress in 1964 to:

1. Collect, analyze, develop and disseminate in the public interest information and recommendations about (a) protection against radiation and (b) radiation measurements, quantities and units, particularly those concerned with radiation protection.
2. Provide a means by which organizations concerned with the scientific and related aspects of radiation protection and of radiation quantities, units and measurements may cooperate for effective utilization of their combined resources, and to stimulate the work of such organizations.
3. Develop basic concepts about radiation quantities, units and measurements, about the application of these concepts, and about radiation protection.
4. Cooperate with the International Commission on Radiological Protection, the International Commission on Radiation Units and Measurements, and other national and international organizations, governmental and private, concerned with radiation quantities, units and measurements and with radiation protection.

The Council is the successor to the unincorporated association of scientists known as the National Committee on Radiation Protection and Measurements and was formed to carry on the work begun by the Committee in 1929.

The participants in the Council's work are the Council members and members of scientific and administrative committees. Council members are selected solely on the basis of their scientific expertise and serve as individuals, not as representatives of any particular organization. The scientific committees, composed of experts having detailed knowledge and competence in the particular area of the committee's interest, draft proposed recommendations. These are then submitted to the full membership of the Council for careful review and approval before being published.

The following comprise the current officers and membership of the Council:

Officers

President	Charles B. Meinhold
Vice President	S. James Adelstein
Secretary and Treasurer	William M. Beckner
Assistant Secretary	Michael F. McBride

Members

Honorary Members

Lauriston S. Taylor Lecturers

Herbert M. Parker (1977) *The Squares of the Natural Numbers in Radiation Protection*
Sir Edward Pochin (1978) *Why be Quantitative about Radiation Risk Estimates?*
Hymer L. Friedell (1979) *Radiation Protection—Concepts and Trade Offs*
Harold O. Wyckoff (1980) *From "Quantity of Radiation" and "Dose" to "Exposure" and "Absorbed Dose"—An Historical Review*
James F. Crow (1981) *How Well Can We Assess Genetic Risk? Not Very*
Eugene L. Saenger (1982) *Ethics, Trade-offs and Medical Radiation*
Merril Eisenbud (1983) *The Human Environment—Past, Present and Future*
Harald H. Rossi (1984) *Limitation and Assessment in Radiation Protection*
John H. Harley (1985) *Truth (and Beauty) in Radiation Measurement*
Herman P. Schwan (1986) *Biological Effects of Non-ionizing Radiations: Cellular Properties and Interactions*
Seymour Jablon (1987) *How to be Quantitative about Radiation Risk Estimates*
Bo Lindell (1988) *How Safe is Safe Enough?*
Arthur C. Upton (1989) *Radiobiology and Radiation Protection: The Past Century and Prospects for the Future*
J. Newell Stannard (1990) *Radiation Protection and the Internal Emitter Saga*
Victor P. Bond (1991) *When is a Dose Not a Dose?*
Edward W. Webster (1992) *Dose and Risk in Diagnostic Radiology: How Big? How Little?*
Warren K. Sinclair (1993) *Science, Radiation Protection and the NCRP*
R.J. Michael Fry (1994) *Mice, Myths and Men*
Albrecht Kellerer (1995) *Certainty and Uncertainty in Radiation Protection*
Seymour Abrahamson (1996) *70 Years of Radiation Genetics: Fruit Flies, Mice and Humans*
William J. Bair (1997) *Radionuclides in the Body: Meeting the Challenge!*
Eric J. Hall (1998) *From Chimney Sweeps to Astronauts: Cancer Risks in the Workplace*
Naomi H. Harley (1999) *Back to Background*
S. James Adelstein (2000) *Administered Radioactivity: Unde Venimus Quoque Imus*
Wesley L. Nyborg (2001) *Assuring the Safety of Medical Diagnostic Ultrasound*

Currently, the following committees are actively engaged in formulating recommendations:

SC 1 Basic Criteria, Epidemiology, Radiobiology and Risk
- SC 1-4 Extrapolation of Risks from Non-Human Experimental Systems to Man
- SC 1-7 Information Needed to Make Radiation Protection Recommendations for Travel Beyond Low-Earth Orbit
- SC 1-8 Risk to Thyroid from Ionizing Radiation

SC 9 Structural Shielding Design and Evaluation for Medical Use of X Rays and Gamma Rays of Energies Up to 10 MeV

SC 46 Operational Radiation Safety

SC 46-8 Radiation Protection Design Guidelines for Particle Accelerator Facilities
SC 46-10 Assessment of Occupational Doses from Internal Emitters
SC 46-13 Design of Facilities for Medical Radiation Therapy
SC 46-14 Radiation Protection Issues Related to Terrorist Activities that Result in the Dispersal of Radioactive Material
SC 46-15 Operational Radiation Safety Program for Astronauts
SC 57-15 Uranium Risk
SC 57-17 Radionuclide Dosimetry Models for Wounds

SC 64 Environmental Issues
SC 64-19 Historical Dose
SC 64-22 Design of Effective Effluent and Environmental Monitoring Programs
SC 64-23 Cesium in the Environment

SC 66 Biological Effects and Exposure Criteria for Ultrasound

SC 72 Radiation Protection in Mammography

SC 85 Risk of Lung Cancer from Radon

SC 87 Radioactive and Mixed Waste
SC 87-1 Waste Avoidance and Volume Reduction
SC 87-2 Waste Classification Based on Risk
SC 87-3 Performance Assessment
SC 87-4 Management of Waste Metals Containing Radioactivity

SC 88 Fluence as the Basis for a Radiation Protection System for Astronauts

SC 89 Nonionizing Electromagnetic Fields
SC 89-3 Biological Effects of Extremely Low-Frequency Electric and Magnetic Fields
SC 89-4 Biological Effects and Exposure Recommendations for Modulated Radiofrequency Fields
SC 89-5 Biological Effects and Exposure Criteria for Radiofrequency Fields

SC 91 Radiation Protection in Medicine
SC 91-1 Precautions in the Management of Patients Who Have Received Therapeutic Amounts of Radionuclides
SC 91-2 Radiation Protection in Dentistry

SC 92 Public Policy and Risk Communication

SC 93 Radiation Measurement and Dosimetry

In recognition of its responsibility to facilitate and stimulate cooperation among organizations concerned with the scientific and related aspects of radiation protection and measurement, the Council has created a category of NCRP Collaborating Organizations. Organizations or groups of organizations that are national or international in scope and are concerned with scientific problems involving radiation quantities, units, measurements and effects, or radiation protection may be admitted to collaborating status by the Council. Collaborating Organizations provide a means by which the NCRP can gain input into its activities from a wider segment of society. At the same time, the relationships with the Collaborating Organizations facilitate wider dissemination of information about the Council's activities, interests and concerns. Collaborating Organizations have the opportunity to comment on draft reports (at the time that these are submitted to the members of the Council). This is intended to capitalize on the fact that Collaborating Organizations are in an excellent position to both contribute to the identification of what needs to be treated in NCRP reports and to identify problems that might result from proposed recommendations. The present Collaborating Organizations with which the NCRP maintains liaison are as follows:

Agency for Toxic Substances and Disease Registry
American Academy of Dermatology
American Academy of Environmental Engineers
American Academy of Health Physics
American Association of Physicists in Medicine
American College of Medical Physics
American College of Nuclear Physicians
American College of Occupational and Environmental Medicine
American College of Radiology
American Dental Association
American Industrial Hygiene Association
American Institute of Ultrasound in Medicine
American Insurance Services Group
American Medical Association
American Nuclear Society
American Pharmaceutical Association
American Podiatric Medical Association
American Public Health Association
American Radium Society
American Roentgen Ray Society
American Society for Therapeutic Radiology and Oncology
American Society of Health-System Pharmacists
American Society of Radiologic Technologists
Association of University Radiologists
Bioelectromagnetics Society
Campus Radiation Safety Officers
College of American Pathologists
Conference of Radiation Control Program Directors, Inc.
Council on Radionuclides and Radiopharmaceuticals
Defense Threat Reduction Agency
Electric Power Research Institute

Electromagnetic Energy Association
Federal Communications Commission
Federal Emergency Management Agency
Genetics Society of America
Health Physics Society
Institute of Electrical and Electronics Engineers, Inc.
Institute of Nuclear Power Operations
International Brotherhood of Electrical Workers
National Aeronautics and Space Administration
National Association of Environmental Professionals
National Electrical Manufacturers Association
National Institute for Occupational Safety and Health
National Institute of Standards and Technology
Nuclear Energy Institute
Office of Science and Technology Policy
Oil, Chemical and Atomic Workers
Radiation Research Society
Radiological Society of North America
Society for Risk Analysis
Society of Nuclear Medicine
U.S. Air Force
U.S. Army
U.S. Coast Guard
U.S. Department of Energy
U.S. Department of Housing and Urban Development
U.S. Department of Labor
U.S. Department of Transportation
U.S. Environmental Protection Agency
U.S. Navy
U.S. Nuclear Regulatory Commission
U.S. Public Health Service
Utility Workers Union of America

The NCRP has found its relationships with these organizations to be extremely valuable to continued progress in its program.

Another aspect of the cooperative efforts of the NCRP relates to the Special Liaison relationships established with various governmental organizations that have an interest in radiation protection and measurements. This liaison relationship provides: (1) an opportunity for participating organizations to designate an individual to provide liaison between the organization and the NCRP; (2) that the individual designated will receive copies of draft NCRP reports (at the time that these are submitted to the members of the Council) with an invitation to comment, but not vote; and (3) that new NCRP efforts might be discussed with liaison individuals as appropriate, so that they might have an opportunity to make suggestions on new studies and related matters. The following organizations participate in the Special Liaison Program:

Atomic Energy Control Board
Australian Radiation Laboratory
Bundesamt für Strahlenschutz (Germany)
Central Laboratory for Radiological Protection (Poland)
Commisariat à l'Energie Atomique
Commonwealth Scientific Instrumentation Research Organization (Australia)
European Commission
Health Council of the Netherlands
International Commission on Non-Ionizing Radiation Protection
Japan Radiation Council
Korea Institute of Nuclear Safety
National Radiological Protection Board (United Kingdom)
Russian Scientific Commission on Radiation Protection
South African Forum for Radiation Protection
World Association of Nuclear Operations

The NCRP values highly the participation of these organizations in the Special Liaison Program.

The Council also benefits significantly from the relationships established pursuant to the Corporate Sponsor's Program. The program facilitates the interchange of information and ideas and corporate sponsors provide valuable fiscal support for the Council's program. This developing program currently includes the following Corporate Sponsors:

3M Corporate Health Physics
Commonwealth Edison
Consolidated Edison
Duke Energy Corporation
Florida Power Corporation
ICN Biomedicals, Inc.
Landauer, Inc.
New York Power Authority
Nuclear Energy Institute
Nycomed Amersham Corporation
Southern California Edison

The Council's activities are made possible by the voluntary contribution of time and effort by its members and participants and the generous support of the following organizations:

3M Health Physics Services
Agfa Corporation
Alfred P. Sloan Foundation
Alliance of American Insurers
American Academy of Dermatology
American Academy of Health Physics
American Academy of Oral and Maxillofacial Radiology
American Association of Physicists in Medicine

American Cancer Society
American College of Medical Physics
American College of Nuclear Physicians
American College of Occupational and Environmental Medicine
American College of Radiology
American College of Radiology Foundation
American Dental Association
American Healthcare Radiology Administrators
American Industrial Hygiene Association
American Insurance Services Group
American Medical Association
American Nuclear Society
American Osteopathic College of Radiology
American Podiatric Medical Association
American Public Health Association
American Radium Society
American Roentgen Ray Society
American Society of Radiologic Technologists
American Society for Therapeutic Radiology and Oncology
American Veterinary Medical Association
American Veterinary Radiology Society
Association of University Radiologists
Battelle Memorial Institute
Canberra Industries, Inc.
Chem Nuclear Systems
Center for Devices and Radiological Health
College of American Pathologists
Committee on Interagency Radiation Research and Policy Coordination
Commonwealth of Pennsylvania
Consumers Power Company
Council on Radionuclides and Radiopharmaceuticals
Defense Nuclear Agency
Eastman Kodak Company
Edison Electric Institute
Edward Mallinckrodt, Jr. Foundation
EG&G Idaho, Inc.
Electric Power Research Institute
Federal Emergency Management Agency
Florida Institute of Phosphate Research
Fuji Medical Systems, U.S.A., Inc.
Genetics Society of America
Health Effects Research Foundation (Japan)
Health Physics Society
Institute of Nuclear Power Operations
James Picker Foundation
Martin Marietta Corporation
Motorola Foundation

National Aeronautics and Space Administration
National Association of Photographic Manufacturers
National Cancer Institute
National Electrical Manufacturers Association
National Institute of Standards and Technology
Picker International
Public Service Electric and Gas Company
Radiation Research Society
Radiological Society of North America
Richard Lounsbery Foundation
Sandia National Laboratory
Siemens Medical Systems, Inc.
Society of Nuclear Medicine
Society of Pediatric Radiology
U.S. Department of Energy
U.S. Department of Labor
U.S. Environmental Protection Agency
U.S. Navy
U.S. Nuclear Regulatory Commission
Victoreen, Inc.
Westinghouse Electric Corporation

Initial funds for publication of NCRP reports were provided by a grant from the James Picker Foundation.

The NCRP seeks to promulgate information and recommen-dations based on leading scientific judgment on matters of radiation protection and measurement and to foster cooperation among organizations concerned with these matters. These efforts are intended to serve the public interest and the Council welcomes comments and suggestions on its reports or activities from those interested in its work.

NCRP Publications

Information on NCRP publications may be obtained from the NCRP website (http://www.ncrp.com), e-mail (ncrppubs@ncrp.com), by telephone (800-229-2652), or fax (301-907-8768). The address is:

NCRP Publications
7910 Woodmont Avenue
Suite 800
Bethesda, MD 20814-3095

Abstracts of NCRP reports published since 1980, abstracts of all NCRP commentaries, and the text of all NCRP statements are available at the NCRP website. Currently available publications are listed below.

NCRP Reports

No.	Title
8	*Control and Removal of Radioactive Contamination in Laboratories* (1951)
22	*Maximum Permissible Body Burdens and Maximum Permissible Concentrations of Radionuclides in Air and in Water for Occupational Exposure* (1959) [Includes Addendum 1 issued in August 1963]
25	*Measurement of Absorbed Dose of Neutrons, and of Mixtures of Neutrons and Gamma Rays* (1961)
27	*Stopping Powers for Use with Cavity Chambers* (1961)
30	*Safe Handling of Radioactive Materials* (1964)
32	*Radiation Protection in Educational Institutions* (1966)
35	*Dental X-Ray Protection* (1970)
36	*Radiation Protection in Veterinary Medicine* (1970)
37	*Precautions in the Management of Patients Who Have Received Therapeutic Amounts of Radionuclides* (1970)
38	*Protection Against Neutron Radiation* (1971)
40	*Protection Against Radiation from Brachytherapy Sources* (1972)
41	*Specification of Gamma-Ray Brachytherapy Sources* (1974)
42	*Radiological Factors Affecting Decision-Making in a Nuclear Attack* (1974)

44 *Krypton-85 in the Atmosphere—Accumulation, Biological Significance, and Control Technology* (1975)

46 *Alpha-Emitting Particles in Lungs* (1975)

47 *Tritium Measurement Techniques* (1976)

49 *Structural Shielding Design and Evaluation for Medical Use of X Rays and Gamma Rays of Energies Up to 10 MeV* (1976)

50 *Environmental Radiation Measurements* (1976)

51 *Radiation Protection Design Guidelines for 0.1 - 10 MeV Particle Accelerator Facilities* (1977)

52 *Cesium-137 from the Environment to Man: Metabolism and Dose* (1977)

54 *Medical Radiation Exposure of Pregnant and Potentially Pregnant Women* (1977)

55 *Protection of the Thyroid Gland in the Event of Releases of Radioiodine* (1977)

57 *Instrumentation and Monitoring Methods for Radiation Protection* (1978)

58 *A Handbook of Radioactivity Measurements Procedures*, 2nd ed. (1985)

59 *Operational Radiation Safety Program* (1978)

60 *Physical, Chemical, and Biological Properties of Radiocerium Relevant to Radiation Protection Guidelines* (1978)

61 *Radiation Safety Training Criteria for Industrial Radiography* (1978)

62 *Tritium in the Environment* (1979)

63 *Tritium and Other Radionuclide Labeled Organic Compounds Incorporated in Genetic Material* (1979)

64 *Influence of Dose and Its Distribution in Time on Dose-Response Relationships for Low-LET Radiations* (1980)

65 *Management of Persons Accidentally Contaminated with Radionuclides* (1980)

67 *Radiofrequency Electromagnetic Fields—Properties, Quantities and Units, Biophysical Interaction, and Measurements* (1981)

68 *Radiation Protection in Pediatric Radiology* (1981)

69 *Dosimetry of X-Ray and Gamma-Ray Beams for Radiation Therapy in the Energy Range 10 keV to 50 MeV* (1981)

70 *Nuclear Medicine—Factors Influencing the Choice and Use of Radionuclides in Diagnosis and Therapy* (1982)

72 *Radiation Protection and Measurement for Low-Voltage Neutron Generators* (1983)

73 *Protection in Nuclear Medicine and Ultrasound Diagnostic Procedures in Children* (1983)

74 *Biological Effects of Ultrasound: Mechanisms and Clinical Implications* (1983)

75 *Iodine-129: Evaluation of Releases from Nuclear Power Generation* (1983)

77 *Exposures from the Uranium Series with Emphasis on Radon and Its Daughters* (1984)
78 *Evaluation of Occupational and Environmental Exposures to Radon and Radon Daughters in the United States* (1984)
79 *Neutron Contamination from Medical Electron Accelerators* (1984)
80 *Induction of Thyroid Cancer by Ionizing Radiation* (1985)
81 *Carbon-14 in the Environment* (1985)
82 *SI Units in Radiation Protection and Measurements* (1985)
83 *The Experimental Basis for Absorbed-Dose Calculations in Medical Uses of Radionuclides* (1985)
84 *General Concepts for the Dosimetry of Internally Deposited Radionuclides* (1985)
85 *Mammography—A User's Guide* (1986)
86 *Biological Effects and Exposure Criteria for Radiofrequency Electromagnetic Fields* (1986)
87 *Use of Bioassay Procedures for Assessment of Internal Radionuclide Deposition* (1987)
88 *Radiation Alarms and Access Control Systems* (1986)
89 *Genetic Effects from Internally Deposited Radionuclides* (1987)
90 *Neptunium: Radiation Protection Guidelines* (1988)
92 *Public Radiation Exposure from Nuclear Power Generation in the United States* (1987)
93 *Ionizing Radiation Exposure of the Population of the United States* (1987)
94 *Exposure of the Population in the United States and Canada from Natural Background Radiation* (1987)
95 *Radiation Exposure of the U.S. Population from Consumer Products and Miscellaneous Sources* (1987)
96 *Comparative Carcinogenicity of Ionizing Radiation and Chemicals* (1989)
97 *Measurement of Radon and Radon Daughters in Air* (1988)
99 *Quality Assurance for Diagnostic Imaging* (1988)
100 *Exposure of the U.S. Population from Diagnostic Medical Radiation* (1989)
101 *Exposure of the U.S. Population from Occupational Radiation* (1989)
102 *Medical X-Ray, Electron Beam and Gamma-Ray Protection for Energies Up to 50 MeV (Equipment Design, Performance and Use)* (1989)
103 *Control of Radon in Houses* (1989)
104 *The Relative Biological Effectiveness of Radiations of Different Quality* (1990)
105 *Radiation Protection for Medical and Allied Health Personnel* (1989)
106 *Limit for Exposure to "Hot Particles" on the Skin* (1989)
107 *Implementation of the Principle of As Low As Reasonably Achievable (ALARA) for Medical and Dental Personnel* (1990)

108 *Conceptual Basis for Calculations of Absorbed-Dose Distributions* (1991)
109 *Effects of Ionizing Radiation on Aquatic Organisms* (1991)
110 *Some Aspects of Strontium Radiobiology* (1991)
111 *Developing Radiation Emergency Plans for Academic, Medical or Industrial Facilities* (1991)
112 *Calibration of Survey Instruments Used in Radiation Protection for the Assessment of Ionizing Radiation Fields and Radioactive Surface Contamination* (1991)
113 *Exposure Criteria for Medical Diagnostic Ultrasound: I. Criteria Based on Thermal Mechanisms* (1992)
114 *Maintaining Radiation Protection Records* (1992)
115 *Risk Estimates for Radiation Protection* (1993)
116 *Limitation of Exposure to Ionizing Radiation* (1993)
117 *Research Needs for Radiation Protection* (1993)
118 *Radiation Protection in the Mineral Extraction Industry* (1993)
119 *A Practical Guide to the Determination of Human Exposure to Radiofrequency Fields* (1993)
120 *Dose Control at Nuclear Power Plants* (1994)
121 *Principles and Application of Collective Dose in Radiation Protection* (1995)
122 *Use of Personal Monitors to Estimate Effective Dose Equivalent and Effective Dose to Workers for External Exposure to Low-LET Radiation* (1995)
123 *Screening Models for Releases of Radionuclides to Atmosphere, Surface Water, and Ground* (1996)
124 *Sources and Magnitude of Occupational and Public Exposures from Nuclear Medicine Procedures* (1996)
125 *Deposition, Retention and Dosimetry of Inhaled Radioactive Substances* (1997)
126 *Uncertainties in Fatal Cancer Risk Estimates Used in Radiation Protection* (1997)
127 *Operational Radiation Safety Program* (1998)
128 *Radionuclide Exposure of the Embryo/Fetus* (1998)
129 *Recommended Screening Limits for Contaminated Surface Soil and Review of Factors Relevant to Site-Specific Studies* (1999)
130 *Biological Effects and Exposure Limits for "Hot Particles"* (1999)
131 *Scientific Basis for Evaluating the Risks to Populations from Space Applications of Plutonium* (2001)
132 *Radiation Protection Guidance for Activities in Low-Earth Orbit* (2000)
133 *Radiation Protection for Procedures Performed Outside the Radiology Department* (2000)
134 *Operational Radiation Safety Training* (2000)
135 *Liver Cancer Risk from Internally-Deposited Radionuclides* (2001)
136 *Evaluation of the Linear-Nonthreshold Dose-Response Model for Ionizing Radiation* (2001)

Binders for NCRP reports are available. Two sizes make it possible to collect into small binders the "old series" of reports (NCRP Reports Nos. 8-30) and into large binders the more recent publications (NCRP Reports

Nos. 32-136). Each binder will accommodate from five to seven reports. The binders carry the identification "NCRP Reports" and come with label holders which permit the user to attach labels showing the reports contained in each binder.

The following bound sets of NCRP reports are also available:

Volume I. NCRP Reports Nos. 8, 22
Volume II. NCRP Reports Nos. 23, 25, 27, 30
Volume III. NCRP Reports Nos. 32, 35, 36, 37
Volume IV. NCRP Reports Nos. 38, 40, 41
Volume V. NCRP Reports Nos. 42, 44, 46
Volume VI. NCRP Reports Nos. 47, 49, 50, 51
Volume VII. NCRP Reports Nos. 52, 53, 54, 55, 57
Volume VIII. NCRP Report No. 58
Volume IX. NCRP Reports Nos. 59, 60, 61, 62, 63
Volume X. NCRP Reports Nos. 64, 65, 66, 67
Volume XI. NCRP Reports Nos. 68, 69, 70, 71, 72
Volume XII. NCRP Reports Nos. 73, 74, 75, 76
Volume XIII. NCRP Reports Nos. 77, 78, 79, 80
Volume XIV. NCRP Reports Nos. 81, 82, 83, 84, 85
Volume XV. NCRP Reports Nos. 86, 87, 88, 89
Volume XVI. NCRP Reports Nos. 90, 91, 92, 93
Volume XVII. NCRP Reports Nos. 94, 95, 96, 97
Volume XVIII. NCRP Reports Nos. 98, 99, 100
Volume XIX. NCRP Reports Nos. 101, 102, 103, 104
Volume XX. NCRP Reports Nos. 105, 106, 107, 108
Volume XXI. NCRP Reports Nos. 109, 110, 111
Volume XXII. NCRP Reports Nos. 112, 113, 114
Volume XXIII. NCRP Reports Nos. 115, 116, 117, 118
Volume XXIV. NCRP Reports Nos. 119, 120, 121, 122
Volume XXV. NCRP Report No. 123I and 123II
Volume XXVI. NCRP Reports Nos. 124, 125, 126, 127
Volume XXVII. NCRP Reports Nos. 128, 129, 130

(Titles of the individual reports contained in each volume are given above.)

NCRP Commentaries

No.	Title
1	*Krypton-85 in the Atmosphere—With Specific Reference to the Public Health Significance of the Proposed Controlled Release at Three Mile Island* (1980)
4	*Guidelines for the Release of Waste Water from Nuclear Facilities with Special Reference to the Public Health*

Significance of the Proposed Release of Treated Waste Waters at Three Mile Island (1987)

5 *Review of the Publication, Living Without Landfills* (1989)

6 *Radon Exposure of the U.S. Population—Status of the Problem* (1991)

7 *Misadministration of Radioactive Material in Medicine—Scientific Background* (1991)

8 *Uncertainty in NCRP Screening Models Relating to Atmospheric Transport, Deposition and Uptake by Humans* (1993)

9 *Considerations Regarding the Unintended Radiation Exposure of the Embryo, Fetus or Nursing Child* (1994)

10 *Advising the Public about Radiation Emergencies: A Document for Public Comment* (1994)

11 *Dose Limits for Individuals Who Receive Exposure from Radionuclide Therapy Patients* (1995)

12 *Radiation Exposure and High-Altitude Flight* (1995)

13 *An Introduction to Efficacy in Diagnostic Radiology and Nuclear Medicine (Justification of Medical Radiation Exposure)* (1995)

14 *A Guide for Uncertainty Analysis in Dose and Risk Assessments Related to Environmental Contamination* (1996)

15 *Evaluating the Reliability of Biokinetic and Dosimetric Models and Parameters Used to Assess Individual Doses for Risk Assessment Purposes* (1998)

Proceedings of the Annual Meeting

No. Title

1 *Perceptions of Risk*, Proceedings of the Fifteenth Annual Meeting held on March 14-15, 1979 (including Taylor Lecture No. 3) (1980)

3 *Critical Issues in Setting Radiation Dose Limits*, Proceedings of the Seventeenth Annual Meeting held on April 8-9, 1981 (including Taylor Lecture No. 5) (1982)

4 *Radiation Protection and New Medical Diagnostic Approaches*, Proceedings of the Eighteenth Annual Meeting held on April 6-7, 1982 (including Taylor Lecture No. 6) (1983)

5 *Environmental Radioactivity*, Proceedings of the Nineteenth Annual Meeting held on April 6-7, 1983 (including Taylor Lecture No. 7) (1983)

6 *Some Issues Important in Developing Basic Radiation Protection Recommendations*, Proceedings of the Twentieth Annual Meeting held on April 4-5, 1984 (including Taylor Lecture No. 8) (1985)

7 *Radioactive Waste*, Proceedings of the Twenty-first Annual Meeting held on April 3-4, 1985 (including Taylor Lecture No. 9) (1986)

8 *Nonionizing Electromagnetic Radiations and Ultrasound,* Proceedings of the Twenty-second Annual Meeting held on April 2-3, 1986 (including Taylor Lecture No. 10) (1988)

9 *New Dosimetry at Hiroshima and Nagasaki and Its Implications for Risk Estimates*, Proceedings of the Twenty-third Annual Meeting held on April 8-9, 1987 (including Taylor Lecture No. 11) (1988)

10 *Radon*, Proceedings of the Twenty-fourth Annual Meeting held on March 30-31, 1988 (including Taylor Lecture No. 12) (1989)

11 *Radiation Protection Today—The NCRP at Sixty Years*, Proceedings of the Twenty-fifth Annual Meeting held on April 5-6, 1989 (including Taylor Lecture No. 13) (1990)

12 *Health and Ecological Implications of Radioactively Contaminated Environments*, Proceedings of the Twenty-sixth Annual Meeting held on April 4-5, 1990 (including Taylor Lecture No. 14) (1991)

13 *Genes, Cancer and Radiation Protection,* Proceedings of the Twenty-seventh Annual Meeting held on April 3-4, 1991 (including Taylor Lecture No. 15) (1992)

14 *Radiation Protection in Medicine,* Proceedings of the Twenty-eighth Annual Meeting held on April 1-2, 1992 (including Taylor Lecture No. 16) (1993)

15 *Radiation Science and Societal Decision Making,* Proceedings of the Twenty-ninth Annual Meeting held on April 7-8, 1993 (including Taylor Lecture No. 17) (1994)

16 *Extremely-Low-Frequency Electromagnetic Fields: Issues in Biological Effects and Public Health,* Proceedings of the Thirtieth Annual Meeting held on April 6-7, 1994 (not published).

17 *Environmental Dose Reconstruction and Risk Implications,* Proceedings of the Thirty-first Annual Meeting held on April 12-13, 1995 (including Taylor Lecture No. 19) (1996)

18 *Implications of New Data on Radiation Cancer Risk*, Proceedings of the Thirty-second Annual Meeting held on April 3-4, 1996 (including Taylor Lecture No. 20) (1997)

19 *The Effects of Pre- and Postconception Exposure to Radiation*, Proceedings of the Thirty-third Annual Meeting held on April 2-3, 1997, Teratology **59**, 181–317 (1999)

20 *Cosmic Radiation Exposure of Airline Crews, Passengers and Astronauts*, Proceedings of the Thirty-fourth Annual Meeting held on April 1-2, 1998, Health Phys. **79**, 466–613 (2000)

21 *Radiation Protection in Medicine: Contemporary Issues*, Proceedings of the Thirty-fifth Annual Meeting held on April 7-8, 1999 (including Taylor Lecture No. 23) (1999)

Lauriston S. Taylor Lectures

No.	Title
1	*The Squares of the Natural Numbers in Radiation Protection by* Herbert M. Parker (1977)

2 *Why be Quantitative about Radiation Risk Estimates?* by Sir Edward Pochin (1978)

3 *Radiation Protection—Concepts and Trade Offs* by Hymer L. Friedell (1979) [Available also in *Perceptions of Risk*, see above]

4 *From "Quantity of Radiation" and "Dose" to "Exposure" and "Absorbed Dose"—An Historical Review* by Harold O. Wyckoff (1980)

5 *How Well Can We Assess Genetic Risk? Not Very* by James F. Crow (1981) [Available also in *Critical Issues in Setting Radiation Dose Limits*, see above]

6 *Ethics, Trade-offs and Medical Radiation* by Eugene L. Saenger (1982) [Available also in *Radiation Protection and New Medical Diagnostic Approaches*, see above]

7 *The Human Environment—Past, Present and Future* by Merril Eisenbud (1983) [Available also in *Environmental Radioactivity*, see above]

8 *Limitation and Assessment in Radiation Protection* by Harald H. Rossi (1984) [Available also in *Some Issues Important in Developing Basic Radiation Protection Recommendations*, see above]

9 *Truth (and Beauty) in Radiation Measurement* by John H. Harley (1985) [Available also in *Radioactive Waste*, see above]

10 *Biological Effects of Non-ionizing Radiations: Cellular Properties and Interactions by Herman P. Schwan (1987) [Available also in Nonionizing Electromagnetic Radiations and Ultrasound*, see above]

11 *How to be Quantitative about Radiation Risk Estimates* by Seymour Jablon (1988) [Available also in *New Dosimetry at Hiroshima and Nagasaki and its Implications for Risk Estimates*, see above]

12 *How Safe is Safe Enough?* by Bo Lindell (1988) [Available also in *Radon*, see above]

13 *Radiobiology and Radiation Protection: The Past Century and Prospects for the Future* by Arthur C. Upton (1989) [Available also in *Radiation Protection Today*, see above]

14 *Radiation Protection and the Internal Emitter Saga* by J. Newell Stannard (1990) [Available also in *Health and Ecological Implications of Radioactively Contaminated Environments*, see above]

15 *When is a Dose Not a Dose?* by Victor P. Bond (1992) [Available also in *Genes, Cancer and Radiation Protection*, see above]

16 *Dose and Risk in Diagnostic Radiology: How Big? How Little?* by Edward W. Webster (1992)[Available also in *Radiation Protection in Medicine*, see above]

17 *Science, Radiation Protection and the NCRP* by Warren K. Sinclair (1993)[Available also in *Radiation Science and Societal Decision Making,* see above]

18 *Mice, Myths and Men* by R.J. Michael Fry (1995)
19 *Certainty and Uncertainty in Radiation Research* by Albrecht M. Kellerer (1995). Health Phys. **69**, 446–453.
20 *70 Years of Radiation Genetics: Fruit Flies, Mice and Humans* by Seymour Abrahamson (1996). Health Phys. **71**, 624–633.
21 *Radionuclides in the Body: Meeting the Challenge* by William J. Bair (1997). Health Phys. **73**, 423–432.
22 *From Chimney Sweeps to Astronauts: Cancer Risks in the Work Place* by Eric J. Hall (1998). Health Phys. **75**, 357–366.
23 *Back to Background: Natural Radiation and Radioactivity Exposed* by Naomi H. Harley (2000). Health Phys. **79**, 121–128.

Symposium Proceedings

No. Title

1 *The Control of Exposure of the Public to Ionizing Radiation in the Event of Accident or Attack*, Proceedings of a Symposium held April 27-29, 1981 (1982)
2 *Radioactive and Mixed Waste—Risk as a Basis for Waste Classification,* Proceedings of a Symposium held November 9, 1994 (1995)
3 *Acceptability of Risk from Radiation—Application to Human Space Flight,* Proceedings of a Symposium held May 29, 1996 (1997)

NCRP Statements

No. Title

1 "Blood Counts, Statement of the National Committee on Radiation Protection," Radiology **63**, 428 (1954)
2 "Statements on Maximum Permissible Dose from Television Receivers and Maximum Permissible Dose to the Skin of the Whole Body," Am. J. Roentgenol., Radium Ther. and Nucl. Med. **84**, 152 (1960) and Radiology **75**, 122 (1960)
3 *X-Ray Protection Standards for Home Television Receivers, Interim Statement of the National Council on Radiation Protection and Measurements* (1968)
4 *Specification of Units of Natural Uranium and Natural Thorium, Statement of the National Council on Radiation Protection and Measurements* (1973)
5 *NCRP Statement on Dose Limit for Neutrons* (1980)
6 *Control of Air Emissions of Radionuclides* (1984)
7 *The Probability That a Particular Malignancy May Have Been Caused by a Specified Irradiation* (1992)
8 *The Application of ALARA for Occupational Exposures* (1999)
9 *Extension of the Skin Dose Limit for Hot Particles to Other External Sources of Skin Irradiation* (2001)

Other Documents

The following documents of the NCRP were published outside of the NCRP report, commentary and statement series:

Somatic Radiation Dose for the General Population, Report of the Ad Hoc Committee of the National Council on Radiation Protection and Measurements, 6 May 1959, Science, February 19, 1960, Vol. 131, No. 3399, pages 482-486

Dose Effect Modifying Factors In Radiation Protection, Report of Subcommittee M-4 (Relative Biological Effectiveness) of the National Council on Radiation Protection and Measurements, Report BNL 50073 (T-471) (1967) Brookhaven National Laboratory (National Technical Information Service Springfield, Virginia)

Index